Encyclopaedia of Mathematical Sciences

Volume 33

Editor-in-Chief: R.V. Gamkrelidze

Yu.V. Egorov M.A. Shubin (Eds.)

Partial Differential Equations IV

Microlocal Analysis and Hyperbolic Equations

Springer-Verlag Berlin Heidelberg GmbH

Title of the Russian edition:
Itogi nauki i tekhniki, Sovremennye problemy matematiki,
Fundamental'nye napravleniya, Vol. 33, Differentsial'nye
uravneniya s chastnymi proizvodnymi 4
Publisher VINITI, Moscow 1988

Mathematics Subject Classification (1991):
35-02, 35Axx, 35A27, 35H05, 35Lxx, 35S30, 46Fxx, 47G30,
58G15, 58G16, 58G17, 70Hxx

ISBN 978-3-642-08099-9

Library of Congress Cataloging-in-Publication Data
Differen͡tsial'nye uravneni͡ia s chastnymi proizvodnymi 4. English
Partial differential equations IV: microlocal analysis and
hyperbolic equations / Yu.V. Egorov, M.A. Shubin.
p. cm.—(Encyclopaedia of mathematical sciences; v. 33)
Includes bibliographical references and indexes.
ISBN 978-3-642-08099-9 ISBN 978-3-662-09207-1 (eBook)
DOI 10.1007/978-3-662-09207-1
1. Mathematical analysis. 2. Differential equations, Hyperbolic. I. Egorov, I͡U. V. (I͡Uriĭ Vladimirovich)
II. Shubin, M.A. (Mikhail Aleksandrovich), 1944- . III. Title. IV. Title: Partial differential equations 4.
V. Title: Partial differential equations four. VI. Series.
QA300.D471513 1993 515'.353—dc20 92-22077

Originally published by Springer-Verlag Berlin Heidelberg New York in 1993
Softcover reprint of the hardcover 1st edition 1993

Typesetting: Asco Trade Typesetting Ltd., Hong Kong
41/3140-543210—Printed on acid-free paper

List of Editors, Authors and Translators

Editor-in-Chief

R.V. Gamkrelidze, Russian Academy of Sciences, Steklov Mathematical Institute, ul. Vavilova 42, 117966 Moscow, Institute for Scientific Information (VINITI), ul. Usievicha 20a, 125219 Moscow, Russia

Consulting Editors

Yu.V. Egorov, Department of Mathematics, Moscow State University, Leninskie Gory, 119899 Moscow, Russia

M.A. Shubin, Department of Mathematics, Moscow State University, Leninskie Gory, 119899 Moscow, Russia

Authors

Yu.V. Egorov, Department of Mathematics, Moscow State University, Leninskie Gory, 119899 Moscow, Russia

V.Ya. Ivrii, Department of Mathematics, Magnitogorsk Mining-Metallurgical Institute, 455000 Magnitogorsk, Russia, and
Department of Mathematics, University of Toronto, 100 St. George Street, Toronto, M5S 1A1, Ontario, Canada

Translator

P.C. Sinha, Jagat Narayan Road, 800003 Patna, India

Contents

I. Microlocal Analysis

Yu.V. Egorov

Translated from the Russian
by P.C. Sinha

Contents

Preface

The microlocal analysis is the local analysis in cotangent bundle space. The remarkable progress made in the theory of linear partial differential equations over the past two decades is essentially due to the extensive applicaton of the microlocalization idea. The Hamiltonian systems, canonical transformations, Lagrange manifolds and other concepts, used in theoretical mechanics for examining processes in the phase space, have in recent years become the central objects of the theory of differential equations. For example, the evolution of singularities of solutions of a differential equation is described most naturally in terms of Lagrange manifolds and Hamiltonian systems, the solvability conditions are formulated in terms of the behaviour of integral curves of the Hamiltonian system whose Hamiltonian function serves as the characteristic form, the class of pseudodifferential equations arises in a natural way from that of differential equations under the action of canonical transformations, the class of subelliptic operators is defined by means of the Poisson brackets, etc. The difficulty faced in the microlocal analysis is connected with the principle of uncertainty which does not permit us to localize a function in any neighbourhood of a point of the cotangent space.

This paper presents a survey of the most interesting results, from our point of view, of the microlocal analysis achieved over the recent years. Unfortunately, due to lack of space many significant results could not be included. Also incomplete is the list of the literature cited; more complete lists can be found in Egorov [1984], Shubin [1978], Hörmander [1963, 1983, 1985], Taylor [1981] and Trèves [1982].

The author expresses his thanks to V.Ya. Ivrii for his useful critical comments.

Chapter 1
Microlocal Properties of Distributions

§ 1. Microlocalization

The study of the singularities of solutions constitutes one of the most important problems in the theory of differential equations. In this theory, just as in other mathematical disciplines, one often examines functions modulo smooth ones, so that the points where a given function is infinitely differentiable may be neglected. This approach reflects physical realities: singular points correspond to those phenomena which are most interesting from the point of view of each physical theory.

In investigating physical processes that take place in a bounded space an extensive use is made of the principle of *locality*. Its essence lies in that by knowing the state of the process at a given moment of time in a fixed region Ω of the physical space one may determine, by means of physical laws, the course of the process in a region Ω', lying strictly inside Ω, for a future time interval. During this time interval the effect of processes taking place outside Ω will have no influence on phenomena in Ω' because the effect is propagated with a fintie velocity.

We can introduce a more general principle, the principle of *microlocality*, by examining the phenomenon in a bounded region of the phase space. If we know the state of the process in this region at a certain moment of time, we can describe this process for future close points lying strictly inside the region. In physical terms, this means that the change in the impulse too takes place with a finite velocity because the acting forces are finite.

The above-mentioned principles are reflected in mathematical physics in the investigation of singularities of solutions of differential equations. Namely, local properties of such solutions are those properties which remain unaltered when the solutions are multiplied by smooth functions with a small support. Microlocal properties of a solution refer naturally to those properties which do not change on "multiplication" of the solution by a smooth function having support in a small neighbourhood of the given point in the phase space. However, this operation is much more complicated. In fact, it consists in multiplying by an ordinary smooth cutoff function with a small support, in applying Fourier transformation, in multiplying successively by a smooth cutoff function of dual coordinates, and in applying inverse Fourier transformation. Instead of Fourier transformation we can also use some other decomposition in plane waves; for example, the Radon transformation. In fact, the microlocal analysis is the local analysis on the cotangent bundle space.

A special feature of the microlocal analysis is the fact that localization in the phase space is possible only to a certain extent: the localization of spatial coordinates obstructs that of impulses. In quantum mechanics, this fact is referred to as the Heisenberg uncertainty principle.

The last two decades have seen immensely fruitful applications of the microlocality principle to the theory of partial differential equations. Every function (ordinary or generalized) can be regarded as an aggregate of linear differential equations which this functon satisfies. The microlocality principle extends in a natural manner this aggregate to the system of pseudodifferential equations derived from differential equations by transforming the phase space without altering its structure. By applying the microlocality principle we not only obtain a more precise description of singular points of a distribution but we also have a simpler description of the propagation process of these singularities. By this principle we can also extend to distributions the operations defined initially for smooth functions only; for example, the operation of taking trace or the operation of multiplication, etc.

Let us explain the idea of microlocalization with the following simple example. Let $n \geqslant 2$ be a natural number and let f be a function in $\mathbb{R}^n$ of the form $f(x) = g(\alpha \cdot x)$, where $\alpha \in \mathbb{R}^n \backslash 0$, $\alpha \cdot x = \sum_{j=1}^n \alpha_j \cdot x_j$ and g is a function of a single variable. If $g(t)$ has a singularity, for example, if $g(t)$ is not differentiable at $t = t_0$, then all the points x lying on the plane $\alpha \cdot x = t_0$ are singular points of f. However, f is a smooth function in each direction lying on this plane so that for it singular will be only the direction of the vector α. Radon's theorem enables us to represent each distribution f in $\mathscr{D}'(\mathbb{R}^n)$ as an integral of plane waves:

$$f(x) = \int_{|\alpha|=1} g_\alpha(\alpha \cdot x)\, d\alpha.$$

Therefore at each point x those directions α will be singular for which the distribution $g_\alpha(t)$ has a singularity at the point $t = x \cdot \alpha$. If, instead of Radon's theorem, we apply Fourier transformation, then f can be represented as an integral of plane waves:

$$f(x) = \int g(\alpha) e^{i\alpha \cdot x}\, d\alpha,$$

where the integration is performed over the whole $\mathbb{R}^n$. Now those directions of α become singular for f in which $g(t\alpha)$ does not decrease, as $t \to \infty$, rapidly enough.

As mentioned earlier, in the modern theory of differential equations the microlocality principle is extensively applied to investigate the singularities of the solution. Many important results achieved in recent years by means of this principle in the theory of boundary-value problems, in the spectral theory, in the theory of functions of several complex variables, and in other branches of mathematics point towards great potentialities of the microlocal analysis.

§2. Wave Front of Distribution. Its Functorial Properties

2.1. Definition of the Wave Front. The notion of a singular point of a distribution does not have only one meaning. Depending on the problem under discussion, a singular point may signify a point of discontinuity, or a point where the function becomes infinite, or a point where the function has an essential singularity in the sense of the theory of complex variables, etc. For the general theory of distributions, the most natural is the following

Definition 2.1. A point x_0 is a non-singular point for a distribution u if there exists a function φ in $C_0^\infty(\mathbb{R}^n)$ such that $\varphi(x_0) \neq 0$ and $\varphi u \in C^\infty(\mathbb{R}^n)$.

It follows from the definition that the singular points of a distribution constitute a closed set. This set is known as the *singular support* of the distribution u and is denoted by sing supp u. We can easily see that this set is invariant under diffeomorphisms of the space; thus the definition of a singular support can be readily extended to distributions on a smooth manifold.

For a smooth manifold X, we denote by $T^*(X)$ the cotangent bundle space and by $T^*(X)\backslash 0$ the same space with the zero section removed (see, for example, Arnol'd [1974], Egorov [1984]). The following definition and examples are due to Hörmander. It should be remarked that in the general theory there are many close concepts that are extensively used, namely, the analytic wave front, Gevrey wave front, oscillation front, etc (see Hörmander [1983, 1985], Trèves [1982], and § 4.2 of Chapter 6).

Definition 2.2. A point $(x_0, \xi_0) \in T^*(\mathbb{R}^n)\backslash 0$ does not belong to the *wave front* of a distribution u in $\mathscr{D}'(\mathbb{R}^n)$ if there are a function φ in $C_0^\infty(\mathbb{R}^n)$, with $\varphi(x_0) \neq 0$, and a cone Γ in $\mathbb{R}^n$, with vertex at the origin of coordinates, which contains in its interior the ray $\{\xi; \xi = t\xi_0, t > 0\}$ such that the relations

$$|\tilde{\varphi u}(\xi)| = 0((1 + |\xi|)^{-N})$$

holds for all $\xi \in \Gamma$ and all integers N.

The wave front of u will be denoted by WF(u).

Example 2.1. If the distribution u is a plane wave, that is, if $u(x) = g(\alpha \cdot x)$, where $\alpha \in \mathbb{R}^n\backslash 0$ and $g \in \mathscr{D}'(\mathbb{R})$, then every direction ξ_0, non-collinear with the vector α, is non-singular for u. That is, WF(u) may only contain points (x_0, ξ_0) for which $\xi_0 = t\alpha$ with $t \in \mathbb{R}\backslash 0$ and $\alpha \cdot x_0 \in$ sing supp g.

Example 2.2. Suppose that $\eta \in \mathbb{R}^n\backslash 0$, $\tilde{\varphi} \in C_0^\infty(\mathbb{R}^n)$, $\tilde{\varphi}(0) = 1$ and $\tilde{\varphi}(0) > 0$. Then for the function

$$u(x) = \sum_{k=1}^{\infty} \frac{\varphi(kx)}{k^2} e^{ik^2(x,\eta)}$$

which is continuous in $\mathbb{R}^n$, the wave front consists of the ray $\{(0, t\eta), t > 0\}$ (Hörmander [1979a]).

Summing over η, we obtain from this example a function whose wave front coincides with an arbitrary conical closed subset of $T^*(\mathbb{R}^n)\backslash 0$.

2.2. Localization of Wave Front. It is comparatively easy to establish the following properties of the wave front (see Hörmander [1971, 1983, 1985], Egorov [1984]):

1. If $u \in \mathscr{D}'(\mathbb{R}^n)$ and $\varphi \in C_0^\infty(\mathbb{R}^n)$, then

$$\mathrm{WF}(\varphi u) \subset \mathrm{WF}(u).$$

2. If $\pi\colon T^*(\mathbb{R}^n) \to \mathbb{R}^n$ is the natural projection, then $\pi \mathrm{WF}(u) = \operatorname{sing\,supp} u$ for every u in $\mathscr{D}'(\mathbb{R}^n)$.

2.3. Wave Front and Singularities of One-Dimensional Distributions. Let X and Y be smooth manifolds, and let $f\colon X \to Y$ be a smooth map. f is called *proper* if the set $f^{-1}(K)$ is compact for every compact set K in Y. For a $\varphi \in C_0^\infty(Y)$ and a proper map f we set

$$f^*\varphi(x) = \varphi(f(x)).$$

Then $f^*\varphi \in C_0^\infty(X)$ and f^* is a continuous map from $\mathscr{D}(Y)$ into $\mathscr{D}(X)$. This enables us to define, by means of duality, the *pushforward* f_*u of every *distribution* $u \in \mathscr{D}'(X)$ by the formula

$$\langle f_*u, \varphi\rangle = \langle u, f^*\varphi\rangle.$$

It is obvious that the same construction remains valid for an arbitrary map f (not necessarily proper) when u has a compact support, that is, when $u \in \mathscr{E}'(X)$.

The *pullback* f^*u of a *distribution* is defined when f is a submersion of X onto Y. This last condition means that for every point $y \in Y$ the set $f^{-1}(y)$ is a smooth submanifold of X and all these submanifolds are diffeomorphic to a fixed k-dimensional smooth manifold. Locally such a map is a projection, and in a suitable local coordinate system it becomes the projection $\mathbb{R}^k \times \mathbb{R}^l \to \mathbb{R}^k$. If $\varphi \in C_0^\infty(X)$ and the support of φ lies in a coordinate neighbourhood, and if $x = (x', x'')$, where $x' \in \mathbb{R}^k$, $x'' \in \mathbb{R}^l$, then $f_*\varphi(x') = \int \varphi(x', x'')\,dx''$. For any $u \in \mathscr{D}'(Y)$ we now set

$$\langle f^*u, \varphi\rangle = \langle u, f_*\varphi\rangle.$$

If f is both proper and submersion (for example, if f is a diffeomorphism), then f^*u and f_*v are simultaneously defined for every $u \in \mathscr{D}'(Y)$, $v \in \mathscr{D}'(X)$.

Example 2.3. Let $\pi\colon \mathbb{R}^n \to \mathbb{R}^1$ be the projection onto the x_1-axis. In this case both π^*u, $u \in \mathscr{D}'(\mathbb{R})$, and π_*v, $v \in \mathscr{E}'(\mathbb{R}^n)$, are defined, and

$$\pi^*v(x^1) = u(x^1) \otimes 1_{x_2,\ldots,x_n};$$

$$\pi_*v(x) = \int v(x_1, x_2, \ldots, x_n)\,dx_2 \ldots dx_n.$$

Using these concepts, we can establish the following

Theorem 2.1. *Let Ω be a domain in $\mathbb{R}^n$, and let $u \in \mathscr{D}'(\Omega)$. A point (x_0, ξ_0) of $T^*(\Omega)\backslash 0$ does not lie in* $\mathrm{WF}(u)$ *if and only if there exist a function φ in $C_0^\infty(\Omega)$, with $\varphi(x_0) \neq 0$, and an $\varepsilon > 0$ such that for every smooth function $f\colon \operatorname{supp}\varphi \to \mathbb{R}$, with $|\operatorname{grad} f(x_0) - \xi_0| < \varepsilon$, the function $f_*(\varphi u)(t)$ is infinitely differentiable on the real line* (Guillemin and Sternberg [1977], Egorov [1985a]).

2.4. Wave Fronts of Pushforwards and Pullbacks of a Distribution

Theorem 2.2. *Let $f\colon X \to Y$ be a submersion, and let $u \in \mathscr{E}'(X)$. Then*

$$\mathrm{WF}(f_* u) \subset \{(f(x), \eta)\colon x \in X, (x, {}^t f_x' \eta) \in \mathrm{WF}(u) \text{ or } {}^t f_x' \eta = 0\},$$

where ${}^t f_x'$ denotes the transpose matrix of the Jacobi matrix f_x' of f (Hörmander [1971]).

Example 2.4. Let $\pi\colon \mathbb{R}^2 \to \mathbb{R}^1$ be the projection onto the x_1-axis, that is, $\pi(x_1, x_2) = x_1$. The Jacobi matrix is of the form $\pi_x' = (1, 0)$ and

$$\mathrm{WF}\left(\int u(x_1, x_2)\, dx_2\right) \subset \{(x_1, \eta);\ \exists x_2, (x_1, x_2, \eta, 0) \in \mathrm{WF}(u)\}.$$

Thus $\mathrm{WF}(\pi_* u)$ is contained in the set of projections of those points of the wave front of the distribution u where singular directions are parallel to the x_1-axis.

The pullback f^*u of a distribution was defined above for submersions f only. Using continuity, we can attempt to define it for the more general case by setting $f^*u = \lim f^*u_j$, where $\{u_j\}$ is a sequence of smooth functions converging to u in $\mathscr{D}'(Y)$ and $f^*u_j(x) = u_j(f(x))$. As Hörmander [1971] has shown, such a path leads to the following

Theorem 2.3. *Let $f\colon X \to Y$ be a smooth map. Set*

$$N_f = \{(y, \eta) \in T^*Y \backslash 0;\ \exists x \in X, y = f(x), {}^t f_x' \eta = 0\}.$$

If $u \in \mathscr{D}'(Y)$ and $\mathrm{WF}(u) \cap \mathrm{N}_f = \varnothing$, then the distribution f^*u can be defined by continuity and the inclusion

$$\mathrm{WF}(f^*u) \subset \{(x, \xi);\ \exists (y, \eta) \in \mathrm{WF}(u), y = f(x), \xi = {}^t f'(x)\eta\}$$

holds. Moreover, the pullback f^*u is unique.

Example 2.5. Let M be a k-dimensional smooth submanifold in $\mathbb{R}^n$ and ρ a smooth function on M. Set $u(x) = \rho \otimes \delta(M)$. Then

$$\mathrm{WF}(u) \subset N(M) \backslash 0,$$

where $N(M)$ is the conormal bundle of M, $N(M) \subset T^*(\mathbb{R}^n)$. If $\rho(x_0) \neq 0$, then $(x_0, \xi) \in \mathrm{WF}(u)$ if and only if the vector ξ is directed along the normal to M. This result follows from Theorem 2.3 if one considers the submersion $\mathbb{R}^n \to M$.

§ 3. Wave Front and Operations on Distributions

3.1. The Trace of a Distribution. Product of Distributions. Let us demonstrate how to define, using theorems of the preceding sections, some of the operations for distributions that have been initially defined for smooth functions only.

Theorem 3.1. *Let X be a smooth manifold and Y a smooth submanifold of X. Let $N(Y)$ be the normal bundle of Y, the normals being tangent to X. Suppose that*

$u \in \mathscr{D}'(Y)$ and $\mathrm{WF}(u) \cap N(Y) = \varnothing$. Then the restriction $u|_y$ can be defined as i^*u, where $i\colon Y \to X$ is the embedding. Furthermore, if $(y, \eta) \in \mathrm{WF}(u|_y)$, then there exists a vector $\xi \in N(Y)$ such that $(y, \eta + \xi) \in \mathrm{WF}(u)$ (Hörmander [1983, 1985]).

This result is an immediate consequence of Theorem 2.3, since $N_i = N(Y)$.

Theorem 3.2 (Hörmander [1983, 1985]). *Let X be a smooth manifold. Suppose that u_1 and $u_2 \in \mathscr{D}'(X)$, with $\mathrm{WF}(u_1) + \mathrm{WF}(u_2) \subset T^*X\backslash 0$. Then the distribution $u_1 u_2 \in \mathscr{D}'(X)$ can be defined as $\Delta^*(u_1 \otimes u_2)$, where $\Delta\colon X \to X \times X$ is the diagonal map. Furthermore,*

$$\mathrm{WF}(u_1, u_2)$$
$$\subset \{(x, \xi + \eta); (x, \xi) \in \mathrm{WF}(u_1), \textit{ or } \xi = 0, (x, \eta) \in \mathrm{WF}(u_2) \textit{ or } \eta = 0; \xi + \eta \neq 0\}$$

Here also Theorem 2.3 is applicable, since $N_\Delta = \{(x, x, \xi, -\xi)\}$.

Example 3.1. If Y and Z are transversally intersecting submanifolds of X and if u and v are smooth densities on Y and Z, respectively, then uv is a smooth density on $Y \cap Z$.

3.2. The Wave Front of the Solution of a Differential Equation

Theorem 3.3. *Let $u \in \mathscr{D}'(\mathbb{R}^n)$, and let $P(x, D_x)$ be a differential operator with smooth coefficients. Let $P(x, D_x)u = f \in C^\infty(\mathbb{R}^n)$. Then*

$$\mathrm{WF}(u) \subset \mathrm{Char}\, P = \{(x, \xi) \in T^*(\mathbb{R}^n)\backslash 0, P_0(x, \xi) = 0\},$$

where P_0 is the characteristic form of P.

In the general case, we have $\mathrm{WF}(u) \subset \mathrm{WF}(Pu) \cup \mathrm{Char}\, P$. In particular, if P is an *elliptic operator*, that is, if $\mathrm{Char}\, P = \varnothing$, then $\mathrm{WF}(Pu) = \mathrm{WF}(u)$ (Hörmander [1983, 1985]).

We recall that P is said to be *hypoelliptic* if sing supp $Pu \supset$ sing supp u.

If $\mathrm{WF}(Pu) \supset \mathrm{WF}(u)$ for each $u \in \mathscr{E}'(X)$, P is called a *microlocally hypoelliptic* operator. The question of singularities of the solution of a differential equation will be examined later in Chapter 4. For the present, we note one of the consequences of Theorems 3.1 and 3.3.

Theorem 3.4 (Hörmander [1983, 1985]). *Suppose that $u \in \mathscr{D}'(\mathbb{R}^{n+1})$, $P(t, x, D_t, D_x)u = f \in C^\infty(\mathbb{R}^{n+1})$, $t \in \mathbb{R}$, $x \in \mathbb{R}^n$. If the plane $t = 0$ is non-characteristic, that is, if $P(0, x; 1, 0) \neq 0$, then the traces $D_t^k u(0, x) \in \mathscr{D}'(\mathbb{R}^n)$ are defined for all $k \geqslant 0$. Moreover,*

$$u \in C^\infty(\mathbb{R}; \mathscr{D}'(\mathbb{R}^n)).$$

This result follows from Theorem 3.1, since the points $(0, x, \tau, 0)$ do not lie in $\mathrm{WF}(u)$ and hence $\mathrm{WF}(u) \cap N(M_0) = \varnothing$, where M_0 is the plane $t = 0$.

3.3. Wave Fronts and Integral Operators. Let Ω_1 and Ω_2 be domains in $\mathbb{R}^n$ and $\mathbb{R}^m$, respectively. Suppose that A is a continuous linear operator from $\mathscr{D}(\Omega_2)$ into $\mathscr{D}'(\Omega_1)$ with the Schwartz kernel K belonging to $\mathscr{D}'(\Omega_1 \times \Omega_2)$. Consider the sets

$$M_1 = \{(x, \xi); \exists y \in \Omega_2, (x, y, \xi, 0) \in \mathrm{WF}(K)\},$$
$$M_2 = \{(y, \eta); \exists x \in \Omega_1, (x, y, 0, -\eta) \in \mathrm{WF}(K)\}.$$

Theorem 3.5. *Suppose that* $K \in \mathscr{D}'(\Omega_1 \times \Omega_2)$, $u \in \mathscr{E}'(\Omega_2)$, *with* $M_1 = \varnothing$, $M_2 \cap \mathrm{WF}(u) = \varnothing$. *Then the distribution*

$$Au(x) = \int K(x, y)u(y)\,dy \in \mathscr{D}'(\Omega_1)$$

is defined and

$$\mathrm{WF}(Au) \subset \{(x, \xi); \exists (y, \eta), (x, y, \xi, -\eta) \in \mathrm{WF}(K), (y, \eta) \in \mathrm{WF}(u)\}.$$

We note that since $M_2 \cap \mathrm{WF}(u) = \varnothing$, the product $K(x, y)u(y)$ is defined, in view of Theorem 3.2. The condition $M_1 = \varnothing$ signifies that $A\colon C_0^\infty(\Omega_2) \to C^\infty(\Omega_1)$, that is, under this mapping smooth functions go over to smooth functions.

Chapter 2
Pseudodifferential Operators

§ 1. Algebra of Pseudodifferential Operators

1.1. Singular Integral Operators. In many questions of the theory of partial differential equations there arises the necessity of using *singular integral operators*. These are operators of the form

$$Au(x) = \int K(x, x - y)u(y)\,dy, \tag{1}$$

where $K(x, z)$ is a function having singularity at $z = 0$ only and $K(x, tz) = t^{-n}K(x, z)$ for $t > 0$. Assuming that

$$\int_{|z|=1} K(x, z)\,dz = 0,$$

we can define A correctly in the sense of the principal value, that is,

$$Au(x) = \lim_{\substack{\varepsilon \to 0 \\ R \to \infty}} \int_{\varepsilon < |x-y| < R} K(x, x - y)u(y)\,dy.$$

It was shown in the works of Giraud, Korn, and Lichtenstein that A is a bounded operator in spaces C^α, $0 < \alpha < 1$, while its boundedness in L_p spaces, $p > 1$, was established in the works of Calderon and Zygmund.

These theorems prove to be crucial in studying smoothness of solutions of elliptic equations. The fundamental solution $F(x, x - y)$ of an elliptic differential operator $L(x, D)$ of order m has a singularity of order $|x - y|^{m-n}$ at $x = y$. Therefore the mth order derivatives of the solution of the equation $Lu = f$ can be expressed in terms of f by means of singular integral operators. Thus $D^\alpha u \in L_p^{\mathrm{loc}}(\Omega)$ for $|\alpha| \leqslant m$ if $Lu \in L_p(\Omega)$, $p > 1$, and $u \in C^{m+\alpha}(\Omega)$ if $Lu \in C^\alpha(\Omega)$, $0 < \alpha < 1$.

The operators of the form $\sum (a_\alpha(x) + A_\alpha) D^\alpha$,where A_α are singular integral operators, are known as *singular integro-differential operators.* It is easy to see that such operators constitute an algebra. In this algebra elliptic differential operators are invertible, up to compact operators.

1.2. The Symbol. In his works S.G. Mikhlin had noted that with every operator A of the form (1) one can associate a function $a(x, \xi)$ defined in the cotangent space. This function, called a *symbol,* plays a leading role in the calculus of singular integro-differential operators. Different variants of this calculus have been developed in the works of Dynin, Agranovich, Vishik, Eskin, Kohn, Hörmander, Nirenberg, Seeley, Unterberger, and Bokobza. At present, the most popular theory of pseudodifferential operators is the one developed first by Kohn and Nirenberg [1965].

By a *pseudodifferential* operator we mean an operator of the form

$$Au(x) = (2\pi)^{-n} \int a(x, \xi) \tilde{u}(\xi) e^{i(x, \xi)}\, d\xi. \tag{2}$$

Here $u \in C_0^\infty(\mathbb{R}^n)$ and $\tilde{u}(\xi) = \int u(y) e^{-i(y, \xi)}\, dy$ is the Fourier transform of u. The function a is called the *symbol* of A. It is assumed that

$$|D_\xi^\alpha D_x^\beta a(x, \xi)| \leqslant C_{\alpha,\beta}(1 + |\xi|)^{m-|\alpha|} \tag{3}$$

for all $\alpha, \beta \in \mathbb{Z}_+^n$. The constants $C_{\alpha,\beta}$ may depend on the compact set K to which the point x belongs. When a is a polynomial in ξ, A turns out to be a differential operator. In particular, $A = I$ if $a = 1$. We denote by S^m the class of symbols satisfying (3).

The number m is called the *order* of the operator. When a satisfies (3) for any $m \in \mathbb{R}$,the order of A is taken to be $-\infty$; $S^{-\infty}$ denotes the set of such symbols. A pseudodifferential operator A with its symbol a in S^m is called *classical* if there is a sequence of symbols $a_j \in S^{m-j}$, $j = 0, 1, 2, \ldots$, such that $a_j(x, t\xi) = t^{m-j} a_j(x, \xi)$ for $t > 1$, $|\xi| > 1$, and

$$\left| D_\xi^\alpha D_x^\beta \left(a(x, \xi) - \sum_{j=0}^{N} a_j(x, \xi) \right) \right| \leqslant C_{\alpha,\beta,N} |\xi|^{m-N-1-|\alpha|}$$

for all α, β, $|\xi| > 1$ and all integers $N \geqslant 0$. In this case the notation $a \sim \sum a_j$ is used. The function a_0 is known as the *principal symbol* of A, and the class of such symbols is denoted by S_{cl}^m.

1.3. Boundedness of Pseudodifferential Operators. We mention a few basic theorems regarding boundedness of the operators under discussion.

1. The operator (2) maps $C_0^\infty(\mathbb{R}^n)$ into $C^\infty(\mathbb{R}^n)$ and can be extended to an operator $A\colon \mathscr{E}'(\mathbb{R}^n) \to \mathscr{D}'(\mathbb{R}^n)$.

2. If $m < -n$, then $Au(x) = \int K(x, x - y)u(y)\,dy$, where $K(x, z) = (2\pi)^{-n}\int a(x, \xi)e^{iz\cdot\xi}\,d\xi \in C^{-m-n}(\mathbb{R}^{2n})$. In particular, $K \in C^\infty$ if A is of order $-\infty$. In this case, $Au(x) = \langle u_y, K(x, x - y)\rangle \in C^\infty(\mathbb{R}^n)$ if $u \in \mathscr{E}'(\mathbb{R}^n)$. Such operators are referred to as *smoothing* operators.

If $m \geqslant -n$, the kernel $K(x, z)$ can be determined by the formula

$$|z|^{2N}K(x, z) = (2\pi)^{-n}\int [(-\Delta_\xi)^N a(x, \xi)]e^{iz\cdot\xi}\,d\xi$$

with $2N > m + n$, which is obtained by integration by parts. This formula shows that $K \in C^\infty$ when $z \neq 0$ and $|K(x, z)| \leqslant C|z|^{-2N}$, that is, the kernel can have only power singularity.

The operator A with the Schwartz kernel $K(x, x - y)$ is called *properly supported* if $K(x, z) = 0$ for $|z| > \rho$, with $\rho > 0$, and $K \in C^\infty(\mathbb{R}^{2n})$. Such an operator maps functions in $C_0^\infty(\mathbb{R}^n)$ into functions in $C_0^\infty(\mathbb{R}^n)$ and distributions in $\mathscr{E}'(\mathbb{R}^n)$ into distributions in $\mathscr{E}'(\mathbb{R}^n)$.

Let a given function $h(z)$ in $C^\infty(\mathbb{R}^n)$ be such that $h(z) = 0$ for $|z| < \rho$ and $h(z) = 1$ for $|z| > 2\rho$, $\rho > 0$. Let A be a pseudodifferential operator with the Schwartz kernel $K(x, x - y)$. We set $K_1(x, y) = h(x - y)$. $K(x, x - y)$ and $K_2(x, y) = [1 - h(x - y)]K(x, x - y)$ and denote by A_1 and A_2 operators with the kernels K_1 and K_2, respectively. Since $K_1 \in C^\infty(\mathbb{R}^{2n})$, it follows that A_1 is a smoothing operator. And A_2 is a properly supported operator. We note that the symbol of a pseudodifferential operator can be obtained from its kernel by the formula

$$a(x, \xi) = (2\pi)^n\int K(x, x - y)e^{i(y-x)\cdot\xi}\,dy.$$

Thus every pseudodifferential operator is the sum of a properly supported operator and a smoothing operator. This situation enables us to define the algebra of pseudodifferential operators up to smoothing operators.

3. Let A be a pseudodifferential operator, and suppose that its symbol vanishes for large values of $|x|$. Then A defines a continuous map

$$A\colon H_s \to H_{s-m}$$

for each real s.

1.4. Composition of Pseudodifferential Operators. If A and B are two pseudodifferential operators of order m and m', respectively, of which B is properly supported, then the composition $C = A \circ B$ is defined. It is a pseudodifferential operator of order $m + m'$ with the symbol

$$C(x, \xi) \sim \sum \frac{i^{|\alpha|}}{\alpha!} D_\xi^\alpha a(x, \xi) D_x^\alpha b(x, \xi). \tag{4}$$

This formula is a generalization of the classical Leibniz formula.

1.5. The Formally Adjoint Operator. The operator A^* is *formally adjoint* to the operator A if

$$\int Au(x)\overline{v(x)}\,dx = \int u(x)\overline{A^*v(x)}\,dx;\ u, v \in C_0^\infty(\mathbb{R}^n).$$

If A is a pseudodifferential operator of order m, then A^* is also a pseudodifferential operator of the same order with the symbol

$$a^*(x, \xi) \sim \sum_\alpha \frac{i^{|\alpha|}}{\alpha!} D_\xi^\alpha D_x^\alpha \overline{a(x, \xi)}. \tag{5}$$

We note that pseudodifferential operators can be defined in the spaces of vector-functions. In this case, their symbols are matrices, and the right side of (5) contains the matrix ${}^t\bar{a}$, the conjugate of a. Other formulae remain valid without any change.

1.6. Pseudolocality. Microlocality. It is known (see Hörmander [1963]) that differential operators are the only operators that are *local* linear operators acting from $C_0^\infty(\mathbb{R}^n)$ into $\mathscr{D}'(\mathbb{R}^n)$, that is, are such that $\operatorname{supp} Pu \subset \operatorname{supp} u$ for all $u \in C_0^\infty(\mathbb{R}^n)$. Pseudodifferential operators possess a weaker property of *pseudolocality*, namely,

$$\operatorname{sing\,supp} Au \subset \operatorname{sing\,supp} u,$$

as well as the property of *microlocality*, namely,

$$\mathrm{WF}(Au) \subset \mathrm{WF}(u).$$

The wave front of a distribution u can be defined by means of pseudodifferential operators as follows: for a distribution u with a compact support in $\mathbb{R}^n$ a point (x_0, ξ_0) does not lie in $\mathrm{WF}(u)$ if and only if there exists a pseudodifferential operator A with the principal symbol $a(x, \xi)$ such that $a(x_0, \xi_0) \neq 0$ and $Au \in C^\infty(\mathbb{R}^n)$.

1.7. Elliptic Operators. An operator A of order m is called *elliptic* if $|a(x, \xi)| \geqslant C_0|\xi|^m - C_1$, where $C_0 = \text{const.} > 0$. Elliptic operators are invertible in the algebra considered up to smoothing operators. More precisely, for each elliptic operator A of order m and for each compact set K there exists an elliptic operator Q of order $-m$ such that $QA = I + T_1$, $AQ = I + T_2$, where T_1 and T_2 are operators of order $-\infty$ on K. This immediately implies that elliptic operators are hypoelliptic, that is, $u \in C^\infty$ on each open set where $Au \in C^\infty$.

An operator A of order m is called elliptic at a point (x_0, ξ_0) of $T^*(\mathbb{R}^n)\backslash 0$ if $|a(x, \xi)| \geqslant C_0|\xi|^m - C_1$ for (x, ξ) lying in a conical neighbourhood of (x_0, ξ_0). By a *conical neighbourhood* we mean, as usual, the set

$$\left\{(x, \xi);\ |x - x_0| < \delta_1, \left|\frac{\xi}{|\xi|} - \frac{\xi_0}{|\xi_0|}\right| < \delta_2\right\}.$$

For an operator A, which is elliptic at a given point, we may construct a microlocal *parametrix*. It is an operator Q of order $-m$ such that

$$Q\varphi A = \varphi I + T_1, \qquad A\varphi Q = \varphi I + T_2,$$

where $\varphi(x, D)$ is a pseudodifferential operator of order zero the symbol of which has a support in a sufficiently small conical neighbourhood of the point (x_0, ξ_0) and which equals 1 in a smaller conical neighbourhood Γ of this point; as to symbols of T_1 and T_2, they vanish in Γ.

1.8. Gårding's Inequality. If A is a classical pseudodifferential operator of order m with the principal symbol $a_0(x, \xi)$ satisfying the condition $\operatorname{Re} a_0(x, \xi) \geqslant C_0|\xi|^m$, $C_0 = \text{const.}$, $|\xi| \geqslant 1$, then

$$\operatorname{Re}(Au, u) \geqslant C_0 \|u\|^2_{m/2} - C_1 \|u\|^2_{(m-1)/2}$$

for all u in $C_0^\infty(K)$, $C_1 = C_1(K)$ (Hörmander [1983, 1985]).

This result can be employed for obtaining a sharper estimate for the norm of the operator in terms of its symbol. Let $B(x, D)$ be a pseudodifferential operator of order m with the principal symbol $b_0(x, \xi)$ satisfying the condition $|b_0(x, \xi)| \leqslant M$ for $|\xi| = 1$. Then for any real s we can find a constant $C = C(s)$ such that

$$\|Bu\|_{s-m} \leqslant M\|u\|_s + C\|u\|_{s-(1/2)}, \qquad u \in C_0^\infty.$$

1.9. Extension of the Class of Pseudodifferential Operators. The operators of the form

$$Au(x) = (2\pi)^{-n} \iint a(x, y, \xi) u(y) e^{i(x-y)\xi}\, dy\, d\xi, \tag{6}$$

where the symbol a satisfies the inequalities

$$|D_\xi^\alpha D_x^\beta D_y^\gamma a(x,y,\xi)| \leqslant C_{\alpha,\beta,\gamma}(1 + |\xi|)^{m-|\alpha|},$$

are also pseudodifferential operators. The symbol b of such an operator has the following asymptotic representation:

$$b(x, \xi) \sim \sum_\alpha \frac{i^{|\alpha|}}{\alpha!} D_y^\alpha D_\xi^\alpha a(x, y, \xi)|_{y=x}.$$

Thus the classes of operators defined by (2) and (6) coincide. The *Weyl operators*

$$Wu(x) = (2\pi)^{-n} \iint a\left(\frac{x+y}{2}, \xi\right) u(y) e^{i(x-y)\xi}\, dy\, d\xi, \tag{7}$$

can also be represented in form (6), where $a \in S^m$.

§ 2. Invariance of the Principal Symbol Under Canonical Transformations

2.1. Invariance Under the Change of Variables. Let us first examine the question of invariance of a pseudodifferential operator under the change of variables. We express a given pseudodifferential operator A in the form (6) and then replace

x by $f(x')$ and y by $f(y')$. We immediately note that

$$x - y = f(x') - f(y') = F(x', y')(x' - y'),$$

where $F(x', y') = \int_0^1 \frac{\partial f(y' + t(x' - y'))}{\partial x} dt$. If ξ is replaced by ${}^tF(x', y')^{-1}$, then

$$\begin{aligned} Au(x) &= (2\pi)^{-n} \iint a(x, \xi) u(y) e^{i(x-y)\xi}\, dy\, d\xi \\ &= (2\pi)^{-n} \iint a(f(x'), {}^tF(x', y')^{-1}\eta) v(y') e^{i(x'-y')\eta} \frac{|f'(y')|\, dy'\, d\eta}{|{}^tF(x', y')|}, \end{aligned}$$

where $v(y') = u(y)$. The resulting operator is of the form (1.6), and is therefore a pseudodifferential operator. Its principal symbol is computed in accordance with § 1.9 and equals $a_0(f(x'), {}^tf'(x')^{-1})$. This implies that the principal symbol of a pseudodifferential operator is invariant on the cotangent space.

Thus under a change of variables the class of pseudodifferential operators is preserved. This enables us to define a pseudodifferential operator on a smooth manifold.

Definition 2.1. Let X be a smooth manifold. A continuous linear map A: $C_0^\infty(X) \to C^\infty(X)$ is called a pseudodifferential operator on X if for each coordinate neighbourhood $U \subset X$, to which there corresponds a coordinate map $\varphi: U \to \omega$, where ω is a domain in $\mathbb{R}^n$, the map $C_0^\infty(\omega) \ni u \to (\varphi^{-1})^* A(\varphi^* u) \in C^\infty(\omega)$ is an operator of the form (2).

Definition 2.2. Let X be a smooth manifold and let E and F be smooth complex bundles on X. A continuous linear map A: $C_0^\infty(X, E) \to C^\infty(X, F)$ is called a pseudodifferential operator if for every open subset $V \subset X$, with the property that $E|_V$ and $F|_V$ are diffeomorphic, respectively, to $V \times \mathbb{C}^{m_1}$ and $V \times \mathbb{C}^{m_2}$, there exists an $m_2 \times m_1$ matrix of pseudodifferential operators A_{ij} on V such that

$$(Au|_V)_i = \sum_{j=1}^{m_1} A_{ij}(u|_V)_j, \qquad u \in C_0^\infty(V, E).$$

2.2. The Subprincipal Symbol. The invariance of the principal symbol under the change of variables enables us to define various classes of pseudodifferential operators on a manifold that are characterized by the principal symbol, namely, elliptic operators, operators of the principal type, etc. In investigating operators with multiple characteristics a significant role is played by the *subprincipal symbol*

$$a_{\text{sub}}(x, \xi) = a_1(x, \xi) - \frac{1}{2i} \sum_{j=1}^n \frac{\partial^2 a_0(x, \xi)}{\partial x_j \partial \xi_j},$$

that is defined in an invariant manner on

$$\Sigma_2 = \{(x, \xi);\, a_0(x, \xi) = 0,\, d_{x,\xi} a_0(x, \xi) = 0\},$$

the set of double characteristics of the operator A.

2.3. Canonical Transformations. We can examine a wider class of transformations under which pseudodifferential operators are mapped into pseudodifferential operators.

Let us set

$$\Phi u(x) = (2\pi)^{-n} \int b(x, \xi)\tilde{u}(\xi)e^{iS(x,\xi)}\, d\xi, \tag{8}$$

where S is a real-valued smooth function on $T^*\mathbb{R}^n$ such that $S(x, t\xi) = tS(x, \xi)$ for $t \geqslant 1$ and $|\xi| \geqslant 1$, and $\det \|\partial^2 S/\partial x_i \partial\xi_j\| \neq 0$ for $|\xi| \geqslant 1$. As to the symbol b, we assume that $b \in S^0$ and that $|b(x, \xi)| \geqslant C_0 > 0$ for $|\xi| \geqslant 1$.

Theorem 2.1 (Egorov [1969, 1984]). *To every pseudodifferential operator A of order m there corresponds a pseudodifferential operator A', of the same order, such that $\Phi A - A'\Phi$ is a smoothing operator. If $a_0(x, \xi)$ is the principal symbol of A, then the principal symbol of A' is defined by the formula*

$$a_0'\left(x, \frac{\partial S(x, \xi)}{\partial x}\right) = a_0\left(\frac{\partial S(x, \xi)}{\partial \xi}, \xi\right).$$

This result implies the validity of the assertion of § 2.1 if one sets $S(x, \xi) = f(x)\cdot\xi$.

The transformation $(x, \xi) \mapsto (y, \eta)$ under which

$$\xi = \frac{\partial S(x, \eta)}{\partial x}, \qquad y = \frac{\partial S(x, \eta)}{\partial \eta}$$

is called a *canonical transformation*, and the function S in this case is referred to as its generating function.

The canonical transformations play a significant role in classical mechanics (see Arnol'd [1974]). In general, a canonical transformation is a transformation of $\mathbb{R}^{2n}$, the space of the points $(x, \xi) = (x_1, \ldots, x_n, \xi_1, \ldots, \xi_n)$, that preserves the Poisson brackets

$$\{f, g\}(x, \xi) = \sum_{j=1}^{n} \left(\frac{\partial f(x, \xi)}{\partial \xi_j}\frac{\partial g(x, \xi)}{\partial x_j} - \frac{\partial f(x, \xi)}{\partial x_j}\frac{\partial g(x, \xi)}{\partial \xi_j}\right)$$

for every pair of functions f and g in $C^\infty(\mathbb{R}^{2n})$. Since the variables x and ξ do not enjoy equal status in the theory of differential equations, we confine our attention only to homogeneous canonical transformations. That is, we consider only those transformations whose generating functions satisfy the homogeneity condition, namely, $S(x, t\xi) = tS(x, \xi)$ for $t \geqslant 1$, $|\xi| \geqslant 1$. However, in the general theory of differential equation arbitrary canonical transformations are also used (see Egorov [1984], Maslov [1976]).

2.4. An Inverse Theorem. It can be shown that the above-mentioned class of homogeneous canonical transformations turns out to be the maximal admissible class that preserves the algebra of the classical pseudodifferential operators. This becomes immediately clear from the fact that the principal symbol of the operator $[A, B] = AB - BA$ is proportional to the Poisson bracket of the principal symbols of the operators A and B. Therefore to every such transformation there corresponds a homogeneous canonical transformation of $T^*\mathbb{R}^n$, and, hence, a homogeneous generating function S and an operator Φ of the form (8).

In fact, a far more strong result is valid (see Beals [1977]; Singer was the first to obtain results of this kind). Let X and Y be smooth manifolds with $H_1(T^*X\backslash 0, C) = 0$. Let $L^m(X)$ be the space of classical pseudodifferential operators of order m whose symbols $a(x, \xi)$ have the asymptotic representation $a \sim \sum a_j$, where $a_j(x, t\xi) = t^{m-j}a_j(x, \xi)$ for $t > 1$, $|\xi| > 1$; thus $L^\infty(X) = \bigcup L^m(X)$. Let i be an algebraic isomorphism $L^\infty(X) \to L^\infty(Y)$ such that i: $L^m(X) \to L^m(Y)$ for all $m \in \mathbb{Z}$. Then there exists an invertible Fourier integral operator A: $C^\infty(Y) \to C^\infty(X)$ such that $A \cdot i(P) = P \cdot A$. The phase function of this operator defines microlocally a canonical transformation that enables us to compute the principal symbol of the operator $i(P)$.

§ 3. Canonical Forms of the Symbol

3.1. Simple Characteristic Points. The canonical transformations make the investigation of differential and pseudodifferential equations considerably simple. Let us remark that such an approach is impossible within the framework of the theory of "pure" differential equations. One obtains the class of pseudodifferential operators in a natural manner by applying canonical transformations to differential operators.

As we mentioned in § 1.7, the microlocal study of the equation $A(x, D)u = f(x)$ at a non-characteristic point (x^0, ξ^0) can be carried out easily after constructing a parametrix in a conical neighbourhood of this point.

In the order of complexity, next are the simple characteristic points. Suppose that the principal symbol a_0 of A is a real-valued function such that $a_0(x^0, \xi^0) = 0$ and $d_\xi a_0(x^0, \xi^0) \neq 0$, $|\xi^0| > 1$. Then in a conical neighbourhood of (x^0, ξ^0) the symbol of A can be reduced to the form $q(x, \xi)\xi_1$, where q is the symbol of an elliptic operator of order $m - 1$. By inverting the operator Q, with symbol q, we can reduce the investigation of the equation $Au = f$ to solving an equation of the form $D_1 v = g$.

To construct the desired canonical transformation, we present a device for finding its generating function. Since the equation $Au = f$ is equivalent to an equation of the form $\Lambda^{1-m}Au = \Lambda^{1-m}f$, where Λ is an operator with symbol $(1 + |\xi|^2)^{1/2}$, we can take, without the loss of generality, the order m of A to be 1. With the aid of the rotation $x' = Tx$, we can achieve that $\partial a_0(x^0, \xi^0)/\partial \xi_1 \neq 0$. The desired canonical transformation Φ: $(x, \xi) \to (y, \eta)$ must be such that $a_0(x, \xi) = \eta_1$. This results in the differential equation

$$a_0\left(x, \frac{\partial S(x, \eta)}{\partial x}\right) = \eta_1.$$

We supplement this equation by the initial conditions

$$S = \sum_{j=2}^{n} x_j \eta_j \qquad \text{for} \quad x_1 = x_1^0.$$

By construction, the plane $x_1 = x_1^0$ is non-characteristic, and hence this Cauchy problem has a unique solution in a certain neighbourhood of (x^0, η^0), where the vector η^0 is found from the condition that $\partial S(x^0, \eta^0)/\partial x = \xi^0$. Thus $\eta_j^0 = \xi_j^0$ for $j = 2, \ldots, n$, and $\eta_1^0 = 0$. It is easy to see that $S(x, t\eta) = tS(x, \eta)$ for $t > 0$ and that $\det \|\partial^2 S(x^0, \eta^0)/\partial x_i \partial \eta_j\| \neq 0$.

In a similar fashion, the canonical transformation, resulting in $a_0(x, \xi) = \eta_1$, can be found also when $d_\xi a_0(x^0, \xi^0) = 0$ but $d_x a_0(x^0, \xi^0) \neq 0$ and the form $d_x a_0(x^0, \xi^0)$ is non-collinear to the form $\xi^0\, dx$. In this case, we seek the generating function $S = S(y, \xi)$. To be definite, we assume that $d_x a_0(x^0, \xi^0) = dx_1$ and $\xi_1^0 = 0$, $|\xi^0| \geqslant 1$. S must satisfy the equation

$$a_0\left(\frac{\partial S(y, \xi)}{\partial \xi}, \xi\right) = \frac{\partial S}{\partial y_1},$$

and the initial conditions

$$S = \sum_{j=2}^{n} y_j \xi_j + \frac{\xi_1^2}{2|\xi|} \qquad \text{for} \quad y_1 = 0.$$

Such a problem has a unique solution in a neighbourhood of (y^0, ξ^0). Here the vector y^0 is determined from the condition that $x^0 = \partial S(y^0, \xi^0)/\partial \xi$; thus $x_j^0 = y_j^0$ for $j = 2, \ldots, n$, and $y_1^0 = 0$. It is easy to see that $S(y, t\xi) = tS(y, \xi)$ for $t > 0$ and that $\det \|\partial^2 S(y^0, \xi^0)/\partial y_i \partial \xi_j\| \neq 0$.

Having constructed the generating function $S(x, \eta)$ or $S(y, \xi)$ in a conical neighbourhood of the point in question, we can extend it to a neighbourhood of the form $\omega \times \mathbb{R}^n$ so that for $|\xi| \geqslant 1$ S becomes a non-degenerate homogeneous function. By applying an operator Φ of the form

$$\Phi v(x) = (2\pi)^{-n} \int \tilde{v}(\xi) e^{iS(x, \xi)}\, d\xi,$$

we can reduce the first-order equation $Au = f$ to the form $Bv = g$; moreover, the principal symbol $b_0(y, \eta)$ coincides with η_1 in the conical neighbourhood of (y^0, η^0). We can then find an elliptic operator Q, of order zero, such that the operator RBQ has the full symbol η_1 in the conical neighbourhood of (y^0, η^0); here R is a parametrix for Q.

In conclusion, we mention

Definition 3.1. An operator $A(x, D)$ is said to be an operator of the *principal type* if the form $d_{x,\xi} a_0(x, \xi)$ is non-proportional to the form $\xi^0\, dx$ at each characteristic point $(x^0, \xi^0) \in T^*\Omega$.

It is clear from the above results that the investigation of a principal-type equation $Au = f$ with a real-valued principal symbol leads to the solution of elliptic equatons and of equations of the form $D_1 v = g$. We mention

Proposition 3.1. *If $A(x, D)$ and $B(x, D)$ are pseudodifferential operators of the principal type with the same principal symbol, $a_0 = b_0$, then there exist elliptic pseudodifferential operators $R(x, D)$ and $S(x, D)$ of order zero such that $AR - SB$ is a smoothing operator* (see Egorov [1984]).

3.2. Double Characteristics. If $a_0(x, \xi) = p^2(x, \xi)$, where p is a first-order real-valued symbol of the principal type, then the operator A with the principal symbol a_0 is microlocally equivalent to every operator of the form $D_1^2 + C(x, D)$. Here $C(x, D)$ is a first-order operator whose principal symbol coincides with the subprincipal symbol of A when $\xi_1 = 0$.

In the general case, when the principal symbol $a_0(x, \xi)$ assume real values, with $a_0(x, \xi) = 0$, $|\xi^0| \geqslant 1$, and $da_0(x^0, \xi^0) = 0$ but $d^2a_0(x^0, \xi^0) \neq 0$, the question of reducing the symbol to the canonical form in a neighbourhood of (x^0, ξ^0) is resolved under the assumption that the resonances are absent (see Sternin [1984]).

3.3. The Complex-Valued Symbol. When the principal symbol a_0 of the operator $A(x, D)$ is a complex-valued function, there arises the question of simultaneous reduction of the pair of functions $\alpha = \operatorname{Re} a_0$ and $\beta = \operatorname{Im} a_0$ to the canonical form. Let us list some of the known results. Two operators A and B are said to be equivalent at a point (x^0, ξ^0) if there exist an elliptic pseudo-differential operator Q and a Fourier integral operator Φ of the form (7) such that the principal symbols of $QA\Phi$ and ΦB coincide in some conical neighbourhood of (x^0, ξ^0).

1. If $a_0(x^0, \xi^0) = 0$, $\xi^0 \neq 0$, and if A is a principal type operator with the principal symbol a_0, then at the point (x^0, ξ^0) A is equivalent to an operator with the symbol $i\xi_1 + b(x, \xi')$. Here $\xi' = (\xi_2, \dots, \xi_n)$, $\xi_1^0 = 0$, and the real-valued smooth function b is independent of ξ_1, and $b \in S^1(\mathbb{R}^{n-1})$ for every fixed value of x_1 (see Egorov [1984]).

2. If $a_0(x^0, \xi^0) = 0$, $\xi^0 \neq 0$, and the forms $d_\xi\alpha(x^0, \xi^0)$ and $d_\xi\beta(x^0, \xi^0)$ are linearly independent, then at the point (x^0, ξ^0) the operator A is equivalent to an operator with the symbol $i\xi_1 + \xi_2 + a(x_1 - x_1^0)|\xi| + \gamma(x, \xi')$, where $a =$ const., $\xi_1^0 = \xi_2^0 = 0$, $\gamma(x^0, \xi^{0\prime}) = 0$, and $d_x\gamma(x^0, \xi^{0\prime}) = 0$ (see Egorov [1984]).

3. If $a_0(x^0, \xi^0) = 0$ and $\{\alpha, \beta\}(x^0, \xi^0) \neq 0$, then at the point (x^0, ξ^0) the operator A is equivalent to an operator with the symbol $\xi_1 + ix_1\xi_2$, where $\xi_1^0 = 0$, $x_1^0 = 0$ and $\xi_2^0 \neq 0$ (see Sato [1970a]).

4. Suppose that $a_0(x^0, \xi^0) = 0$ and let the forms $d\alpha(x^0, \xi^0)$,$d\beta(x^0, \xi^0)$ and $\xi^0\, dx$ be linearly independent. Furthermore, let $\{\alpha, \beta\}(x, \xi) = 0$ at all characteristic points (x, ξ) lying in a conical neighbourhood of (x^0, ξ^0).Then at the point (x^0, ξ^0) the operator $A(x, D)$ is equivalent to the operator $D_1 + iD_2$ (see Sjöstrand [1972]).

3.4. The Canonical Form of the Symbol in a Neighbourhood of the Boundary. Let X be a symplectic manifold, and F and G be two hypersurfaces in X defined, respectively, by the equations $f = 0$ and $g = 0$. F and G are said to be glancing surfaces at a point $\rho \in F \cap G$ if at this point df and dg are linearly independent and $\{f, g\}(\rho) = 0$ but $\{f, \{f, g\}\}(\rho) \neq 0$, $\{g, \{f, g\}\}(\rho) \neq 0$. Melrose [1976] has proved that by means of a canonical transformation every pair of glancing surfaces can be reduced to the form $f = x_1$ and $g = \xi_1^2 - x_1 - \xi_2$, and to the form $f = x_1$ and $g = \xi_1^2 - x_1\xi_2^2 - \xi_2\xi_3$ if one restricts to only those canonical transformations that are homogeneous in ξ.

Let A be a strictly hyperbolic operator of second order in the domain $\Omega = \{x \in \mathbb{R}^n, x_1 > 0\}$, $n \geqslant 3$, and let $a(x, \xi)$ be its principal symbol. Let ρ be a point in $T^*(\partial\Omega)\backslash 0$ such that $a(\rho) = 0$ and $\{a, x_1\}(\rho) = 0$. Melrose's theorem enables us to reduce A to its canonical form in a neighbourhood of ρ, with the plane $x_1 = 0$ preserved. If $\{a, \{a, x_1\}\}(\rho) < 0$, then A is equivalent to an operator with the symbol $\xi_1^2 + x_1\xi_n^2 + \xi_2\xi_n$.

If $\{a, \{a, x_1\}\}(\rho) > 0$ (in this case ρ is referred to as a point of diffraction),then A is microlocally equivalent to the operator $D_1^2 - x_1 D_n^2 - D_2 D_n$. This enables us to construct microlocally a parametrix for the boundary-value problem for A in the form of an Airy-Fourier integral operator which in the local coordinates is of the form

$$Au(x) = (2\pi)^{-n} \int a(x, \xi)\tilde{u}(\xi)e^{iS(x,\xi)} \frac{A(S_1(x, \xi)|\xi|^{-1/3})}{A(S_2(x, \xi)|\xi|^{-1/3})}\, d\xi,$$

where S, S_1, S_2 are the phase functions and $A(s)$ the Airy functions

$$A(s) = \frac{1}{2\pi}\int_{-\infty}^{\infty} e^{i(ts + t^3/3)}\, dt.$$

§4. Various Classes of Pseudodifferential Operators

4.1. The $L^m_{\rho,\delta}$ classes. We have considered hitherto pseudodifferential operators of the form (2) from the class L^m with the symbols belonging to S^m, that is, with symbols satisfying the inequalities (3). The most important property of L^m is that there exists a parametrix for an elliptic operator in L^m. If $L^{-\infty} = \bigcap L^m$, the elliptic operators are invertible in the classes $L^m/L^{-\infty}$, and no other operator does have an inverse in this class. From this point of view more interesting are $L^m_{\rho,\delta}$, the spaces of operators of the form (2) with symbols belonging to the classes $S^m_{\rho,\delta}$. These classes are characterized by the estimates

$$|D_\xi^\alpha D_x^\beta a(x, \xi)| \leqslant C_{\alpha,\beta}(1 + |\xi|)^{m - \rho|\alpha| + \delta|\beta|},$$

where $\rho \geqslant 0$ and $\delta \geqslant 0$. As in (3), the constants $C_{\alpha,\beta}$ may depend on R for $|x| < R$.

Examples. 1. Let $P(D)$ be a hypoelliptic differential operator of order m, with constant coefficients. It is known (Hörmander [1963]) that such operators are characterized by the relations

$$\lim_{|\xi| \to \infty} |D^\alpha P(\xi) \cdot P(\xi)^{-1}| = 0, \qquad \forall \alpha \neq 0.$$

It is clear that, in particular, $P(\xi) \neq 0$ for all sufficiently large $|\xi|$. For a parametrix of $P(D)$ one can take an operator $Q(D)$ with the symbol $q(\xi) = h(\xi)P(\xi)^{-1}$, where $h \in C^\infty$, $h(\xi) = 0$ for $|\xi| \leqslant a$ and $h(\xi) = 1$ for $|\xi| \geqslant b$; here a is so large that $P(\xi) \neq 0$ for $|\xi| \geqslant a$. It can be shown (Hörmander [1983, 1985]) that $Q \in L^{m'}_{\rho,0}$ with some m', $\rho > 0$. If $A(x)$ is an invertible matrix depending on x, then $q(A(x)\xi) \in S^{m'}_{\rho,1-\rho}$.

2. Let $\sum \varphi_\alpha(x) = 1$, $\alpha \in \mathbb{Z}^n$, where $\varphi_\alpha \in C_0^\infty(\mathbb{R}^n)$ and $\varphi_\alpha(x) = 0$ for $|x - \alpha| \geqslant 1$, and $|D^\beta \varphi_\alpha(x)| \leqslant C_\beta$ for all α and β. Let $a \in S^m_{\rho,\delta}$. Then a function of the form

$$\sum_{\alpha,\beta} \varphi_\alpha(\xi|\xi|^{-\rho})\varphi_\beta(x|\xi^\alpha|^\delta)a(x_{\alpha\beta}, \xi^\alpha)$$

belongs to the class $S^m_{\rho,\delta}$; here ξ^α is an arbitrary point in supp $\varphi_\alpha(\xi|\xi|^{-\rho})$ and $x_{\alpha\beta}$ is an arbitrary point in supp $\varphi_\beta(x|\xi^\alpha|^\delta)$. This function may be regarded as an approximation to the symbol $a(x, \xi)$. Similar properties are also possessed by the function

$$\sum_{\alpha,\beta} \varphi_\alpha(\xi|\xi|^{-\rho})\varphi_\beta(x|\xi^\alpha|^\delta) \sum_{|\delta+\gamma|\leqslant N} a^{(\delta)}_{(\gamma)}(x_{\alpha\beta}, \xi^\alpha)(x - x_{\alpha\beta})^\gamma(\xi - \xi^\alpha)^\delta \frac{1}{\gamma!\delta!},$$

where $a^{(\delta)}_{(\gamma)}(x, \xi) = (iD_\xi)^\delta(iD_x)^\gamma a(x, \xi)$.

The calculus developed above for operators of the class L^m can be extended to those belonging to the class $L^m_{\rho,\delta}$ with $\rho > \delta$. This last requirement is imposed so that the asymptotic series corresponding to the symbols of the product or of the adjoint operator are series in descending order of $|\xi|$. By a change of variables $x = \varphi(y)$, the symbol $a(x, \xi)$ goes over to $a(\varphi(y), {}^t\varphi'(y)^{-1}\eta) + \cdots$. Thus the class $S^m_{\rho,\delta}$ is preserved if $1 - \rho \leqslant \delta < \rho$. Under these conditions, a class of operators $L^m_{\rho,\delta}(X)$ is defined on a smooth manifold X.This condition is sufficient for an elliptic operator to have a parametrix.

The operators from the class $L^0_{\rho,\delta}$ are bounded in L_2 if $\delta < \rho$. For an operator A, with symbol $a \in S^0_{\rho,\delta}$, to be compact, it is necessary and sufficient that $a(x, \xi) \to 0$ as $\xi \to \infty$, and the convergence is uniform if x lies in a compact subset (see Kohn and Nirenberg [1965], Hörmander [1983, 1985]).

Calderon and Vaillancourt have proved that the operator A with symbol in $S^0_{\rho,\rho}$ is bounded in L_2 when $\rho < 1$. More precisely, they showed that the operator A defined by

$$Au(x) = (2\pi)^{-n} \int\int e^{i(x-y)\cdot\xi} a(x, y, \xi)u(y)\, dy\, d\xi$$

is bounded in L_2 if $0 \leqslant \rho \leqslant \delta < 1$, $m \leqslant (\rho - \delta)n$ and if

$$|D^\alpha_x D^\beta_y D^\gamma_\xi a(x, y, \xi)| \leqslant C(1 + |\xi|)^{m+\delta|\alpha+\beta|-\rho|\gamma|}$$

for $0 \leqslant |\gamma| \leqslant 2[n/2] + 2$, $0 \leqslant |\alpha + \beta| \leqslant 2m_j$, where m_j is the smallest value of the integer r for which $r(1 - \delta) \geqslant 5n/4$ (see Calderon and Vaillancourt [1971]).

Hörmander has established that the operator A with symbol a and having compact support with respect to x and y is bounded in L_2 when $\rho > 0$, $m \leqslant 0$ and $2m < (\rho - \delta)n$. He also demonstrated that this result fails to be valid if $m > 0$ or if $2m > (\rho - \delta)n$ (see Hörmander [1983, 1985]). Chin-Hung-Chin has shown that A may not be bounded if $\delta = \rho = 1$ and $m = 0$.

Properties close to the above mentioned ones are possessed by operators whose symbols belong to $S^m_{\lambda,\rho,\delta}$, the classes introduced by Kumano-go and Taniguchi [1976].

A symbol $a \in S^m_{\lambda,\rho,\delta}$ if

$$|D^\alpha_\xi D^\beta_x a(x,\xi)| \leqslant C_{\alpha,\beta}\lambda(x,\xi)^{m+\delta|\beta|-\rho|\alpha|}.$$

Here λ is a weight function that satisfies the following conditions:

$$1 \leqslant \lambda(x,\xi) \leqslant A_0(1+|x|)^{\tau_0}(1+|\xi|), \qquad \tau_0 \geqslant 0;$$

$$|D^\alpha_\xi D^\beta_x \lambda(x,\xi)| \leqslant A_{\alpha,\beta}\lambda(x,\xi)^{1+\delta|\beta|-|\alpha|}, \qquad 0 \leqslant \delta < 1;$$

$$\lambda(x+y,\xi) \leqslant A_1(1+|y|)^{\tau_1}\lambda(x,\xi), \qquad \tau_1 \geqslant 0.$$

The classes close to the above mentioned ones were also studied by Vishik and Grushin [1969] and by Grushin [1971a, 1972]. Within the framework of this calculus, one may construct, for example, a parametrix for the operator $-\Delta + |x|^{2k}$ in the whole space. For this construction, it is convenient to use a weighted Sobolev space, for example, the space with a norm of the form

$$\|u\|_m = \left(\sum_{|\alpha| \leqslant m} \||x|^{(m-|\alpha|)k} D^\alpha u\|^2_{L_2}\right)^{1/2}.$$

Let us also mention the class of *operators of variable order*, introduced by Unterberger and Bokobza [1964]. Symbols belonging to this class are homogeneous functions in ξ but $m(x)$, the degree of homogeneity, depends on x, and, moreover,

$$|D^\alpha_\xi D^\beta_x a(x,\xi)| \leqslant C_{\alpha\beta}(1+|\xi|)^{m(x)-|\alpha|}[\ln(1+|\xi|)]^{|\beta|}.$$

4.2. The $L^m_{\Phi,\varphi}$ Classes. In their works Beals and Fefferman [1975, 1977] proposed a wide-ranging extension of the classes of symbols of pseudodifferential operators. The calculus developed by them enables us to invert operators with symbols of the form

$$\xi_1 + ip_1(x)x_1^{2k}\xi_2 + p_2(x), \qquad p_1(0) \neq 0, \qquad p_1\colon \mathbb{R}^n \to \mathbb{R}^1;$$

$$\xi_1^2 + p_1(x)x_1^{2k}\xi_2^2 + p_2(x)x_1^{k-1}\xi_2 + p_3(x)\xi_1 + p_4(x), \qquad p_1(0) > 0;$$

$$x_1\xi_2^2 + ip_1(x)\xi_1 + p_2(x)\xi_2 + p_3(x), \qquad p_1(0) > 0.$$

This was not possible in other variants of the calculus of pseudodifferential operators.

The definition of the classes of symbols in this calculus starts by describing the *weight functions*. The continuous positive functions φ and Φ defined on the space of variables $(x,\xi) \in \mathbb{R}^{2n}$ constitute a pair of weight functions if there are positive constants c, C and δ such that

(A_1) $\varphi \leqslant C$;

(A_2) $\Phi\varphi \geqslant c$;

$$(A_3)\ c \leqslant \Phi(x, \xi)\Phi(y, \eta)^{-1} \leqslant C, c \leqslant \varphi(x, \xi)\varphi(y, \eta)^{-1} \leqslant C,$$

$$\text{if} \quad |x - y| \leqslant c\varphi(x, \xi), |\xi - \eta| \leqslant c\Phi(x, \xi);$$

$$(A_4)\ R(x, 0) \leqslant C(1 + |x|)^C, \qquad \text{where} \quad R = \Phi/\varphi;$$

$$(A_5)\ c \leqslant R(x, \xi)R(y, \eta)^{-1} \leqslant C,$$

$$\text{if} \quad |x - y| \leqslant cR(x, \xi)^{\delta}R(y, \eta)^{-1/2}, |\xi - \eta| \leqslant cR(x, \xi)^{\delta+1/2}.$$

The pair Φ, φ is said to be localizable if for every compact set $K \subset \mathbb{R}^n$ a constant $\varepsilon > 0$ can be found for which $\Phi \geqslant \varepsilon(1 + |\xi|^2)$ on $K \times \mathbb{R}^n$.

The functions φ and Φ serve as distinctive units of length in the directions of x and ξ in the cotangent bundle space. The conditions (A_3) signify that these units do not change very much with an increase in the distance. The condition (A_2) replaces the condition $\rho \geqslant \delta$ for the classes $S^m_{\rho,\delta}$. To these classes there correspond the functions $\varphi = (1 + |\xi|)^{-\delta}$, $\Phi = (1 + |\xi|)^{\rho}$, $0 \leqslant \delta \leqslant \rho \leqslant 1, \delta \neq 1$.

Let φ and Φ be weight functions. The order of an operator is determined by the function λ belonging to the class $0(\Phi, \varphi)$. This class is characterized by the following conditions:

$$(0_1)\ |\lambda(x, \xi) - \lambda(y, \eta)| \leqslant C, \qquad \text{if} \quad |x - y| \leqslant c\varphi(x, \xi),$$

$$|\xi - \eta| \leqslant \Phi(x, \xi);$$

$$(0_2)\ c(\Phi\varphi)^{-m} \leqslant e^{\lambda}\Phi^{-K}\varphi^{-K} \leqslant C(\Phi\varphi)^m$$

for some real numbers k, K and m.

Definition. A function $a \in S^{\lambda}_{\Phi,\varphi}$ if

$$|D^{\alpha}_{\xi}D^{\beta}_{x}a(x, \xi)| \leqslant C_{\alpha,\beta}e^{\lambda} \cdot \Phi^{-|\alpha|}\varphi^{-|\beta|}$$

for all α and β.

Examples. 1. If

$$\Phi = (1 + |\xi|)^{\rho}, \qquad \varphi = (1 + |\xi|)^{-\delta}, \qquad \lambda = m \ln |\xi|,$$

we obtain the Hörmander classes $S^m_{\rho,\delta}$.

2. The Kumano-go-Taniguchi classes $S^m_{\lambda,\rho,\delta}$ are obtained if $\Phi = \lambda^0$, $\varphi = \lambda^{-\delta}$ and $\mu = m \ln \lambda$.

3. If we take $\Phi = 1 + |\xi|$, $\varphi = [1 + \ln(1 + |\xi|)]^{-1}$, and $\lambda = \rho(x) \ln(1 + |\xi|)$, we obtain the Unterberger-Bokobza classes.

The operators

$$u \mapsto a(x, D)u(x) = (2\pi)^{-n} \int a(x, \xi)\tilde{u}(\xi)e^{ix\xi}\, d\xi$$

whose symbols a belong to $S^{\lambda}_{\Phi,\varphi}$ constitute a class of operators $L^{\lambda}_{\Phi,\varphi}$. These operators map continuously the Schwartz space S into itself. Let us mention the basic facts regarding the calculus of such operators.

I. If $A \in L^{\lambda}_{\Phi,\varphi}$ and $B \in L^{\mu}_{\Phi,\varphi}$, then $AB \in L^{\lambda+\mu}_{\Phi,\varphi}$. The symbol $a \circ b$ of the operator AB has an asymptotic expansion

$$a \circ b \sim \sum \frac{i^{|\alpha|}}{\alpha!} D^{\alpha}_{\xi} a(x, \xi) D^{\alpha}_{x} b(x, \xi)$$

in the sense that

$$a \circ b - \sum_{|\alpha|<N} \frac{i^{|\alpha|}}{\alpha!} D^{\alpha}_{\xi} a \cdot D^{\alpha}_{x} b \in S^{\lambda+\mu-N \ln \Phi - N \ln \varphi}$$

for each natural number N.

II. If $A \in L^{\lambda}_{\Phi,\varphi}$, then the formally adjoint operator A^* belongs to $L^{\lambda}_{\Phi,\varphi}$ and its symbol a^* admits an asymptotic expansion

$$a^*(x, \xi) \sim \sum \frac{i^{|\alpha|}}{\alpha!} D^{\alpha}_{\xi} D^{\alpha}_{x} \overline{a(x, \xi)}.$$

III. Suppose that $\lambda_j \in 0(\Phi, \varphi)$ for $j = 0, 1, \ldots$, with

$$\lambda_{j+1} \leqslant \lambda_j + c_j,$$

$$\lambda_j - \lambda_{j+1} \leqslant d_j(\lambda_{j-1} - \lambda_j) + c_j,$$

where c_j and d_j are positive constants. Let $a_j \in S^{\lambda_j}$. Then there exists a symbol $a \in S^{\lambda_0}$ such that $a \sim \sum a_j$, that is,

$$a - \sum_{j<N} a_j \in S^{\lambda_N}.$$

IV. If $A \in L^0$,then $A: L_2 \to L_2$ is a continuous operator.

V. Let $\lambda \in 0(\Phi, \varphi)$. Put

$$H^{\lambda} = H^{\lambda}_{\Phi,\varphi} = \{Au; u \in L_2, A \in L^{-\lambda}\}.$$

Then $H^0 = L_2$ and $S \subset H^{\lambda} \subset S'$. Moreover, these embeddings are continuous and the image sets are everywhere dense; also $(H^{\lambda})' = H^{-\lambda}$ as topological spaces. If $A \in L^{\mu}$, $\mu \in 0(\Phi, \varphi)$ and $\lambda \in 0(\Phi, \varphi)$, then the map $A: H^{\lambda+\mu} \to H^{\lambda}$ is continuous. There exists an operator $A \in L^{\mu}$ such that the operator $A: H^{\lambda+\mu} \to H^{\lambda}$ performs a topological isomorphism. In particular, H^{λ} is a Hilbert space.

If $a \in S^{\mu}$ and $e^{-\mu}\Phi^{\alpha}\varphi^{\beta}D^{\alpha}_{\xi}D^{\beta}_{\xi}a(x, \xi) \to 0$ as $|x| + |\xi| \to \infty$ for all α and β, then the operator $a(x, D): H^{\lambda+\mu} \to H^{\lambda}$ is compact. In particular, the embedding $H^{\mu}_{K} \subset H^{\lambda}$ is compact if $\mu - \lambda \to +\infty$ as $|\xi| \to \infty$ uniformly for x lying in a neighbourhood of a compact set K. Here H^{μ}_{K} is a subspace in H^{μ} and consists of functions having supports in K.

Beals has also examined more general spaces where Φ and φ are the vectors $\Phi = (\Phi_1, \ldots, \Phi_n)$, $\varphi = (\varphi_1, \ldots, \varphi_n)$, and Φ_j and φ_j are weight functions of the variables x_j and ξ_j. What is more, $\Phi_j(x, \xi) \leqslant C\Phi_k(x, \xi)^C$ for all j and k, and if $\pi(x, \xi) = \min_k \Phi_k(x, \xi)\varphi_k(x, \xi)$, then

$$\Phi_j(x, \xi)\Phi_j(y, \eta)^{-1} \leqslant C[\pi(x, \xi) + \pi(y, \eta)]^C,$$

$$\varphi_j(x, \xi)\varphi_j(y, \eta)^{-1} \leqslant C[\pi(x, \xi) + \pi(y, \eta)]^C$$

with

$$|x_k - y_k| \leqslant cR_k(x,\xi)^\delta R_k(y,\eta)^{-1/2}, \qquad |\xi_k - \eta_k| \leqslant cR_k(x,\xi)^{\delta+(1/2)}.$$

Example. Let $a(x,\xi) = \xi_1 + ix_1^{2k}p_1(x,\xi) + p_0(x,\xi)$, where $p_1 \in S^1$, $p_0 \in S^0$, p_1 is a real-valued function and where

$$|p_1(x,\xi)| \geqslant c|\xi'| \text{ for } |\xi'| \geqslant C, \qquad \xi' = (\xi_2, \dots, \xi_n).$$

We set $g(x,\xi) = |\xi_1| + x_1^{2k}|\xi'| + (1+|\xi|)^\delta$, where $\delta = 1/(2k+1)$

$$\Phi_1 = g^{1/2k}(1+|\xi|)^{\delta(2k-1)/2k}, \qquad \varphi_1 = g^{1/2k}(1+|\xi|)^{-1/2k},$$

$$\Phi_j = 1 + |\xi|, \qquad \varphi_j = 1 \qquad \text{for} \quad j = 2, 3, \dots, n.$$

The operator A with the symbol a has a right parametrix and a left parametrix belonging to the class $S^{-\mu}$,where $\mu = \ln g$. In particular, the estimate

$$\|u\|_\mu \leqslant C(\|Au\|_{\mathscr{L}_2} + \|u\|_{\mathscr{L}_2}), \qquad u \in C_0^\infty(K).$$

holds. This implies the subelliptic estimate, because $e^\mu \geqslant c(1+|\xi|)^\delta$ (see Beals [1977]).

4.3. The Weyl Operators. Hörmander [1979] has introduced more general classes of pseudodifferential operators. He proposed that they be expressed in a form generalizing the Weyl's form:

$$a(x,D)u(x) = (2\pi)^{-n} \int\int a\left(\frac{x+y}{2}, \xi\right) e^{i(x-y)\xi} u(y)\, dy\, d\xi. \tag{9}$$

If $a \in S^m$, then such an operator differs from the usual operator with symbol a by an operator of order $m-1$. However, the operators of the form (9) have the important property that they remain invariant under linear symplectic coordinate transformatons: $U^{-1}a(x,D)U = (a\cdot\chi)(x,D)$. Here $\chi\colon \mathbb{R}^{2n} \to \mathbb{R}^{2n}$ is a transformation that preserves the form $\sigma = \sum d\xi_j \wedge dx_j$ and the operator U is determined uniquely by χ to within the constant multiplier.

Let V be a finite-dimensional vector space and let $g_x(y)$ be a positive definite quadratic form in this space for each $x \in V$. For a function $u\colon \omega \to \mathbb{C}$ belonging to the class $C^k(\omega)$, where ω is a neighbourhood of x in V, we define the norm of its kth differential by the formula

$$|u|_k^g(x) = \sup_{t_j \in V} |u^{(k)}(x; t_1, \dots, t_k)| \prod_1^k g(t_j)^{-1/2}.$$

The metric g is said to vary slowly if there exist positive constants c and C such that

$$g_x(y) \leqslant c \Rightarrow g_{x+y}(t) \leqslant Cg_x(t), \qquad \forall t \in V.$$

A function m, defined on V and assuming positive real values, is said to be g-continuous if there are positive constants c and C such that

$$g_x(y) \leqslant c \Rightarrow m(x)/C \leqslant m(x+y) \leqslant Cm(x).$$

If g varies slowly and m is g-continuous, one can introduce the class $S(m, g)$ of functions u that are smooth on V and for which

$$\sup |u|_k^g(x)/m(x) < \infty \qquad \text{for each} \quad k \in \mathbb{Z}_+.$$

Clearly, $C_0^\infty(V) \subset S(m, g)$. Moreover, $uv \in S(mm', g)$ if $u \in S(m, g)$ and $v \in S(m', g)$. If $|u| > cm$, then $u^{-1} \in S(1/m, g)$.

Let B be a quadratic form, with real coefficients, on V', the dual space of V. The differential operator $B(D)$ is defined by the formula $B(D)e^{ix\xi} = B(\xi)e^{ix\xi}$ on functions belonging to $C^2(V)$. We set

$$g_y^B(x) = \sup_\xi (x, \xi)^2/g(B\xi).$$

Suppose that the Riemannian metric g varies slowly in V and that m is g-continuous. In this case, g and m are said to be compatible with B if there are constants C and N such that the inequalities

$$g_y(t) \leqslant Cg_x(t)[1 + g_y^B(x - y)]^N,$$
$$m(y) \leqslant Cm(x)[1 + g_y^B(x - y)]^N$$

holds for all $y, t \in V$.

Proposition 4.1. *If g and m are compatible with B and if $g_x(y) \leqslant g_x^B(y)$ for all x and y, then the continuous function $e^{iB(D)}u(x)$ is defined for $u \in S(m, g)$, and*

$$|e^{iB(D)}u(x)| \leqslant m(x)\|u\|,$$

where $\|\cdot\|$ is some seminorm in $S(m, g)$ that depends on C and N.

If $u \in S$, then $v = e^{iB(D)}u = U(1, x)$ is the value at $t = 1$ of the solution to the problem

$$\frac{\partial U}{\partial t} = iB(D)U, \qquad U = u \quad \text{at} \quad t = 0.$$

It is easy to see that the function

$$K(x, y) = (2\pi)^{-n} \int a\left(\frac{x + y}{2}, \xi\right) e^{i(x-y)\xi}\, d\xi$$

serves as a Schwartz kernel for the operator (9), and, moreover,

$$a(x, \xi) = \int K\left(x + \frac{t}{2}, x - \frac{t}{2}\right) e^{-it\xi}\, dt.$$

We set

$$W = V \oplus V', \quad \sigma(x, \xi; y, \eta) = (y, \xi) - (x, \eta), \quad (x, \xi) \in W, \quad (y, \eta) \in W.$$

It is immediately clear from the definition that the formally adjoint operator $a(x, D)^*$ has the symbol $\overline{a(x, \xi)}$. The symbol $a(x, \xi)$ of the composotion $a_1(x, D)$. $a_2(x, D)$ can be found by applying Proposition 4.1 to the space $W \oplus W$ and the form σ, regarded as a quadratic form on $W \oplus W$. This yields

$$a(x, \xi) = e^{i\sigma(D_x, D_\xi; D_y, D_\eta)/2} a_1(x, \xi) a_2(y, \eta)|_{(y,\eta)=(x,\xi)}.$$

Here it is assumed that the hypotheses of Proposition 4.1 hold, that is,

$$g_w^\sigma \leqslant C g_{w_1}^\sigma (1 + g_{w_1}^\sigma(w_1 - w))^N,$$

$$m(w_1) \leqslant C m(w)(1 + g_{w_1}^\sigma(w_1 - w))^N, \qquad w, w_1 \in W.$$

Moreover, if $a_1 \in S(m_1, g)$ and $a_2 \in S(m_2, g)$, then $a \in S(m_1 m_2, g)$, and

$$a(x, \xi) - \sum_{j<N} \frac{1}{j!} \left[\frac{i}{2} \sigma(D_x, D_\xi; D_y, D_\eta) \right]^j a_1(x, \xi) a_2(y, \eta)|_{(y,\eta)=(x,\xi)} \in S(h^N m_1 m_2, g)$$

for all N. Here $h(x, \xi)^2 = \sup g_{x,\xi}/g_{x,\xi}^\sigma$.

Examples. 1. The space $S_{\rho,\delta}^m$ of symbols a satisfying the inequalities

$$|D_\xi^\alpha D_x^\beta a(x, \xi)| \leqslant C_{\alpha,\beta}(1 + |\xi|)^{\mu - \rho|\alpha| + \delta|\beta|}$$

is obtained as $S(m, g)$ if one sets

$$g_{x,\xi}(y, \eta) = |y|^2(1 + |\xi|)^{2\delta} + |\eta|^2(1 + |\xi|)^{-2\rho}, \quad m(x, \xi) = (1 + |\xi|)^\mu.$$

Moreover, g varies slowly if and only if $\rho \leqslant 1$.Since

$$g_{x,\xi}^\sigma(y, \eta) = |y|^2(1 + |\xi|)^{2\rho} + |\eta|^2(1 + |\xi|)^{-2\delta},$$

it follows that the inequality $g_{x,\xi} \leqslant g_{x,\xi}^\sigma$ holds if and only if $\delta \leqslant \rho$. The compatibility of g with σ implies that

$$\left(\frac{1 + |\eta|}{1 + |\xi|}\right)^\delta + \left(\frac{1 + |\xi|}{1 + |\eta|}\right)^\rho \leqslant C[1 + |x - y|(1 + |\eta|)^\rho + |\xi - \eta|(1 + |\eta|)^{-\delta}]^N,$$

that is, $\delta < 1$. Thus the calculus is developed for $0 \leqslant \delta \leqslant \rho \leqslant 1$ and $\delta < 1$. These conditions are necessary for the operators belonging to $S_{\rho,\delta}^0$ to be bounded in L_2 (Hörmander [1983, 1985]).

2. If we take

$$g_{x,\xi}(y, \eta) = |y|^2 \varphi(x, \xi)^{-2} + |\eta|^2 \Phi(x, \xi)^{-2},$$

then the operators belonging to $S(m, g)$ become operators of the Beals-Fefferman class. In this case, the condition $\Phi\varphi \geqslant 1$ coincides with the condition $g \leqslant g^\sigma$. The remaining conditions may be reduced to the condition

$$R(y, \eta) \leqslant C R(x, \xi), \qquad \text{where} \quad R(x, \xi) = \Phi(x, \xi)\varphi(x, \xi)^{-1}$$

and

$$R(x, \xi)|x - y|^2 + |\xi - \eta|^2 R^{-1}(y, \eta) \leqslant C[R(y, \eta)R(x, \xi)^{-1}]^\delta$$

for some $\delta > 0$.

Theorem 4.1 (Hörmander [1979]). *Let $g \leqslant g^\sigma$, and let the metric g and the function m be compatible with the form σ. The operator $a(x, D)$, with the symbol a in $S(m, g)$, is bounded in L_2 if and only if m is bounded. This operator is compact in L_2 if and only if $m(x) \to 0$ as $x \to \infty$.*

Using the calculus developed by him, Hörmander [1979] obtained an explicit formula for the index of an elliptic operator that generalizes the results of Atiyah and Singer and of Fedosov [1974].

Theorem 4.2 (Hörmander [1979]). *Assume that the metric g is compatible with σ and that* $\dfrac{g(x, \xi)}{g^\sigma(x, \xi)} \leqslant (1 + |x| + |\xi|)^{-\delta}$ *for some* $\delta > 0$. *Let* $a \in S(1, g)$, *where* $a(x, \xi) \in L(\mathbb{C}^\nu, \mathbb{C}^\nu)$. *Let the matrix* $a^{-1}(x, \xi)$ *exist and be bounded outside some ball* B. *Then the operator* $a(x, D)$ *is Nöether and*

$$\operatorname{ind} a = -(2\pi i)^{-n} \frac{(n-1)!}{(2n-1)!} \int_{\partial B} \operatorname{tr}(a^{-1}\, da)^{2n-1}.$$

Here an orientation in $\mathbb{R}^{2n}$ *has been so chosen that the condition* $d\xi_1 \wedge dx_1 \wedge \cdots \wedge d\xi_n \wedge dx_n > 0$ *is satisfied.*

We also note the following result which generalizes Theorem 2.1.

Theorem 4.3 (Hörmander [1979]). *Suppose that the metric g varies slowly and that m and g are compatible with σ. Suppose that for each point* $(x, \xi) \in \mathbb{R}^n \times \mathbb{R}^n$ *there is a value* $b = b(x, \xi) \leqslant (1 + |\xi|^2)^{1/3}$ *such that*

$$b(|dx|^2 + |d\xi|^2(1 + |\xi|^2)^{-1}) \leqslant g_{x,\xi}(dx, d\xi) \leqslant b^2(|dx|^2 + |d\xi|^2(1 + |\xi|^2)^{-1}).$$

Suppose that $a \in S(m, g)$ *and* $a = 0$ *for large* $|x|$. *Let* $\varphi \in S_1^0$ *and* $\varphi = 0$ *for large* $|x| + |\xi|$. *Let* $A_1 = \Phi A \Phi^*$, *where* Φ *is the Fourier integral operator*

$$\Phi u(x) = \int \varphi(x, \xi) \tilde{u}(\xi) e^{iS(x, \xi)}\, d\xi$$

with a non-degenerate phase function S that defines the canonical transformation

$$\chi\colon \left(\frac{\partial S(x, \xi)}{\partial \xi}, \xi\right) \mapsto \left(x, \frac{\partial S(x, \xi)}{\partial x}\right).$$

Then $a_1 \in S_{1,0}^{-\infty}$ *outside each conical neighbourhood of the set*

$$\{(x, \partial S(x, \xi)/\partial x)\colon (x, \xi) \in \text{cone supp}\, \varphi\}$$

and $a_1 \circ \chi \in S(m, g)$ *in a conical neighbourhood of the set*

$$\{(\partial S(x, \xi)/\partial \xi, \xi)\colon (x, \xi) \in \text{cone supp}\, \varphi\}.$$

Moreover, $a_1 \circ \chi$ *can be expressed explicitly in terms of* a, *and if* $b(x, \xi) < (1 + |\xi|)^\delta$ *for* $\delta < 2/3$, *then the principal term of* $a_1 \circ \chi$ *equals* $|(2\pi)^n \varphi(x, \zeta)|^2 |\det S''_{x\zeta}(x, \zeta)|^{-1} a(z, \zeta)$, *where* $(z, \zeta) = \chi^{-1}(x, \xi)$.

§ 5. Complex Powers of Elliptic Operators

5.1. The Definition of Complex Powers. Let us consider an elliptic pseudodifferential operator A of order m on an n-dimensional closed manifold M. We shall define its power A^z for $z \in \mathbb{C}$ and show in the sequel how these operators can be used in the theory of index and in the spectral theory of elliptic operators (see Seeley [1967] and Shubin [1978]).

Assume that the principal part of the symbol a_m of A does not take negative values. Then it can be shown that the operator $A - \lambda I$ is invertible for those negative λ that are sufficiently large in absolute value and that for them the inequality $\|(A - \lambda I)^{-1}\| \leqslant C|\lambda|^{-1}$ holds. We assume further that $A - \lambda I$ is invertible for all $\lambda \leqslant 0$. Then in the λ-plane there exists a contour Γ which comes from $-\infty$ from below the real axis, goes around $\lambda = 0$ along a given curve and then again goes off to $-\infty$ but now above the real axis, and is such that the spectrum of A lies to the right of Γ. For $\operatorname{Re} z < 0$, we set

$$A_z = \frac{1}{2\pi i}\int_\Gamma \lambda^z (A - \lambda I)^{-1}\, d\lambda,$$

where λ^z is defined as a holomorphic function of λ that equals $e^{z \ln \lambda} = |\lambda|^z e^{iz \arg \lambda}$ for $\lambda \in \Gamma$.

Using the Hilbert identity

$$(\lambda - \mu)(A - \lambda I)^{-1}(A - \mu I)^{-1} = (A - \lambda I)^{-1} - (A - \mu I)^{-1},$$

we can easily verify that the semigroup property

$$A_z A_w = A_{z+w}$$

holds if $\operatorname{Re} z < 0$ and $\operatorname{Re} w < 0$. In particular, the relation $A_{-k} = (A^{-1})^k$ holds for non-negative integers k. Moreover, for $\operatorname{Re} z < 0$ the operators A_z are bounded in each space $H_s(M)$.

If $z \in \mathbb{C}$, we can define A^z by the formula $A^z = A^k \cdot A_{z-k}$, where k is an integer such that $k > \operatorname{Re} z$. It is easy to verify that $A^0 = I$, $A^1 = A$, A^{-1} is the inverse of A and that A^z coincides with A_z for $\operatorname{Re} z < 0$. For $\operatorname{Re} z < k$, A^z is a bounded linear operator from the space $H^s(M)$ into the space $H^{s-mk}(M)$, and depends holomorphically on $z \in \mathbb{C}$.The family of operators A^z possesses the group property: $A^z A^w = A^{z+w}$ for all z and $w \in \mathbb{C}$.

5.2. The Construction of the Symbol for the Operator A^z. The theory of pseudodifferential operators enables us to construct the symbol of A^z in an explicit form.

We start by constructing the symbol for a parametrix of the operator $A - \lambda I$. Let X be a coordinate neighbourhood on M that can be identified with a domain in $\mathbb{R}^n$ by fixing a coordinate system. We represent the symbol $a(x, \xi)$ of A in X in the form of the sum

$$a(x, \xi) = \sum_{j=0}^{m} a_j(x, \xi), \qquad \text{where} \quad a_j(x, \xi) = \sum_{|\alpha|=j} a_\alpha(x)\xi^\alpha, j = 0, 1, \ldots, m.$$

It is natural to seek the symbol of the parametrix of $A - \lambda I$ in the form of a sum of functions that are homogeneous in $(\xi, \lambda^{1/m})$. We denote these functions by $b^0_{-m-j}(x, \xi, \lambda), j = 0, 1, \ldots,$ where

$$b^0_{-m-j}(x, t\xi, t^m\lambda) = t^{-m-j} b^0_{-m-j}(x, \xi, \lambda) \qquad \text{for} \quad t > 0 \text{ and } |\xi| + |\lambda| \neq 0.$$

These functions are determined from the recurrence formulae

$$[a_m(x, \xi) - \lambda] b^0_{-m}(x, \xi, \lambda) = 1,$$

$$[a_m(x, \xi) - \lambda] b^0_{-m-j}(x, \xi, \lambda) + \sum_{\substack{|\alpha|+k+l=j \\ l<j}} \frac{1}{\alpha!} (iD_\xi)^\alpha a_{m-k}(x, \xi) D^\alpha_x b^0_{-m-l}(x, \xi, \lambda) = 0,$$

$$j = 1, 2, \ldots$$

The homogeneous components $b^{(z),0}_{mz-j}(x, \xi)$ of the symbol of A^z are now constructed by means of the above homogeneous components of the parametrix in exactly the same way as the powers A^z themselves were constructed with the aid of the resolvent $(A - \lambda I)^{-1}$. It is clear from the construction that $b^0_{-m-j}(x, \xi, \lambda)$ are holomorphic functions of λ on and inside Γ. For Re $z < 0$, we now put

$$b^{(z),0}_{mz-j}(x, \xi) = \frac{1}{2\pi i} \int_\Gamma \lambda^z b^0_{-m-j}(x, \xi, \lambda)\, d\lambda, \qquad j = 0, 1, \ldots,$$

defining the values of λ^z as above. In particular, for $j = 0$ the Cauchy formula gives

$$b^{(z),0}_{mz}(x, \xi) = \frac{1}{2\pi i} \int_\Gamma \lambda^z [a_m(x, \xi) - \lambda]^{-1}\, d\lambda = a^z_m(x, \xi).$$

It is not difficult to see that the functions $b^{(z),0}_{mz-j}(x, \xi)$ are positive homogeneous in ξ of degree $mz - j$, and that the semigroup property

$$\sum_{|\alpha|+p+q=j} (iD_\xi)^\alpha b^{(z),0}_{mz-p}(x, \xi) \cdot D^\alpha_x b^{(w),0}_{mw-q}(x, \xi) \frac{1}{\alpha!} = b^{(z+w),0}_{m(z+w)-j}(x, \xi)$$

holds for Re $z < 0$ and Re $w < 0$.

For an arbitrary $z \in \mathbb{C}$ we can set

$$b^{(z),0}_{mz-j}(x, \xi) = \sum_{p+q+|\alpha|=j} \frac{1}{\alpha!} (iD_\xi)^\alpha a^{(k)}_{mk-p}(x, \xi) \cdot D^\alpha_x b^{(z-k),0}_{m(z-k)-q}(x, \xi)$$

if $k \in \mathbb{Z}$ is chosen so that Re $z < k$. It can be verified that these functions $b^{(z),0}_{mz-j}$ do not depend on k, and for Re $z < 0$ they coincide with the functions defined initially, and that for integers k the relation

$$b^{(k),0}_{mk-j}(x, \xi) = a^{(k)}_{mk-j}(x, \xi)$$

holds. Here and above, $a^{(k)}_j(x, \xi)$ are the homogeneous components, of degree j, of the symbol of A^k.

Having constructed the symbols of A^z in a coordinate neighbourhood X, we can construct these symbols on the whole M by means of the decomposition of unity. Any singularity which these symbols may have for $|\xi| = 0$ can be removed by multiplying by a smooth cutoff function.

5.3. The Construction of the Kernel of the Operator A^z. We recall that the Schwartz kernel of the pseudodifferential operator

$$Bu(x) = \int\int e^{i(x-y)\xi} b(x, \xi) u(y)\, dy\, d\xi,$$

where $b \in S^m$, is defined for $m < -n$ by the formula

$$K(x,y) = \int e^{i(x-y)\xi} b(x, \xi)\, d\xi.$$

The Schwartz kernel K_z of the operator A^z is defined by this formula for $\operatorname{Re} z < -n/m$ and is a holomorphic function of z. It can be shown (Seeley [1967]) that the function $K_z(x, y)$, with $x \neq y$, can be extended to an entire function of z that vanishes for $z = 0, 1, 2, \ldots$. Moreover, $K_z(x, x)$ can be extended to a meromorphic function having no more than simple poles at the points $z_j = (j - n)/m; j = 0, 1, 2, \ldots$. The residues $\gamma_j(x)$ at these points equal

$$-\frac{(2\pi)^n}{m} \int_{|\xi|=1} b^{(z_j),0}_{-n}(x, \xi)\, d\xi = -\frac{(2\pi)^{n-1}}{mi} \int_{|\xi|=1} d\xi \int_\Gamma \lambda^{(j-n)/m} b^0_{-m-j}(x, \xi, \lambda)\, d\lambda.$$

If $z_j = l \in Z_+$, then $\gamma_j(x) = 0$ and

$$K_l(x, x) = \frac{(-1)^l (2\pi)^{-n}}{m} \int_{|\xi|=1} d\xi \int_0^\infty \lambda^l b^0_{-m-ml-n}(x, \xi, \lambda)\, d\lambda.$$

We note that at the extreme left pole $z_0 = -n/m$, the residue γ_0 is given by

$$\gamma_0(x) = -\frac{(2\pi)^{-n}}{m} \int_{|\xi|=1} a_m^{-n/m}(x, \xi)\, d\xi,$$

and the kernel K_0 is regular at $x = y$, and

$$K_0(x, x) = \frac{(2\pi)^{-n}}{m} \int_{|\xi|=1} d\xi \int_0^{+\infty} b^0_{-m-n}(x, \xi, -\lambda)\, d\lambda.$$

It is clear that $b_{-m} = (a_m - \lambda)^{-1}$, and by the recurrence relations defining b_{-m-j}, we have

$$b_{-m-j}(x, \xi, \lambda) = \sum_{k=1}^{2j} \gamma_{k,j}(x, \xi)[a_m(x, \xi) - \lambda]^{-k-1},$$

where $\gamma_{k,j}$ are polynomials in $a_m, \ldots, a_{m-j}$ and their derivatives of orders $\leqslant j$. Therefore

$$\gamma_j(x) = -\frac{(2\pi)^{-n}}{m} \int_{|\xi|=1} \sum_{k=1}^{2j} \gamma_{k,j}(-1)^k s(s-1)\ldots(s-k+1) a_m^{s-k}/k!\, d\xi,$$

where $s = (j - n)/m$, and

$$K_0(x, x) = \frac{(2\pi)^{-n}}{m} \int_{|\xi|=1} \sum_{k=1}^{2\pi} k^{-1} \gamma_{k,n}(x, \xi) a_m^{-k}(x, \xi)\, d\xi.$$

5.4. The ζ-Function of an Elliptic Operator. In the spectral theory a significant role is played by the ζ-function which is defined for $\operatorname{Re} z < -n/m$ by the formula

$$\zeta_A(z) = \int_M K_z(x, x)\, dx.$$

An analytic continuaton of this function yields a meromorphic function that has no more than simple poles at the points $z_j = (j - n)/m$, $j = 0, 1, \ldots$, excluding the points $z_j = l = 0, 1, \ldots$. Moreover, it is easy to derive formulae for residues and for values of $\zeta(l)$ with the aid of the above formulae for the kernels.

Let us assume that a smooth positive density dx is fixed on M and that A is a self-adjoint operator with respect to this density. Then $\zeta_A(z) = \sum_{j=1}^{\infty} \lambda_j^z$, and the series converges absolutely if $\operatorname{Re} z < -n/m$ and uniformly in z for each $\varepsilon > 0$ if $\operatorname{Re} z < -n/m - \varepsilon$.

The following arguments are due to Atiyah and Bott. Let A be a normal operator that has a complete set of eigenfunctions ψ_j corresponding to the eigenvalues λ_j. Then the operator A^z has a kernel $K_z(x, y)$ which is such that

$$K_z(x, x) = \sum \lambda_j^z |\psi_j(x)|^2, \qquad \int_M K_z(x, x)\, dx = \sum \lambda_j^z.$$

If ψ is an eigenfunction of the operator L^*L corresponding to the eigenvalue $\lambda > 0$, then $L\psi$ is an eigenfunction of LL^* corresponding to the same eigenvalue. Thus L maps the eigensubspace $E_\lambda(L^*L)$ in a one-to-one manner into $E_\lambda(LL^*)$ and $\dim E_\lambda(LL^*) \geqslant \dim E_\lambda(L^*L)$. Similarly, we find that $\dim E_\lambda(L^*L) \geqslant \dim E_\lambda(LL^*)$. Thus for L^*L and LL^* the non-zero eigenvalues and their multiplicities coincide. It is clear that if $\zeta_1(z)$ is the trace of $(I + L^*L)^z$ and $\zeta_2(z)$ the trace of $(I + LL^*)^z$, then

$$\zeta_1(z) - \zeta_2(z) = \dim \ker L - \dim \ker L^* = \operatorname{ind} L.$$

Taking now $A = I + LL^*$, we obtain $K_0(x, x)$ explicitly in terms of the symbol of L. The same is true for $I + L^*L$. Thus we can obtain an explicit analytical formula for the index of L.

Let us note, in particular, the connection with the classical Riemann zeta function $\zeta(z) = \sum_{n=1}^{\infty} n^{-z}$. To do this, we take for M the unit circle and for A the operator $-D_x^2 + P$, where P is the projection operator onto the subspace of constant functions. The eigenvalues of A are n^2, $n = 1, 2, \ldots$, and their multiplicity is 2 if $n > 1$ and 3 if $n = 1$. Hence

$$\zeta_A(z) = \sum_{j=1}^{\infty} \lambda_j^z = 1 + 2\zeta(-2z).$$

The symbol of A is ξ^2. Therefore $b_{-2} = (\xi^2 - \lambda)^{-1}$ and $b_{-2-j} = 0$ for $j > 0$. These facts and the above formulae show that $\gamma_j(x) = 0$ for $j > 0$ and that

$$\gamma_0(x) = -\frac{1}{4\pi} \int_{|\xi|=1} |\xi|^{-1}\, d\xi = -\frac{1}{2}.$$

Thus when $s = 1$, $\zeta(s)$ has a pole with the residue 1. What is more, $\zeta(0) = -\frac{1}{2}$.

5.5. The Asymptotics of the Spectral Function and Eigenvalues. We shall mention some of the simple results regarding the asymptotics of the spectral function of a self-adjoint elliptic operator A on a smooth manifold M (see Shubin [1978] and Seeley [1967]). Assume that $a_m(x, \xi) > 0$ for $\xi \neq 0$. Then A is a semibounded operator, and its eigenfunctions $\varphi_1, \varphi_2, \ldots$ constitute a complete orthogonal system. We shall assume that these functions have been normalised and the corresponding eigenvalues $\lambda_1, \lambda_2, \ldots$ have been arranged in the increasing order.

Let E_t be the spectral projector of A, that is, let

$$E_t u = \sum_{\lambda_j \leqslant t} (u, \varphi_j)\varphi_j.$$

The kernel of this operator is known as the spectral function of A, and is of the form

$$e(x, y, t) = \sum_{\lambda_j \leqslant t} \varphi_j(x)\overline{\varphi_j(y)}.$$

We observe that

$$\int_M e(x, x, t)\, dx = \sum_{\lambda_j \leqslant t} 1 = N(t),$$

where $N(t)$ denotes the number of eigenvalues of A not exceeding t. It is evident that $N(t)$ is a decreasing function and that

$$\zeta_A(z) = \int_0^\infty t^z\, dN(t).$$

Assume that, as $t \to +\infty$,

$$N(t) = ct^\alpha + 0(t^\beta),$$

where $\operatorname{Re}\beta < \operatorname{Re}\alpha$. Then

$$\zeta_A(z) = c\int_1^\infty t^z\, dt^\alpha + f(z) = -\frac{c\alpha}{z+\alpha} + f(z)$$

and f is a holomorphic functon for $\operatorname{Re} z < -\operatorname{Re}\beta$. Thus if we know the poles of $\zeta_A(z)$ and the residues at these points, we can find the asymptotics of $N(t)$.

Similarly, the formula

$$K_z(x, y) = \sum_{j=1}^\infty \lambda_j^z \varphi_j(x)\overline{\varphi_j(y)}$$

implies that

$$K_z(x, x) = \int_0^\infty t^z\, de(x, x, t).$$

Moreover, if $e(x, x, t) = c_0(x)t^{\alpha_0} + 0(t^{\beta_0})$ and $\operatorname{Re}\beta_0 < \operatorname{Re}\alpha_0$, then

$$K_z(x, x) = -\frac{c_0(x)\alpha_0}{z+\alpha_0} + f_0(z),$$

where the function f_0 is holomorphic for $\operatorname{Re} z < -\operatorname{Re}\beta$. Since we know the poles of $K_z(x, x)$ and the residues at these points, we can find, as above, the asymptotics of $e(x, x, t)$.

Thus we obtain the formulae

$$e(x, x, t) \sim \frac{1}{n(2\pi)^n} \int_{|\xi|=1} a_m^{-n/m}(x, \xi)\, d\xi \cdot t^{n/m},$$

$$N(t) \sim \frac{1}{(2\pi)^n} \int_{a_m(x,\xi)<t} dx\, d\xi,$$

$$\lambda_k \sim a k^{m/n} \text{ as } t \to \infty, k \to \infty, \qquad \text{where} \quad a = (2\pi)^m \left(\int_{a_m(x,\xi)<1} dx\, d\xi \right)^{-m/n}.$$

5.6. Complex Powers of an Elliptic Operator with Boundary Conditions. The above constructions can be extended to the far more complicated case of a boundary-value problem for a system of elliptic equatons (Seeley [1969]).

Let A be an elliptic system of smooth differential operators in a domain G and let B be a system of differential operators on the boundary ∂G. Let us consider an unbounded operator A_B defined in $L_p(G)$, $1 < p < \infty$, on the space of functions satisfying the conditions $Bu = 0$ on ∂G.

We assume that (1) the matrix $A_0(x, \xi) - \lambda I$ is invertible for $\lambda \leqslant 0$, $(\xi, \lambda) \neq 0$, and (2) the boundary-value problem

$$A_0(x', 0; \xi', D_t)u(t) = \lambda u(t), \qquad t > 0;$$

$$B_{j0}(x', \xi', D_t)u(0) = c_j, j = 1, \ldots, k,$$

$$u(+\infty) = 0$$

has a unique solution with arbitrary constants $c_j \in \mathbb{C}$, $\lambda \leqslant 0$, $(\xi', \lambda) \neq 0$. Here (x', t) are local coordinates in a neighbourhood of some boundary point and $t > 0$ in G.

These assumptions guarantee the existence of the resolvent $R_\lambda = (A_B - \lambda I)^{-1}$ for $\lambda < 0$ (see Seeley [1969]). What is more, if $|\lambda|$ is large, then

$$(A_B - \lambda I)^{-1} \sim \sum Op(C_{-m-j}) + \sum Op'(d_{-m-j}),$$

where

$$Op(c)f(x) = (2\pi)^{-n} \iint e^{i(x-y)\xi} C(x, \xi) f(y)\, dy\, d\xi,$$

$$Op'(d)f(x', t) = (2\pi)^{1-n} \int_0^\infty ds \int_{R^{n-1}} \int_{R^{n-1}} e^{i(x'-y')\xi'} d(x', t, \xi', s) f(y', s)\, dy'\, d\xi',$$

$C_{-m} = [A_0(x, \xi) - \lambda I]^{-1}$, and where $C_{-m-j}(x, \xi, \lambda)$ are homogeneous in $(\xi, \lambda^{1/m})$ of degree $-m-j$ while $d_{-m-j}(x', t, \xi', s, \lambda)$ are homogeneous in $(\xi, \lambda^{1/m}, t^{-1}, s^{-1})$ of degree $1 - m - j$. The functions C_{-m-j} and d_{-m-j} can be computed explicitly in terms of the coefficients of A and B.

If A_B has a complete system of orthonormal eigenfunctions φ_j, we can put

$$A_B^z f = \sum \lambda_j^z (f, \varphi_j)\varphi_j.$$

For Re $z < n/m$, A_B^z has the kernel

$$K_z(x, y) = \sum \lambda_j^z \varphi_j(x)\overline{\varphi_j(y)}.$$

The trace of this operator is given by

$$\int_G \text{Trace } K_z(x, x)\, dx = \int_G \sum \lambda_j^z |\varphi_j(x)|^2\, dx = \sum \lambda_j^z.$$

If we represent A_B^z in the form

$$A_B^z = \frac{1}{2\pi i} \int_\Gamma \lambda^z (A_B - \lambda I)^{-1}\, d\lambda,$$

we can extend the function $K_z(x, x)$ for $x \notin \partial G$ to a meromorphic function having no more than simple poles at the points $z_j = (j - n)/m$, $j = 0, 1, \ldots$, $z \notin \mathbb{Z}_+$. The residue $\gamma_j(x)$ of this function at z_j depends only on c_{-m-j}. If $z_j = (j - n)/m = l \in \mathbb{Z}_+$, the value of $K_l(x, x)$ can be expressed explicitly in terms of c_{-m-j}.

The function $\int_G K_z(x, x)\, dx$ has similar singularities. We can show that the residues at $z_j = (j - n)/m$ equal

$$\int_G \gamma_j(x)\, dx + \int_{\partial G} \delta_j(x')\, dx',$$

where $\delta_j(x')$ are determined explicitly by the functions d_{-m-j+1}. The same is true also for the values of this function at the points $z_j = l \in Z_+$.

In recent years Ivrii [1980b, 1982], Vasil'ev [1984, 1986] and Melrose have obtained more precise results regarding the asymptotics, as $t \to \infty$, of $N(t)$ in the form $N(t) = a_0 t^{n/m} + a_1 t^{(n-1)/m} + 0(t^{(n-1)/m})$ for boundary-value problems for elliptic equations. These works are based on the study of wave fronts of solutions to the boundary-value problems for suitable hyperbolic equations.

§ 6. Pseudodifferential Operators in $\mathbb{R}^n$ and Quantization

6.1. The Analogy Between the Microlocal Analysis and the Quantization. Let us recall what constitutes the process of quantization of a non-relativistic classical dynamical system with n degrees of freedom. The state of such a system can be uniquely determined by a set of values of the generalized coordinates $(x_1, \ldots, x_m)$ and the generalized impulses $(\xi_1, \ldots, \xi_n)$, where $\xi_j = m_j \dot{x}_j$. In classical mechanics, the quantities under observation are the values of the functions $F(x, \xi)$. The result of observation of F in a state (x_0, ξ_0) is the value $F(x_0, \xi_0)$. In the classical theory, a significant role is played by the function

$$H(x, \xi) = \sum_{i=1}^n \frac{1}{2m_i} \xi_i^2 + V(x),$$

which is referred to as the total energy of the system or as the Hamiltonian. The function V is the potential of the field where the particles move. By the Newton's Second law, $m_i\ddot{x}_i = -\partial V/\partial x_i$, and this implies that

$$\dot{x}_i = \frac{\partial H(x, \xi)}{\partial \xi_i}, \qquad \dot{\xi}_i = -\frac{\partial H(x, \xi)}{\partial x_i}, \qquad i = 1, \dots, n.$$

Evidently, the evolution with respect to time of each observable $F(x, \xi)$ is described by the equation

$$\frac{dF}{dt} = \sum_{i=1}^{n} \left(\frac{\partial H}{\partial \xi_i} \frac{\partial F}{\partial x_i} - \frac{\partial H}{\partial x_i} \frac{\partial F}{\partial \xi_i} \right).$$

The last expression is known as the *Poisson bracket* of the functions H and F, and is denoted by $\{H, F\}$. It follows from this equation that all the laws of classical mechanics are invariant under canonical transformations that preserve the values of the Poisson bracket for any pair of functions.

In the quantum theory, the observable quantities are the self-adjoint operators in a Hilbert space $\mathscr{H}$. The state of the system is determined by a wave function ψ in $\mathscr{H}$. The role of coordinates here is played by the multiplication operator $\psi \to x_j\psi$ while those of impulses by the differentiation operator $\psi \to D_j\psi$. The result of measuring the quantity A in the state ψ is a random quantity having mathematical expectation $\langle A\psi, \psi \rangle$. The Hamilton operator H determines the evolution of the state of the system ψ in accordance with the Schrödinger equation

$$\frac{d\psi}{dt} = iH\psi.$$

In the equivalent Heisenberg theory, the state of the system ψ remains unchanged while every observable A is subjected to the equation

$$\frac{dA}{dt} = i(HA - AH) \equiv i[H, A].$$

Therefore all the laws of quantum mechanics remain invariant under unitary transformations $U: \mathscr{H} \to \mathscr{H}$.

By *quantization* we mean the transition from a classical mechanical system to a quantum mechanical system that is carried out according to the following rules. We take for $\mathscr{H}$ the space $\mathscr{L}_2(\mathbb{R}^n)$ of complex-valued functions $u(x)$ whose squares are integrable with respect to the Lebesgue measure. We associate with each coordinate x_j the operator of multiplication by x_j and with each impulse the operator of differentiation, namely, $u \to hD_ju$. The quantity h is known as the Planck constant. The problem of quantization consists in associating with each real-valued function $a(x, \xi)$ a self-adjoint operator A in $\mathscr{H}$. The following two conditions must be satisfied.

1. The operator A depends on the parameter h, and the commutator $[A, B]$ is of the form $ihC + 0(h)$, where C is an operator associated with the Poisson bracket $\{a, b\}$;

2. In a certain sense, $\lim_{h\to 0} A = a(x, \xi)$, that is, the original classical system must be derivable, as $h \to 0$, from the quantum mechanical system thus constructed.

These conditions do not define the quantization operation uniquely. We shall not discuss here various ways of quantization, but only note the analogy with the microlocal analysis where with the function $a(x, \xi)$ is associated a pseudodifferential operator $a(x, D)$ in such a manner that the Poisson bracket $\{a, b\}$ of the two functions a and b becomes the principal symbol of the commutator $[A, B]$.

The theory of pseudodifferential operators meets the needs of quantum mechanics specially well if to a real symbol there corresponds (even if formally) a self-adjoint operator and if the transition from the coordinate representation to the impulse representation does not present difficulties, that is, if the variables x and ξ play analogous roles.

6.2. Pseudodifferential Operators in $\mathbb{R}^n$. We remark that the quantization rules do not define uniquely the operator that corresponds to the function $a(x, \xi)$ even in the simplest case of differential operators. Thus with the function $x_j\xi_j$ one can associate the following operators:

$$u \to hx_jD_ju; \qquad u \to hD_j(x_ju); \qquad u \to \frac{h}{2}x_jD_ju + \frac{h}{2}D_j(x_ju).$$

The last of these is to be preferred because it is formally self-adjoint. Therefore with the function $a(x, \xi)$ of the general form it is natural to associate the Weyl operator in the form (9).

The function a is not necessarily homogeneous in ξ, and, in general, the variables x and ξ may enjoy equal status. This is especially convenient in the investigation of some problems of quantum mechanics where the transition from the coordinate representation to the impulse representation is natural.

Definition 6.1. A function $a(x, \xi)$ belongs to the class T^m if the following conditions hold:

1. $a \in C^\infty(\mathbb{R}^n_x \times \mathbb{R}^n_\xi)$.
2. For any multi-indices α and β

$$|D^\alpha_\xi D^\beta_x a(x, \xi)| < C_{\alpha\beta}(1 + |x|)^m(1 + |\xi|)^m, \qquad (x, \xi) \in \mathbb{R}^n \times \mathbb{R}^n.$$

To construct the calculus of pseudodifferential operators in this class, it is natural to introduce a small parameter ε and then consider the symbols $a(\varepsilon, x, \xi)$ with $\varepsilon > 0$.

Definition 6.2. A function $a(\varepsilon, x, \xi)$ belongs to the class T^m_+ if

1. $a \in C^\infty(\mathbb{R}^+_\varepsilon \times \mathbb{R}^n_x \times \mathbb{R}^n_\xi)$.
2. For any integer $N > 0$

$$a(\varepsilon, x, \xi) = \sum_{k=0}^{N} \varepsilon^k a_k(x, \xi) + \varepsilon^{N+1} R_N(\varepsilon, x, \xi),$$

where $a_k \in T^m$, and

$$|D_\xi^\alpha D_x^\beta D_\varepsilon^\gamma R_N(\varepsilon, x, \xi)| \leqslant C_{\alpha,\beta,\gamma}(1 + |x|)^m(1 + |\xi|)^m$$

for any multi-indices α, β, and for any integer $\gamma \geqslant 0$. Clearly,

$$D_x^\alpha a \in T_+^m, \qquad D_\xi^\alpha a \in T_+^m, \qquad x^\alpha a \in T_+^{m+|\alpha|}, \qquad \xi^\alpha a \in T_+^{m+|\alpha|}$$

if $a \in T_+^m$.

These relations enable us to construct an algebra of pseudodifferential operators with symbols belonging to T_+^m (Maslov and Fedoryuk [1976]).

Examples 1. The Schrödinger equation

$$ih\frac{\partial\psi}{\partial t} = -\frac{h^2}{2m}\Delta\psi + V(x)\psi$$

describes the motion of a non-relativistic quantum particle of mass m in a field with potential energy $V(x)$. To this equation there corresponds the symbol

$$\varepsilon\tau + \frac{\varepsilon^2}{2m}|\xi|^2 + V(x), \qquad \varepsilon = h,$$

belonging to T_+^2 if $|V(x)| \leqslant C(1 + |x|)^2$.

2. The Helmholtz equation

$$(\Delta + k^2 n^2(x))u(x) = 0$$

describes the propagation of waves of frequency k in a medium with refractive index $n^2(x)$. To this equation there corresponds the symbol

$$\varepsilon^2|\xi|^2 + n^2(x), \qquad \varepsilon = k^{-1},$$

belonging to T_+^2 if $|n(x)| \leqslant C(1 + |x|)$.

3. The difference equation

$$\frac{u_m^{n+1} - 2u_m^n + u_m^{n-1}}{h^2} = a^2\frac{u_{m+1}^n - 2u_m^n + u_{m-1}^n}{h^2}$$

is associated with the wave equation $u_{tt} = a^2 u_{xx}$, where $u_m^n = u(hn, hm)$. Using the formulae

$$e^{hd/dx}f(x) = f(x + h), \qquad e^{hd/dt}f(t) = f(t + h),$$

we obtain the equation

$$[\sin^2(hD_t) - a^2\sin^2(hD_x)]u(t, x) = 0.$$

This equation coincides with the above difference equation if the values of $u(t, x)$ are taken at the points of the mesh. To the given equation there corresponds the symbol

$$\sin^2\varepsilon\tau - a^2\sin^2\varepsilon\xi, \qquad \varepsilon = h,$$

belonging to T_+^0 (Maslov and Fedoryuk [1976]).

Chapter 3
Fourier Integral Operators

§ 1. The Parametrix of the Cauchy Problem for Hyperbolic Equations

1.1. The Cauchy Problem for the Wave Equation. Let us first examine the solution to the Cauchy problem for the wave equation

$$\frac{\partial^2 u}{\partial t^2} = a^2 \Delta u \qquad \text{for} \quad t > 0,$$

$$u = \varphi, \qquad \frac{\partial u}{\partial t} = \psi \qquad \text{for} \quad t = 0,$$

where $a = \text{const.}$ and φ, ψ are given functions from $C_0^\infty(\mathbb{R}^n)$. If we put

$$v(t, \xi) = \int u(t, x) e^{-ix\xi} \, dx,$$

then this problem reduces to the Cauchy problem for an ordinary differential equation:

$$\frac{d^2 v}{dt^2} = -a^2 |\xi|^2 v \qquad \text{for} \quad t > 0,$$

$$v = \tilde{\varphi}, \qquad \frac{dv}{dt} = \tilde{\psi} \qquad \text{for} \quad t = 0,$$

This yields

$$v(t, \xi) = \tilde{\varphi}(\xi) \cos at|\xi| + \tilde{\psi}(\xi) \frac{\sin at|\xi|}{a|\xi|},$$

so that

$$u(t, x) = (2\pi)^{-n} \int \frac{1}{2}\left[\tilde{\varphi}(\xi) + \frac{1}{ia|\xi|}\tilde{\psi}(\xi)\right] e^{iat|\xi| + ix\xi} \, d\xi$$
$$+ (2\pi)^{-n} \int \frac{1}{2}\left[\tilde{\varphi}(\xi) - \frac{1}{ia|\xi|}\tilde{\psi}(\xi)\right] e^{-iat|\xi| + ix\xi} \, d\xi.$$

Thus the solution of the Cauchy problem has been represented as a sum of two integrals that define Fourier integral operators with the phase functions $\pm at|\xi| + x\xi$.

In general, by a Fourier integral operator we mean an operator that maps a function u into the function

$$Au(x) = (2\pi)^{-n} \int a(x, \xi) \tilde{u}(\xi) e^{iS(x, \xi)} \, d\xi.$$

Here $a \in S^m$ is the symbol of the operator and S is the phase function. It is usually assumed that the phase function satisfies the non-degeneracy condition $\det \left\| \frac{\partial^2 S}{\partial x_i \partial \xi_j} \right\| \neq 0$ and that $S(x, t\xi) = tS(x, \xi)$ for $t > 0$. Thus a Fourier integral operator may be regarded as a generalization of a pseudodifferential operator.

1.2. The Cauchy Problem for the Hyperbolic Equation of an Arbitrary Order. Let $P(t, x, D_t, D_x)$ be a strictly hyperbolic differential operator of order m, with smooth coefficients, in the strip $\Omega = \{(t, x): 0 < t < T, x \in \mathbb{R}^n\}$. This implies that the roots $\lambda_1, \ldots, \lambda_m$ of the characteristic equation

$$P_0(t, x, \lambda, \xi) = 0$$

are real and distinct if $\xi \neq 0$, and, moreover,

$$|\lambda_j(t, x, \xi) - \lambda_k(t, x, \xi)| \geqslant c_0 = \text{const} > 0$$

if $(t, x) \in \Omega$, $|\xi| = 1$, $j \neq k$.

We consider the Cauchy problem

$$P(t, x, D_t, D_x)u = 0 \quad \text{for} \quad (t, x) \in \Omega,$$

$$D_t^j u(0, x) = \varphi_j(x) \quad \text{for} \quad x \in \mathbb{R}^n, j = 0, 1, \ldots, m-1.$$

The solution of this problem for a sufficiently small t is given by a formula of the form

$$u(t, x) = (2\pi)^{-n} \int \sum_{j,k=0}^{m-1} a_{jk}(t, x, \xi) \tilde{\varphi}_j(\xi) e^{iS_k(t, x, \xi)}\, d\xi, \tag{1}$$

where the functions S_k satisfy the eikonal equations

$$\frac{\partial S_k}{\partial t} = \lambda_k\left(t, x, \frac{\partial S_k}{\partial x}\right) \quad \text{for} \quad t > 0$$

and the initial conditions

$$S_k = x \cdot \xi \quad \text{for} \quad t = 0.$$

In physical terms, S_k signifies the distance covered in time t by a ray of light issuing from the point x in the direction of the vector ξ in a medium the optical properties of which are characterized by the function λ_k. It is not difficult to see that S_k satisfies the non-degeneracy condition if T is sufficiently small, because for $t = 0$ we have

$$\det \|\partial^2 S_k / \partial x_i \partial \xi_j\| = 1, \qquad k = 0, \ldots, m-1.$$

The functions a_{jk} (more precisely, their asymptotic series) can be easily found from the linear transport equations which are obtained by equating to zero the functions b_{jk} in the following equality obtained from (1):

$$Pu(t, x) = (2\pi)^{-n} \int \sum_{j,k=0}^{m-1} b_{jk}(t, x, \xi) \tilde{\varphi}_j(\xi) e^{iS_k(t, x, \xi)}\, d\xi.$$

To solve the Cauchy problem on a large segment $[0, T]$, it is necessary to use the Maslov canonical operator (see § 2.3 below).

1.3. The Method of Stationary Phase. If $\operatorname{Re} a \geqslant 0$, $a \neq 0$, then the Fourier transform of e^{-ay^2} is $(2\pi/a)^{-1/2}e^{-\xi^2/2a}$. Here, and in what follows, $z^{1/2}$ is defined in such a way that $1^{1/2} = 1$. Therefore the Fourier transform of $e^{i\sum a_j y_j^2/2}$, $a_j \in \mathbb{R}\backslash 0$, coincides with the function

$$(2\pi)^{n/2} \prod |a_j|^{-1/2} e^{\pi i\sigma/4} e^{-i\sum \eta_j^2/2a_j},$$

where $\sigma = \sum \operatorname{sgn} a_j$. Similarly, if A is a non-degenerate symmetric matrix, the Fourier transform of $e^{i(Ay,y)/2}$ is given by

$$(2\pi)^{n/2} |\det A|^{-1/2} e^{\pi i \operatorname{sgn} A/4} e^{-i(A^{-1}\eta,\eta)/2},$$

where $\operatorname{sgn} A$ denotes the signature of A.

If $f \in C_0^\infty(\mathbb{R}^n)$, the Parseval theorem gives

$$\int f(y)e^{i\lambda(Ay,y)/2}\,dy = (2\pi\lambda)^{-n/2} |\det A|^{-1/2} e^{\pi i \operatorname{sgn} A/4} \int \tilde{f}(\eta) e^{-i(A^{-1}\eta,\eta)/2}\,d\eta.$$

Let us expand the exponential function on the right-hand side of this equation in a Taylor's series. Since

$$\left| e^{iz} - \sum_{k=0}^{N-1} \frac{(iz)^k}{k!} \right| \leqslant \frac{|z|^N}{N!}, \qquad z \in \mathbb{R},$$

it follows that

$$\left| \int f(y)e^{i\lambda(Ay,y)/2}\,dy - S_N(\lambda) \right| \leqslant C\lambda^{-N-(n/2)} \sum_{|\alpha| \leqslant 2N+n+1} \int |D^\alpha f(y)|\,dy,$$

where

$$S_N(\lambda) = \left(\frac{2\pi}{\lambda}\right)^{n/2} |\det A|^{-1/2} e^{\pi i \operatorname{sgn} A/4} \sum_{|\alpha|<2N} C_\alpha D^\alpha f(0) \lambda^{-|\alpha|/2}.$$

The coefficients C_α here do not depend on λ or f, and are of the form

$$C_\alpha = \frac{1}{\alpha!} \frac{\partial^{|\alpha|}}{\partial \eta^\alpha} e^{-i(A^{-1}\eta,\eta)/2} \Big|_{\eta=0}.$$

These coefficients vanish if $|\alpha|$ is odd.

We now consider the integral

$$I(\lambda) = \int f(y) e^{i\lambda S(y)}\,dy,$$

where $S \in C^\infty$, $f \in C_0^\infty$ and S has a unique critical point x_0 in $\operatorname{supp} f$. We shall assume that this point is a point of non-degeneracy, that is, we shall assume that $\det \partial^2 S(x_0)/\partial x^2 \neq 0$. Then by a non-degenerate coordinate transformation $S(x)$ can be transformed to the form $S = \sum a_j y_j^2/2$ in a neighbourhood of x_0. Applying the above estimate, we have

$$|I(\lambda) - S_N(\lambda)| \leqslant C\lambda^{-N-(n/2)} \sum_{|\alpha|\leqslant 2N+n+1} \int |D^\alpha f(y)|\, dy,$$

where

$$S_N(\lambda) = \lambda^{-n/2} e^{i\lambda S(x_0)} \sum_{j=0}^{N-1} a_j(\varphi, S)\lambda^{-j}.$$

The principal term of the asymptotics is

$$S_1(\lambda) = \left(\frac{2\pi}{\lambda}\right)^{n/2} \left|\det \frac{\partial^2 S(x_0)}{\partial x^2}\right|^{-1/2} e^{i\lambda S(x_0)} e^{(\pi/4)i \operatorname{sgn}(\partial^2 S(x_0)/\partial x^2)} \varphi(x_0).$$

§2. The Maslov Canonical Operator

2.1. The Maslov Index. To construct a Maslov canonical operator or a global Fourier integral operator we need some topological concepts introduced in the works of Maslov [1965a] and Arnol'd [1967].

We consider the space $\mathbb{C}^n$ of points $z = x + i\xi$, $x \in \mathbb{R}^n$, $\xi \in \mathbb{R}^n$, where the unitary structure is defined by the scalar product

$$(z^1, z^2) = \sum_{j=1}^{n} (z^1)_j \overline{(z^2)_j} = \sum_{j=1}^{n} (x_j^1 x_j^2 + \xi_j^1 \xi_j^2) + i \sum_{j=1}^{n} (x_j^1 \xi_j^2 - x_j^2 \xi_j^1).$$

An n-dimensional linear subspace Λ in $\mathbb{C}^n$ is called a Lagrange subspace if $\operatorname{Im}(z^1, z^2) = 0$ for any two points $z^1, z^2 \in \Lambda$.

For example, $x = 0$ or $\xi = 0$ are Lagrange subspaces. It is natural to refer to the Lagrange subspaces as real-similar because every such subspace can be transformed to the real subspace $\xi = 0$ by an unitary transformaton U. The transformation U is defined to within an orthogonal transformation of a real subspace. Therefore the manifold $\Lambda(n)$ of Lagrange subspaces is isomorphic to the factor group $U(n)/0(n)$.

Since the determinant of an orthogonal transformation is ± 1, the quantity $(\det U)^2$ depends only on Λ and, since $|\det U| = 1$, determines a point on $S_1 \subset \mathbb{C}$. Thus if we take a closed curve $\gamma\colon S^1 \to \Lambda(n)$ to it there corresponds a mapping $S^1 \to S^1$ that is characterized by its index. It is possible to deduce from here that $H^1(\Lambda(n), \mathbb{Z}) \simeq \mathbb{Z}$, and for α, a generator of the group of cohomologies $H^1(\Lambda(n), \mathbb{Z})$, one can take the value of this index.

Definition 2.1. A submanifold M^n in $\mathbb{C}^n$ is called Lagrange if at each of its points the tangent space is Lagrange.

Let M be a Lagrange manifold, and let $\tau\colon M \to \Lambda(n)$ be a map that associates a tangent space with each point $x \in M$. This map induces another map $\tau^*\colon H^1(\Lambda(n), \mathbb{Z}) \to H^1(M, \mathbb{Z})$ under which a class α^* of one-dimensional cohomologies of M is associated with α.

Definition 2.2. The class $\alpha^* \in H^1(M, \mathbb{Z})$ is known as the *Maslov index.*

The Maslov index can also be defined as the index of intersection with a certain two-sided $(n-1)$-dimensional cycle on M, namely, the cycle of singularities.

Let M be an n-dimensional Lagrange manifold and $f: M \to \mathbb{R}^n$ the projection of M onto the plane $\xi = 0$, that is, $f(x, \xi) = x$. The set Σ of points of M, where the rank of df is less than n, is known as the set of singularities of f. This set consists, in general, of an open $(n-1)$-dimensional manifold Σ^1, where the rank of df equals $n-1$, and the boundary $\Sigma - \Sigma^1$, the dimension of which is strictly less than $n-2$. Thus Σ defines an $(n-1)$-dimensional cycle in M. In a neighbourhood of any point $x \in \Sigma^1$ the Lagrange manifold M is defined by n equations of the form

$$x_k = x_k(\xi_k, x_{\hat{k}}), \qquad \xi_{\hat{k}} = \xi_{\hat{k}}(\xi_k, x_{\hat{k}}),$$

where $\hat{k} = 1, \ldots, k-1, k+1, \ldots, n$ for a certain k, $1 \leqslant k \leqslant n$. It turns out that the manifold Σ^1 can be made orientable by labelling the side where $\partial x_k / \partial \xi_k > 0$ as positive. The expression "in general" used above signifies that the statement becomes valid after an unitary rotation, no matter how small.

If on a Lagrange manifold M in "generic state" we prescribe a curve γ that is transversal to the cycle Σ and whose end points are not on Σ, then by the Maslov index we shall mean the index of its intersection with Σ, that is, the difference between the number of transition points from the negative side to the positive side and the number of transition points in the reverse direction.

2.2. Pre-canonical Operator. Let M be a Lagrange manifold in the phase space $\mathbb{R}^{2n}$ of points (x, ξ). We shall denote by α a subset of the set $(1, 2, \ldots, n)$ and by $|\alpha|$ the number of elements in α. We set $\bar{\alpha} = (1, 2, \ldots, n) \backslash \alpha$.

Proposition 2.1. *For each point $A \in M$ there exists an α such that in some neighbourhood of this point the coordinates $(x_\alpha, \xi_{\bar{\alpha}})$ may be chosen as the local coordinates for M.*

The atlas of M where the local coordinates have been chosen in accordance with this proposition is known as *canonical* and the coordinates themselves are known as the *local canonical* coordinates.

Proposition 2.2. *Let $(x_\alpha, \xi_{\bar{\alpha}})$ be the local canonical coordinates on M in a neighbourhood of A. Then there exists a smooth scalar function $S(x_\alpha, \xi_{\bar{\alpha}})$ such that*

$$x_{\bar{\alpha}} = -\frac{\partial S}{\partial \xi_{\bar{\alpha}}}, \qquad \xi_\alpha = \frac{\partial S}{\partial x_\alpha}.$$

in this neighbourhood.

S is called a generating function. What is more,

$$S = -x_\alpha \cdot \xi_\alpha + \int_{(x^0, \xi^0)}^{(x, \xi)} \xi \, dx$$

and the integral is independent of the path joining the points (x^0, ξ^0) and (x, ξ) because M is a Lagrange manifold.

We first assume that M is uniquely projected onto the x-space. We introduce the operator

$$K\colon C_0^\infty(M) \to C^\infty(\mathbb{R}_x^n)$$

by the formula

$$(K\varphi)(x) = \sqrt{\left|\frac{d\sigma}{dx}\right|}\, e^{i\lambda S(x)} \varphi(x),$$

where S is the generating function and $d\sigma$ is the volume density on M that is defined by the standard density dx by means of projection.

Note that K preserves the norm in L_2:

$$\int |K\varphi(x)|^2\, dx = \int |\varphi|^2\, d\sigma.$$

Suppose now that M is uniquely projected onto the ξ-space. Then we may assume that the generating function S_1 depends only on ξ and that the equation $x = -\partial S_1/\partial \xi$ defines M. In this case, we set

$$(K_1\varphi)(x) = \lambda^{n/2}(2\pi)^{-n} \int \sqrt{\left|\frac{d\sigma(\xi)}{d\xi}\right|}\, e^{i\lambda S_1(\xi)} \varphi(x(\xi)) e^{i\lambda x\cdot\xi}\, d\xi.$$

Next, suppose that M is uniquely projected onto both x- and ξ-spaces. Let us compare the values of the operators K and K_1 so that $x = x(\xi)$ and $\xi = \xi(x)$ on M.

We compute the last integral by the method of the stationary phase (see § 1.3) noting that on M

$$S(x) = \int_{(x^0,\xi^0)}^{(x,\xi)} \xi\, dx, \qquad S_1(\xi) = -x\xi + \int_{(x^0,\xi^0)}^{(x,\xi)} \xi\, dx = -x\xi + S(x).$$

We obtain, with the assumption that $|\partial^2 S_1(\xi)/\partial\xi^2| \neq 0$,

$$\begin{aligned}(K_1\varphi)(x) &= \sqrt{\left|\frac{d\sigma}{d\xi}\right|} \left|\frac{\partial^2 S_1(\xi)}{\partial \xi^2}\right|^{-1/2} e^{(i\pi/4)\,\mathrm{sgn}\,(\partial^2 S_1/\partial\xi^2)} e^{i\lambda S(x)} \varphi(x) + r(x) \\ &= e^{(i\pi/4)\,\mathrm{sgn}\,(\partial^2 S_1/\partial\xi^2)} K\varphi(x) + r(x),\end{aligned}$$

where $r(x)$ is a function such that $\|r\|_{L_2} = 0(\lambda^{-1})$.

Thus if M is uniquely projected onto the ξ-space, then we must set

$$K\varphi(x) = e^{-i(\pi/4)\,\mathrm{sgn}\,(\partial^2 S_1/\partial\xi^2)} \lambda^{n/2}(2\pi)^{-n} \int \sqrt{\left|\frac{d\sigma(\xi)}{d\xi}\right|}\, e^{i\lambda S_1(\xi) + i\lambda x\xi} \varphi(x(\xi))\, d\xi,$$

where $\mathrm{sgn}\, \partial^2 S_1/\partial\xi^2$ denotes the signature of the matrix $\partial^2 S_1/\partial\xi^2$ (that is, the difference of the number of positive and negative eigenvalues).

When x_α, $\xi_{\bar{\alpha}}$ serve as canonical coordinates on M, we set

$$K\varphi(x) = e^{-(i\pi/4)\,\mathrm{sgn}\,(\partial^2 S/\partial\xi_{\bar{\alpha}}^2)}\lambda^{|\bar{\alpha}|/2}(2\pi)^{-|\bar{\alpha}|}\int\sqrt{\left|\frac{d\sigma}{dx_\alpha\,d\xi_{\bar{\alpha}}}\right|}\,e^{i\lambda S(x_\alpha,\xi_{\bar{\alpha}})+i\lambda x_{\bar{\alpha}}\xi_{\bar{\alpha}}}$$

$$\times\ \varphi(x_\alpha,\xi_{\bar{\alpha}})\,d\xi_{\bar{\alpha}}.$$

2.3. The Canonical Operator. Let M be a simply connected Lagrange manifold, and let $\sum\varphi_j = 1$ be the decomposition of unity on M subject to a fixed canonical atlas. By the *Maslov canonical operator* we mean the operator

$$K_M(\varphi)\colon C_0^\infty(M) \to C^\infty(\mathbb{R}_x^n)$$

defined by the formula

$$K_M(\varphi) = \sum_{j=1}^{N} K_j(\varphi_j\varphi).$$

Here K_j is the operator constructed for the jth neighbourhood in accordance with the above-mentioned rule.

The quantity sgn $\partial^2 S/\partial\xi_{\bar{\alpha}}^2$, the signature of $\partial^2 S/\partial\xi_{\bar{\alpha}}^2$, is not defined uniquely, because the local coordinates can be chosen, in general, in more than one way. However, arguments similar to the above show that the value of the operator $K(\varphi)$ for functions φ having support in such a coordinate neighbourhood is well defined to within functions which have the norm of the order $0(\lambda^{-1})$ in L_2. At any rate, this is true for a simply connected manifold M on which the generating function is chosen in every neighbourhood by means of integration from one common point (x_0, ξ_0).

In the general case, when M is not simply connected, the canonical operator may be well defined on M (to within $0(\lambda^{-1})$) if and only if

1. $\int_\gamma \xi\,dx = 0$ for every closed curve γ on M,
2. for every closed curve γ on M its Maslov index is zero.

2.4. Some Applications

1. The Bohr's Quantization Rule. Let V be a real-valued continuous function on R such that $V(x) \to +\infty$ as $|x| \to \infty$. Consider the one-dimensional Schrödinger equation

$$-h^2\psi'' + V(x)\psi = E\psi.$$

For each fixed $h > 0$ this operator has a pure discrete spectrum and it consists of simple eigenvalues

$$E_0(h) < E_1(h) < \cdots < E_m(h) < \cdots,$$

and $E_m(h) \to +\infty$ as $m \to +\infty$. Moreover,

$$E_m(h) = \lambda_m(h) + 0(h^2),$$

where λ_m is determined from the equation

$$\int_{x_-}^{x_+} \sqrt{\lambda - V(x)}\, dx = \pi h \left(m + \frac{1}{2} \right)$$

in which x_+, x_-, $x_- < x_+$, are the roots of the equation $V(x_\pm) = \lambda$. This is the condition, referred to as the Bohr's quantization formula, in accordance with which the Maslov canonical operator can be constructed on the one-dimensional Lagrange manifold $\xi^2 + V(x) = \lambda$, that is, conditions 1 and 2 of § 2.3 above are satisfied.

2. The Quasi-Classical Asymptotics of the Cauchy Problem for the Schrödinger Equation. Consider the equation

$$ih\frac{\partial\psi}{\partial t} = -\frac{h^2}{2}\frac{\partial^2\psi}{\partial x^2} + V(x)\psi, \qquad t > 0, x \in \mathbb{R},$$

with the initial condition

$$\psi = \varphi(x)e^{(i/h)f(x)} \quad \text{for} \quad t = 0.$$

Let M be the Lagrange manifold $\{(x, \xi), \xi = f'(x)\}$. Let us consider the classical dynamical system

$$\dot{x} = \frac{\partial H}{\partial \xi}, \qquad \dot{\xi} = -\frac{\partial H}{\partial x}, \qquad H = \frac{1}{2}\xi^2 + V(x).$$

This system defines a phase flow g_t: $\mathbb{R}^2 \to \mathbb{R}^2$ by means of the translations along its trajectories. The curves $M_t = g_t M$ are Lagrange manifold for each $t > 0$. We shall assume that these curves and the manifold $M_T = \bigcup_{0<t<T} M_t$ are smooth. It is not difficult to see that M_T is a two-sided Lagrange manifold. Maslov [1965a] has shown that

$$\psi(t, x, h) = K \sum_{j=0}^{N} (ih)^k \psi_k + R_{N+1}(t, x, h)$$

for $0 < t < T$, with $\max \|R_{N+1}(t, x, h)\|_{L_2(R)} \leqslant C_N h^{N+1}$. Here K is the canonical operator constructed for M_T.

§ 3. Fourier Integral Operators

3.1. The Oscillatory Integrals. Let Ω be a domain in $\mathbb{R}^n$, and let $u \in C_0^\infty(\Omega)$. Consider the integral

$$I_S(au) = \iint e^{iS(x,\theta)} a(x, \theta) u(x)\, dx\, d\theta, \tag{2}$$

where $\Theta \in \mathbb{R}^N$, $x \in \Omega$. We shall assume that there exists a constant $C_{\alpha,\beta,K}$ such that

$$|\partial_\theta^\alpha \partial_x^\beta a(x, \theta)| \leqslant C_{\alpha,\beta,K}(1 + |\theta|)^{m - \rho|\alpha| + \delta|\beta|}$$

for all α, β and for any compact set $K \subset \Omega$. These inequalities define the class $S^m_{\rho,\delta}(\Omega \times \mathbb{R}^N)$ of symbols introduced by Hörmander [1971]. The class $S^m_{1,0}$ is usually denoted by S^m, and one simply writes S^m_ρ instead of $S^m_{\rho,1-\rho}$, and $S^{-\infty} = \bigcap S^m$.

The function $S(x, \theta)$ is known as the *phase function.* It is assumed that $S \in C^\infty(\Omega \times \mathbb{R}^N \backslash 0)$, and that S is positive, real-valued and homogeneous in θ of degree 1 and $\operatorname{grad}_{x,\theta} S \neq 0$. S is said to be *non-degenerate* if the differentials $d(\partial S/\partial \theta_k)$ are linearly independent, $k = 1, 2, \ldots, N$, at those points where $\partial S/\partial \theta = 0$.

The integral (2) in which $a \in S^m_{\rho,\delta}$ and S is a phase function is referred to as an *oscillatory integral.*

Proposition 3.1. *There exists an operator*

$$L = \sum_{j=1}^{N} a_j(x, \theta)\frac{\partial}{\partial \theta_j} + \sum_{k=1}^{n} b_k(x, \theta)\frac{\partial}{\partial x_k} + C(x, \theta),$$

with $a_j \in S^0$, $b_k \in S^{-1}$ *and* $C \in S^{-1}$, *such that*

$${}^tLe^{iS} = e^{iS},$$

where the operator ${}^tL$ *is the transpose of* L, *that is,*

$${}^tLu = -\sum \frac{\partial}{\partial \theta_j}(a_j u) - \sum \frac{\partial}{\partial x_k}(b_k u) + Cu.$$

If $m < -N$, then the integral (2) converges absolutely. In this case,

$$I_s(au) = \iint e^{iS(x,\theta)} L^k(a(x, \theta)u(x))\, dx\, d\theta \tag{3}$$

for any integer $k \geqslant 0$. Moreover, $L^k(au) \in S^{m-k\sigma}_{\rho,\delta}$, where $\sigma = \min(\rho, 1 - \delta)$. Assuming that $\sigma > 0$, we can define the integral (2) by the formula (3) if k is so large that $m - k\sigma < -N$; thus the integral (3) converges absolutely.

The integral (2) can be defined in an alternative way also. If $h \in C_0^\infty(\mathbb{R}^N)$, with $h(\theta) = 1$ in a neighbourhood of $0 \in \mathbb{R}^N$, we set

$$I_{S,\varepsilon}(au) = \iint h(\varepsilon\theta) e^{iS(x,\theta)} a(x, \theta) u(x)\, dx\, d\theta, \qquad \varepsilon > 0.$$

Since

$$I_{S,\varepsilon}(au) = \iint e^{iS(x,\theta)} L^k(h(\varepsilon\theta) a(x, \theta) u(x))\, dx\, d\theta$$

for any k, we can choose k so that $m - k\sigma < -N$ and then pass on to limit as $\varepsilon \to 0$. Hence

$$\lim_{\varepsilon \to 0} I_{S,\varepsilon}(au)$$

exists and is independent of h. It is natural to take this limit as $I_S(au)$.

3.2. The Local Definition of the Fourier Integral Operator. Let Ω and Ω' be domains in $\mathbb{R}^n$ and $\mathbb{R}^m$, respectively. Consider the integral

$$Au(x) = \iint e^{iS(x,y,\theta)} a(x, y, \theta) u(y)\, dy\, d\theta. \tag{4}$$

Here $u \in C_0^\infty(\Omega')$, $x \in \Omega$, S is the phase function on $\Omega \times \Omega' \times \mathbb{R}^N$ and $a \in S^m_{\rho,\delta}(\Omega \times \Omega' \times \mathbb{R}^N)$, with $\rho > 0$, $\delta < 1$.

If $v \in C_0^\infty(\Omega)$, the integral

$$(Au, v) = \iiint e^{iS(x,y,\theta)} a(x, y, \theta) u(y) v(x)\, dx\, dy\, d\theta$$

makes sense and is an oscillatory integral. Hence A is a linear operator acting from $C_0^\infty(\Omega)$ into $\mathscr{D}'(\Omega)$. This is the operator that is known as a *Fourier integral operator*. The kernel of A is the distribution $K_A \in \mathscr{D}'(\Omega \times \Omega')$ defined by the oscillatory integral

$$(K_A, w) = \iiint e^{iS(x,y,\theta)} a(x, y, \theta) w(x, y)\, dx\, dy\, d\theta, \qquad w \in C_0^\infty(\Omega \times \Omega').$$

It is easy to see that K_A is a kernel of A in the sense of Schwartz and that A can be extended to a continuous linear map

$$A\colon \mathscr{E}'(\Omega') \to \mathscr{D}'(\Omega).$$

Examples of Fourier integral operators have been cited in § 3.1 in connection with the solution of the Cauchy problem for hyperbolic equations.

The simplest variant of a Fourier integral operator is provided by an operator of the form

$$Au(x) = (2\pi)^{-n} \int a(x, \xi) \tilde{u}(\xi) e^{iS(x,\xi)}\, d\xi, \tag{5}$$

which was examined above in § 2.3.

A particular case of the Fourier integral operator is the pseudodifferential operator corresponding to the phase function $S(x, \xi) = (x, \xi)$ in (5) or $S(x, y, \theta) = (x - y, \theta)$ in (4). By a change of variables, one can transform to a pseudodifferential operator the Fourier integral operator with the phase function $S(x, y, \theta)$ if this function is linear in θ and if $\partial S/\partial\theta = 0$ implies $x = y$.

3.3. The Equivalence of Phase Functions. Consider the oscillatory integral

$$I_S(au) = \iint a(x, \theta) u(x) e^{iS(x,\theta)}\, dx\, d\theta,$$

where S is a phase function and $a \in S^m_{\rho,\delta}$. If we perform a change of variables

$$(x, \theta) \to (x, \tilde{\theta}(x, \theta))$$

that preserves fibres in $T^*(\Omega)$, we obtain

$$I_S(au) = \iint e^{i\tilde{S}(x,\tilde{\theta})} \tilde{a}(x, \tilde{\theta}) u(x)\, dx\, d\tilde{\theta},$$

where

$$S(x, \tilde{\theta}(x, \theta)) = S(x, \theta), \qquad \tilde{a}(x, \tilde{\theta}(x, \theta))|D\tilde{\theta}/D\theta| = a(x, \theta).$$

If the above transformation is diffeomorphic, then $\tilde{S}$ is a phase function and $\tilde{a} \in S^m_{\rho,\delta}$. In this case S and $\tilde{S}$ are said to be locally equivalent.

Consider the map

$$C_S \ni (x, \theta) \mapsto \left(x, \frac{\partial S}{\partial x}\right) \in T^*(\Omega)\backslash 0,$$

where

$$C_S = \left\{(x, \theta): \frac{\partial S(x, \theta)}{\partial \theta} = 0\right\}.$$

It is easy to see that the range of this map does not change when one goes over to the equivalent phase function and that Λ, the set of points lying in this range, constitutes a conical Lagrange manifold in ξ.

Example 3.1. If the function S is linear in θ, then $C_S = Y \times \mathbb{R}^N$, where Y is a submanifold of codimension N.

Example 3.2. Let $H(\xi)$ be a positive homogeneous function of degree one that is smooth for $\xi \neq 0$, and let

$$S(x, \xi) = (x, \xi) - H(\xi).$$

Then the condition $\partial S/\partial \xi = 0$ is equivalent to the condition that $x = H'(\xi)$ and

$$\Lambda = \{H'(\xi), \xi\}.$$

This example is of a general nature, as shown by the next theorem.

Theorem (Maslov and Fedoryuk [1976]). *Let $\Lambda \subset T^*(X)$ be a conical Lagrange manifold in ξ. For each point $\lambda_0 \in \Lambda$ we can choose local coordinates $x_1, \ldots, x_n$ such that in some neighbourhood of this point Λ is defined by means of the phase function $S(x, \xi) = x \cdot \xi - H(\xi)$, where H is a smooth function for $\xi \neq 0$ that is positive and homogeneous of degree one.*

The following theorem of Hörmander's answers the question posed above regarding the equivalence of phase functions.

Theorem (Hörmander [1971]). *Let S and $\tilde{S}$ be non-degenerate functions in conical neighbourhoods of the points $(x_0, \theta_0) \in \Omega \times (\mathbb{R}^N\backslash 0)$ and $(x_0, \tilde{\theta}_0) \in \Omega \times (\mathbb{R}^{\tilde{N}}\backslash 0)$, respectively. S and $\tilde{S}$ are equivalent if and only if*

(I) *they define at corresponding points one and the same Lagrange manifold;*

(II) $N = \tilde{N}$;

(III) *the signatures of the matrices $\partial^2 S(x_0, \theta_0)/\partial\theta^2$ and $\partial^2 \tilde{S}(x_0, \tilde{\theta}_0)/\partial\theta^2$ coincide.*

3.4. The Connection with the Lagrange Manifold. We shall demonstrate that the values of the Fourier integral operators are connected with the Lagrange manifold in an invariant manner. To do this, consider the integral

$$\iint e^{iS(x,\theta)}a(x,\theta)u(x)\,dx\,d\theta.$$

A diffeomorphic change of variables

$$x = x(\tilde{x}), \qquad \theta = \theta(\tilde{x}, \tilde{\theta})$$

transforms this integral to the form

$$\iint e^{i\tilde{S}(\tilde{x},\tilde{\theta})}\tilde{a}(\tilde{x},\tilde{\theta})\tilde{u}(\tilde{x})\,d\tilde{x}\,d\tilde{\theta},$$

where

$$\tilde{S}(\tilde{x},\tilde{\theta}) = S(x(\tilde{x}), \theta(\tilde{x},\tilde{\theta})), \qquad \tilde{u}(\tilde{x}) = |Dx/D\tilde{x}|^{1/2}u(x),$$

$$\tilde{a}(\tilde{x},\tilde{\theta}) = a(x(\tilde{x}), \theta(\tilde{x},\tilde{\theta}))\left|\frac{Dx}{D\tilde{x}}\right|^{1/2}\left|\frac{D\theta}{D\tilde{\theta}}\right|.$$

These formulae show that it is more convenient to consider, in place of distributions u, densities of order $\frac{1}{2}$ that are transformed according to the formulae just mentioned.

Let $C = \{(x,\theta), \partial S(x,\theta)/\partial\theta = 0\}$. Let d_C be a density on C which is a pullback of the Dirac measure in $\mathbb{R}^N$ under the map $(x,\theta) \to \partial S(x,\theta)/\partial\theta$. Finally, let $\lambda_1, \dots, \lambda_n$ be local coordinates on C that are extended to a neighbourhood of this manifold. Then

$$d_C = |D(\lambda_1, \dots, \lambda_n, \partial S/\partial\theta_1, \dots, \partial S/\partial\theta_n)/D(x,\theta)|^{-1}\,d\lambda_1 \dots d\lambda_n.$$

It can be shown (Hörmander [1971]) that

$$|D(\lambda, \partial S/\partial\theta)/D(x,\theta)|^{-1/2}a(x,\theta) = |D(\tilde{\lambda}, \partial\tilde{S}/\partial\tilde{\theta})/D(\tilde{x},\tilde{\theta})|^{-1/2}\tilde{a}(\tilde{x},\tilde{\theta}),$$

that is, the density of order $\frac{1}{2}$ on Λ, which is the pushforward of the density $a\sqrt{d_C}$ under the map $C \ni (x,\theta) \to (x, \partial S/\partial x) \in \Lambda$ coincides with the density obtained in the same manner from $\tilde{S}$ and $\tilde{a}$.

Thus the integral under consideration is connected in an invariant manner with the Lagrange manifold, at least when the number of independent variables N does not change under the coordinate transformation. That the integral is invariant under transition to another parametrization $\theta' = (\theta_1, \dots, \theta_{N'})$ can be checked by the method of stationary phase. We sum up the above discussion in the following.

Theorem (Hörmander [1971]). *Let* $S(x,\theta)$ *and* $\tilde{S}(x,\tilde{\theta})$ *be non-degenerate phase functions in neighbourhoods of the points* $(x_0,\theta_0) \in \Omega \times \mathbb{R}^N$ *and* $(x_0,\tilde{\theta}_0) \in \Omega \times \mathbb{R}^{\tilde{N}}$ *defining elements of a common Lagrange manifold, and let* $\partial S(x_0,\theta_0)/\partial x = \partial\tilde{S}(x_0,\tilde{\theta}_0)/\partial x = \xi_0$.

(I) *The difference* $\sigma = \operatorname{sgn} \partial^2 S(x,\theta)/\partial\theta^2 - \operatorname{sgn} \partial^2\tilde{S}(x,\tilde{\theta})/\partial\tilde{\theta}^2$ *for the points* θ *and* $\tilde{\theta}$ *that are connected by the relations*

$$\partial S(x,\theta)/\partial\theta = 0, \qquad \partial S(x,\tilde{\theta})/\partial\tilde{\theta} = 0,$$
$$\partial S(x,\theta)/\partial x = \partial\tilde{S}(x,\tilde{\theta})/\partial x = \xi \in T_x^*,$$

is a constant in some neighbourhood of (x_0, ξ_0) *on* Λ.

(II) *Every distribution A of the form*

$$(A, u) = (2\pi)^{-(n+2N)/4} \iint e^{iS(x,\theta)} a(x, \theta) u(x)\, dx\, d\theta, \qquad u \in C_0^\infty(\mathbb{R}^n),$$

where $a \in S_\rho^{\mu+(n-2N)/4}$, $\rho > \frac{1}{2}$, and where the conical support of the distribution a lies in a sufficiently small conical neighbourhood of the point (x_0, θ_0), can be written in the same form by replacing S by $\tilde{S}$ and a by $\tilde{a} \in S_\rho^{\mu+(n-2\tilde{N})/4}$. Moreover, the conical support of $\tilde{a}$ lies in some small neighbourhood of $(x_0, \tilde{\theta}_0)$, and

$$e^{\pi i\sigma/4} a(x, \theta)\sqrt{d_C} - \tilde{a}(x, \tilde{\theta})\sqrt{d_{\tilde{C}}} \in S_\rho^{\mu+n/4+1-2\rho}(\Lambda, \Omega_{1/2}).$$

3.5. The Global Definition of the Fourier Distribution. Let Ω be a smooth manifold and Λ a closed conical Lagrange submanifold in $T^*(\Omega)\backslash 0$. Let $\bigcup U_j$ be a finite local cover of Ω by coordinate neighbourhoods. Suppose that in the conical neighbourhood V_j of each domain $U_j \times (\mathbb{R}^N\backslash 0)$, where $N = N(j)$ is an integer, the non-degenerate phase function $S_j(x, \theta)$ is defined, and that the map

$$(x, \theta) \to (x, \partial S/\partial x)$$

is a diffeomorphism of the domain $\{(x, \theta) \in V_j, \partial S_j(x, \theta)/\partial\theta = 0\}$ on an open subset U^Λ of the manifold Λ.

Consider the distribution $A = \sum A_j \in \mathscr{D}'(\Omega, \Omega_{1/2})$, where

$$(A_j, u) = (2\pi)^{-(n+2N(j))/4} \iint e^{i(S_j(x,\theta) - \pi N(j))/4} a_j(x, \theta) u(x)\, dx\, d\theta,$$

$u \in C^\infty(\Omega)$, dx is the Lebesgue measure in U_j, $\theta \in \mathbb{R}^{N(j)}$, and where

$$a_j \in S_\rho^{m+(n-2N(j))/4}(\mathbb{R}^n \times \mathbb{R}^{N(j)}), \qquad \operatorname{supp} a_j \subset \{(x, t\theta); t \geqslant 1, (x, \theta) \in K\},$$

in which K is a compact subset of the image set of V_j in $\mathbb{R}^n \times \mathbb{R}^{N(j)}$ under the above diffeomorphism.

We denote the set of such distributions by $I_\rho^m(\Omega, \Lambda)$. Let

$$\sigma_{jk} = \frac{1}{2}\left[\left(\operatorname{sgn} \frac{\partial^2 S_k(x, \theta_k)}{\partial\theta^2} - N(k)\right) - \left(\operatorname{sgn} \frac{\partial^2 S_j(x, \theta_j)}{\partial\theta^2} - N(j)\right)\right],$$

where the points θ_k and θ_j are such that

$$\frac{\partial S_k(x, \theta_k)}{\partial\theta} = 0, \qquad \frac{\partial S_j(x, \theta_j)}{\partial\theta} = 0, \qquad \frac{\partial S_k(x, \theta_k)}{\partial x} = \frac{\partial S_j(x, \theta_j)}{\partial x} = \xi \in T_x^* \cap \Lambda.$$

It is not difficult to see that $\sigma_{jk} \in \mathbb{Z}$ and that this function is locally constant in $U_j^\Lambda \cap U_k^\Lambda$, in view of the above-mentioned theorem. We thus obtain an integral cochain that defines the element $\sigma \in H^1(\Lambda, \mathbb{Z})$.

Consider the corresponding one-dimensional linear complex bundle L on Λ that is obtained from this cochain in such a way that the multiplication by σi, i is the imaginary unit, corresponds to σ in $\mathbb{C}$. Evidently, L is trivial as a complex vector bundle and is defined by the image of σ in $H^1(\Lambda, \mathbb{Z}_4)$. It is clear that L is a $\mathbb{R}_+$-bundle because $\mathbb{R}_+$ acts on $\mathbb{C}$ trivially. Therefore the spaces $S_\rho^m(\Lambda, \Omega_{1/2} \otimes L)$ are well defined.

We can establish the existence of the isomorphism

$$S_\rho^{m+n/4}(\Lambda, \Omega_{1/2} \otimes L)/S_\rho^{m+n/4+1-2\rho}(\Lambda, \Omega_{1/2} \otimes L) \to I_\rho^m(\Omega, \Lambda)/I_\rho^{m+1-2\rho}(\Omega, \Lambda).$$

To do this, we proceed as follows. If $s \in S_\rho^{m+n/4}(\Lambda, \Omega_{1/2} \otimes L)$, then for each j an element

$$s_j \in S_\rho^{m+n/4}(U_j^\Lambda, \Omega_{1/2} \otimes L)$$

is defined such that

$$s_j = i^{\sigma_{jk}} s_k \quad \text{in} \quad U_j^\Lambda \cap U_k^\Lambda.$$

If cone supp $s \subset U_j^\Lambda$, then we can define, as above, the element $A_j = A_j(s) \in I_\rho^m(\Omega, \Lambda)$, corresponding to the symbol $a_j \in S_\rho^{m+(n-2N_j)/4}(U_j)$, and $a_j\sqrt{d_{C_j}}$ goes over to s_j under the map

$$C_j = \left\{(x, \theta) \in U_j; \frac{\partial S_j(x, \theta)}{\partial \theta} = 0\right\} \ni (x, \theta) \to \left(x, \frac{\partial S_j}{\partial x}\right) \in U_j^\Lambda.$$

This condition defines a_j as an element of $S_\rho^{m+(n-2N_j)/4}(C_j)$. Finally, a_j is determined by multiplication by a homogeneous cutoff function in U_j. By the theorem of § 3.4 a change in the distribution a_j leads to a change in A by an element belonging to $I_\rho^{m+n/4+(1-2\rho)}(\Omega)$.

3.6. The Global Fourier Integral Operators. Let X and Y be manifolds of dimension n_X and n_Y, respectively.The points of X will be denoted by x, those of Y by y and of $X \times Y$ by (x, y).

Definition 3.1. The submanifold $C \subset T^*(X \times Y)\backslash 0$ is called a *canonical homogeneous relation* from T^*Y to T^*X if it is, first, closed and conical, second, is contained in $(T^*X\backslash 0) \times (T^*Y\backslash 0)$, and, finally, if it is Lagrange with respect to the 1-form $\alpha_X - \alpha_Y$, where $\alpha_X = \xi\, dx$ and $\alpha_Y = \eta\, dy$.

Example 3.3. Let $n_X = n_Y$. Suppose that f is a C^∞-map from $T^*Y\backslash 0$ to $T^*X\backslash 0$ that is positive and homogeneous in η of degree 1. The set $\{(f(y, \eta), y, \eta); (y, \eta) \in T^*Y\backslash 0\}$ is a canonical relation if and only if $\sigma_Y = f^*\sigma_X$, that is, if the map f is symplectic.

Example 3.4. Let $X = Y$, and let f be the identity map. Its graph is the set $\Delta_X^* = \{(x, \xi, x, \xi), x \in X, \xi \neq 0\}$, that is, is a diagonal in $(T^*X\backslash 0) \times (T^*X\backslash 0)$. Clearly, Δ_X^* is a canonical homogeneous relation.

Let C be a canonical homogeneous relation from T^*Y into T^*X. The image of C under a map that transforms the point (x, ξ, y, η) into $(x, \xi, y, -\eta)$ is a conical Lagrange submanifold in $T^*(X \times Y)\backslash 0$ with respect to the standard symplectic form $\sigma_X + \sigma_Y$. We denote this image by C'. With this manifold we can associate, as in § 3.5, Fourier distributions. They are defined microlocally by integrals of the form

$$(2\pi)^{-(n_X+n_Y)/4-N/2} \int e^{iS(x,y,\theta)} a(x, y, \theta)\, d\theta.$$

The phase function is defined in some open conical set $\Gamma \subset X \times Y \times (\mathbb{R}_1^N \backslash 0)$ and the map

$$(x, y, \theta) \to \left(x, \frac{\partial S(x, y, \theta)}{\partial x}, y, -\frac{\partial S(x, y, \theta)}{\partial y}\right)$$

transforms Γ smoothly into an open conical subset $\Gamma^0 \subset (T^*X \backslash 0) \times (T^*Y \backslash 0)$ and the set C_S of zeros of $\partial S/\partial\theta$ in Γ onto $C \cap \Gamma^0$.

Proposition 3.2. *The differential of the map $C \to T^*(X)$ is bijective if and only if $n_X = n_Y$ and*

$$D(S) = \det \begin{bmatrix} \dfrac{\partial^2 S}{\partial\theta^2} & \dfrac{\partial^2 S}{\partial\theta\partial x} \\ \dfrac{\partial^2 S}{\partial y\partial\theta} & \dfrac{\partial^2 S}{\partial y\partial x} \end{bmatrix} \neq 0 \quad \text{in} \quad \Gamma.$$

The map $C_S \to C \to T^*(X)$ induces local coordinates (x, ξ) in some neighbourhood on C_S, and the density $d_{C_S} = |D(S)|^{-1}\, dx\, d\xi$.

Definition 3.2. A canonical homogeneous relation C from $T^*(Y)$ into $T^*(X)$ is called *canonical local graph* if the projection $C \to T^*(X)$ (and hence also the projection $C \to T^*(Y)$) is a local diffeomorphism; thus C is a local graph of the canonical transformation. (Clearly, $n_X = n_Y$ in this case.)

The density μ on a local canonical graph can be defined by means of the standard density in $T^*(X)$ (or $T^*(Y)$) and the map $C \to T^*(X)$ (or $C \to T^*(Y)$). Evidently, $\sqrt{\mu} \in S_1^{n/2}(C, \Omega_{1/2})$, where $n = n_X = n_Y$. If L is a linear bundle on C that is obtained from the linear bundle associated with the Lagrange manifold $\Lambda = C'$, then the map

$$S_\rho^m(C, L) \to S_\rho^{m+n/2}(C, \Omega_{1/2} \otimes L),$$

defined by means of multiplication by $\sqrt{\mu}$, is bijective. Hence we obtain the isomorphism

$$S_\rho^m(C, L)/S_\rho^{m+1-2\rho}(C, L) \to I_\rho^m(X \times Y, C')/I_\rho^{m+1-2\rho}(X \times Y, C').$$

Let S be a non-degenerate phase function in a conical neighbourhood U of the point (x_0, y_0, θ_0) from $X \times Y \times (\mathbb{R}^N \backslash 0)$ to which corresponds a neighbourhood of the point c_0 in C. Suppose that $a \in S_\rho^{m+(n-N)/2}$ and $a = 0$ outside a sufficiently small conical neighbourhood of (x_0, y_0, θ_0). Consider an operator A with the Schwartz kernel K_A defined by the formula

$$(K_A, u) = (2\pi)^{-(n+N)/2} \iiint e^{iS(x,y,\theta)} a(x, y, \theta) u(x, y)\, dx\, dy\, d\theta,$$

where $u \in C_0^\infty(X \times Y)$. Then the symbol of A coincides with the image of $a\sqrt{d_C}$ on C under the map $C_S \to \Lambda \to C$. By Proposition 3.2, under this map the density $d_{C_S} = |D(s)|^{-1}\, dx\, d\xi$ corresponds to the density d_C, and

$$b(x, y, \theta) = a(x, y, \theta)|D(S)|^{-1/2} \in S_\rho^m(X \times Y \times (\mathbb{R}^N \backslash 0)).$$

Hence

$$(K_A, u) = (2\pi)^{-(n+N)/2} \iiint e^{iS(x,y,\theta)} b(x, y, \theta) |D(S)|^{1/2} u(x, y)\, dx\, dy\, d\theta,$$

or, equivalently,

$$Au(x) = (2\pi)^{-(n+N)/2} \iint e^{iS(x,y,\theta)} b(x, y, \theta) |D(S)|^{1/2} u(y)\, dy\, d\theta$$

for $u \in C_0^\infty(Y)$. With this operator is associated a symbol which is obtained from b by a composition with the inverse of the map

$$C_S \ni (x, y, \theta) \mapsto \left(x, \frac{\partial S}{\partial x}, y, -\frac{\partial S}{\partial y}\right) \in C,$$

and which may be regarded as an element of the space $S_\rho^m(C, L)$. As shown above, this symbol is invariantly defined to within an element from $S_\rho^{m+1-2\rho}(C, L)$. Therefore, it is natural to refer to this as the principal symbol.

Using Proposition 3.1, we can show that $A: \mathscr{E}'(Y) \to \mathscr{D}'(X)$ if $d_{x,\theta} S \neq 0$. A becomes a continuous operator from $C_0^\infty(Y)$ into $C^\infty(X)$ if $d_{y,\theta} S \neq 0$.

§ 4. The Calculus of Fourier Integral Operators

4.1. The Adjoint Operator. If u and v are two densities of order $\frac{1}{2}$ on a manifold X and if the set $\operatorname{supp} u \cap \operatorname{supp} v$ is compact, then

$$(u, v) = \langle u, \bar{v} \rangle = \int u\bar{v}.$$

The adjoint to A whose kernel is in $I_\rho^m(X \times Y, C')$, where C is a canonical homogeneous relation from $T^*(Y)$ into $T^*(X)$, is defined by the equation

$$(Au, v) = (u, A^*v), \qquad u \in C_0^\infty(Y, \Omega_{1/2}), \qquad v \in C_0^\infty(X, \Omega_{1/2}).$$

If, in local coordinates, the kernel $K_A(x, y)$ of A is of the form

$$\langle K_A, u \rangle = (2\pi)^{-(n_X + n_Y + 2N)/4} \iiint e^{iS(x,y,\theta)} a(x, y, \theta) u(x, y)\, dx\, dy\, d\theta,$$

then

$$\langle K_{A^*}, u \rangle = (2\pi)^{-(n_X + n_Y + 2N)/4} \iiint e^{-iS(x,y,\theta)} \overline{a(x, y, \theta)} u(x, y)\, dx\, dy\, d\theta$$

for $u \in C_0^\infty$. The function $-S$ is a phase function in $X \times Y \times (\mathbb{R}^N \backslash 0)$ and the corresponding canonical relation is defined by the image of the map

$$C_S \ni (x, y, \theta) \to \left(y, -\frac{\partial S}{\partial y}, x, \frac{\partial S}{\partial x}\right).$$

We thus have the following result.

Theorem 4.1 (Hörmander [1971]). *Let C be a canonical homogeneous relation from $T^*(Y)$ into $T^*(X)$ and let $K_A \in I_\rho^m(XxY, C)$, $\rho > \frac{1}{2}$. Then the kernel K_{A^*} of the adjoint operator lies in $I_\rho^m(YxX, C'_S)$, where C_S is the inverse image of C under the map $S: T^*YxT^*X \to T^*XxT^*Y$ that interchanges the factors. If $a \in S_\rho^{m+(n_X+n_Y)/4}(C, \Omega_{1/2} \otimes L_C)$ is the principal symbol of A, then the principal symbol of A^* is $S^*\bar{a} \in S_\rho^{m+(n_X+n_Y)/4}(C_S, \Omega_{1/2} \otimes L_{C_S})$, because there exists a natural isomorphism between the bundles L_{C_S} and $S^*L_C^{-1}$.*

4.2. The Composition of Fourier Integral Operators. Let A_1 and A_2 be Fourier integral operators with the Schwartz kernels K_1 and K_2, respectively. Suppose that $K_1 \in I_\rho^{m_1}(X \times Y, C'_1)$, where C_1 is the canonical relation between $T^*(Y)$ and $T^*(X)$, and $K_2 \in I_\rho^{m_2}(Y \times Z, C'_2)$, where C_2 is the canonical relation between $T^*(Z)$ and $T^*(Y)$. Let $A_2: C_0^\infty(Z) \to C_0^\infty(Y)$ and $A_1: C_0^\infty(Y) \to C^\infty(X)$, where A_2 is a proper map. Let us show that the operator A_1A_2 has the Schwartz kernel $K \in I_\rho^{m_1+m_2}(X \times Z, C')$, where C is obtained from the composition of C_1 and C_2.

The direct product $C_1 \times C_2$ is a symplectic manifold with respect to the symplectic form $\sigma_X - \sigma_{Y_1} + \sigma_{Y_2} - \sigma_Z$, where Y_1 and Y_2 are replicas of the manifold Y. Assume that $C_1 \times C_2$ intersects the diagonal Δ in $T^*(X) \times T^*(Y_1) \times T^*(Y_2) \times T^*(Z)$, where the diagonal consists of the elements whose components coincide transversally in $T^*(Y_1)$ and $T^*(Y_2)$. This implies that no non-zero vector exists that would be normal to the tangent plane to $C_1 \times C_2$ and to Δ relative to the symplectic form. Or, equivalently, there exists no non-zero vector that would be tangent to $C_1 \times C_2$ and to Δ with zero components in $T(T^*(X))$ and $T(T^*(Z))$. Under this condition, the projection of $(C_1 \times C_2) \cap \Delta$ into $T^*(X) \times T^*(Z)$ is a local manifold, with dimension $\dim X + \dim Z$, on which $\sigma_X - \sigma_Z = 0$, because $\sigma_{Y_1} - \sigma_{Y_2} = 0$ on Δ. We now define the composition of C_1 and C_2 as a projection of $(C_1 \times C_2) \cap \Delta$ onto $T^*(X) \times T^*(Z)$.

It is clear that $\dim(C_1 \times C_2) \cap \Delta = \dim C_1 \times C_2 - \operatorname{codim} \Delta = n_X + n_Y$. Assume that the projection $(C_1 \times C_2) \cap \Delta \to T^*(X) \times T^*(Z)\backslash 0$ is a proper injective map.

Let us express the transversality condition in terms of phase functions. Let S_1 be a phase function defined in a neighbourhood of the point $(x_0, y_0, \theta_0) \in X \times Y \times (\mathbb{R}^{N_1}\backslash 0)$ and S_2 a phase function defined in a neighbourhood of the point $(y_0, z_0, \sigma_0) \in Y \times Z \times (\mathbb{R}^{N_2}\backslash 0)$. Assume that these functions are non-degenerate and that they define C_1 locally by the equation $\partial S_1(x, y, \theta)/\partial\theta = 0$ and C_2 by $\partial S_2(y, z, \sigma)/\partial\sigma = 0$. Assume further that

$$\left(x, \frac{\partial S_1(x, y, \theta)}{\partial x}, y, -\frac{\partial S_1(x, y, \theta)}{\partial y}, y, \frac{\partial S_2(y, z, \sigma)}{\partial y}, z, -\frac{\partial S_2(y, z, \sigma)}{\partial z}\right) \in \Delta$$

for $x = x_0, \ldots, \sigma = \sigma_0$, that is, that

$$\partial S_1(x_0, y_0, \theta_0)/\partial y + \partial S_2(y_0, z_0, \sigma_0)/\partial y = 0.$$

As we stated, transversality implies that there is no non-zero vector $(0, 0, t, \tau t, \tau, 0, 0)$ orthogonal to the tangent plane to $C_1 \times C_2$ with respect to the symplectic form $\sigma_X - \sigma_{Y_1} + \sigma_{Y_2} - \sigma_Z$. It is easy to see that the equation

$$d\left\langle \frac{\partial S_1(x, y, \theta)}{\partial \theta}, a\right\rangle + d\left\langle \frac{\partial S_2(y, z, \sigma)}{\partial \sigma}, b\right\rangle + d\left\langle \frac{\partial S_1(x, y, \theta)}{\partial y} + \frac{\partial S_2(y, z, \sigma)}{\partial y}, t\right\rangle = 0$$

implies that $a = b = t = 0$ (here $d = d_{x,y,\theta,\sigma}$).

We now consider the function

$$S(x, z, \omega) = S_1(x, y, \theta) + S_2(y, z, \sigma),$$

where

$$\omega = ((|\theta|^2 + |\sigma|^2)^{1/2} y, \theta, \sigma) \in \mathbb{R}^{N_1+N_2+n_Y} \backslash 0.$$

It is clear that S is homogeneous in ω of degree 1 and that the equation $\partial S/\partial \omega = 0$ implies that

$$\frac{\partial S_1}{\partial \theta} = \frac{\partial S_2}{\partial \sigma} = \frac{\partial S_1}{\partial y} + \frac{\partial S_2}{\partial y} = 0. \tag{6}$$

Therefore the transversality condition means precisely that S is a non-degenerate function in a neighbourhood of the point (x_0, z_0, ω_0). The corresponding canonical relation is of the form

$$\left\{\left(x, \frac{\partial S_1}{\partial x}, z, -\frac{\partial S_2}{\partial z}\right); \frac{\partial S_1}{\partial \theta} = \frac{\partial S_2}{\partial \sigma} = \frac{\partial S_1}{\partial y} + \frac{\partial S_2}{\partial y} = 0\right\},$$

that is, S defines $C_1 \circ C_2$ locally.

If we put

$$A_1 v(x) = (2\pi)^{-(n_X+n_Y+2N_1)/4} \iint e^{iS_1(x,y,\theta)} a_1(x, y, \theta) v(y)\, dy\, d\theta,$$

$$A_2 u(y) = (2\pi)^{-(n_Y+n_Z+2N_2)/4} \iint e^{iS_2(y,z,\sigma)} a_2(y, z, \sigma) u(z)\, dz\, d\sigma,$$

where

$$a_1 \in S_\rho^{m_1+(n_X+n_Y-2N_1)/4}(X \times Y \times \mathbb{R}^{N_1}),$$

$$a_2 \in S_\rho^{m_2+(n_Y+n_Z-2N_2)/4}(Y \times Z \times \mathbb{R}^{N_2}),$$

and where S_1 and S_2 are non-degenerate phase functions, a_1 and a_2 vanish in a neighbourhood of the sections $\theta = 0$ and $\sigma = 0$, respectively, with cone supp $a_1 \subset \Gamma_1$, cone supp $a_2 \subset \Gamma_2$, and where S_1, S_2 define the manifolds C_1 and C_2 locally, then

$$A_1 A_2 u(x) = (2\pi)^{-(n_X+n_Z+2N)/4} \iiiint e^{i(S_1(x,y,\theta)+S_2(y,z,\sigma))}$$

$$\times\, a_1(x, y, \theta) a_2(y, z, \sigma) u(z)\, dz\, dy\, d\theta\, d\sigma.$$

Here $N = N_1 + N_2 + n_Y$. It is important to observe that $a_1(x, y, \theta) \times a_2(y, z, \sigma)$ is a symbol in a neighbourhood of points satisfying (6). In fact, there exist positive constants c_1 and c_2 such that $c_1|\sigma| < |\theta| < c_2|\sigma|$ provided that $\partial S_1(x, y, \theta)/\partial y + \partial S_2(y, z, \sigma)/\partial y = 0$. Let $\chi(\theta, \sigma)$ be a homogeneous function of degree zero that equals 1 for $c_1|\sigma|/2 < |\theta| < 2c_2|\sigma|$ and vanishes for $c_1|\sigma|/3 > |\theta|$ and for $|\theta| > 3c_2|\sigma|$. Then the operator $A_1A_2 - B$ has a C^∞-kernel if

$$Bu(x) = (2\pi)^{-(n_X+n_Z+2N)/4} \iiiint e^{i(S_1(x,y,\theta)+S_2(y,z,\sigma))}$$

$$\times\, b(x, z, y, \theta, \sigma)u(z)\, dz\, dy\, d\theta\, d\sigma,$$

where

$$b(x, z, y, \theta, \sigma) = \chi(\theta, \sigma)a_1(x, y, \theta)a_2(y, z, \sigma) \in S_\rho^m$$

with

$$m = m_1 + m_2 + (n_X + n_Z - 2N)/4 + n_Y.$$

The principal symbol of the composition $A_1 \circ A_2$ is obtained by restricting the principal symbols of A_1 and A_2 to the diagonal of the direct product.

It can be verified that the corresponding $\frac{1}{2}$ density on $C_1 \circ C_2$ is obtained from the product of densities on C_1 and C_2 by means of the projection of $C_1 \times C_2$ onto the diagonal.

4.3. The Boundedness in L_2. Let A be a proper Fourier integral operator with the principal symbol $a \in S_\rho^0(C, L)$. Then the adjoint operator A^* is associated with the homogeneous canonical relation C^* obtained from C by interchanging X and Y (or T^*X and T^*Y). Clearly, $C \circ C^*$ is the graph of the identity map in $T^*Y\backslash 0$. Since in this case the transversality condition is always fulfilled, it follows that the operator A^*A is defined and is a pseudodifferential operator with principal symbol $|a|^2$. Therefore A is a bounded operator from $L^2_{\text{comp}}(Y, \Omega_{1/2})$ into $L^2_{\text{comp}}(X, \Omega_{1/2})$ and from $L^2_{\text{loc}}(Y, \Omega_{1/2})$ into $L^2_{\text{loc}}(X, \Omega_{1/2})$ provided that C is a canonical local graph. If the principal symbol of A tends to zero as the point tends to ∞ along C over each compact subset in $X \times Y$, then A is a compact operator (see Hörmander [1971]).

Suppose that now A is a proper Fourier integral operator with the principal symbol $a \in S^m(C, L)$ and that B and B_1 are pseudodifferential operators with the principal symbols $b \in S^{m'}(X)$ and $b_1 \in S^{m'}(Y)$, respectively. Since these operators are associated with the graph of the identity map, it follows that the transversality conditions are fulfilled and the compositions BA and AB_1 are defined. They are Fourier integral operators with kernels from the class $I^{m+m'}(X \times Y, C, \Omega_{1/2})$. The principal symbol of BA is of the form

$$b(x, x, \partial S(x, y, \theta)/\partial x)a(x, y, \theta)(\sqrt{d_C})^{-1},$$

and that of AB_1 of the form

$$a(x, y, \theta)b_1(y, y, -\partial S(x, y, \theta)/\partial y)(\sqrt{d_C})^{-1}.$$

It is clear from this and the preceding fact that if A is a proper Fourier integral operator with kernel from $I^m_\rho(X \times Y, C')$, then it is a bounded operator from $H^s_{\text{comp}}(Y, \Omega_{1/2})$ into $H^{s-m}_{\text{comp}}(X, \Omega_{1/2})$ and from $H^s_{\text{loc}}(Y, \Omega_{1/2})$ into $H^{s-m}_{\text{loc}}(X, \Omega_{1/2})$ for all real s.

Definition 4.1. A Fourier integral operator A belonging to $I^m(X \times Y, C, \Omega_{1/2})$ is said to be *elliptic* in an open conical subset $\Gamma_X \times \Gamma_Y \subset (T^*X\backslash 0) \times (T^*Y\backslash 0)$ if its principal symbol vanishes nowhere on $C \cap (\Gamma_X \times \Gamma_Y)$.

Clearly, the ellipticity condition implies the existence of constants K_1 and K_2 such that

$$|a(x, y, \theta)| \geqslant K_1 |\theta|^{m+(n-N)/2} - K_2.$$

For an elliptic operator A we can construct an approximate inverse operator E in $\Gamma_Y \times \Gamma_X$. Indeed, AA^* is an elliptic pseudodifferential operator of order $2m$ in $\Gamma_X \times \Gamma_Y$ and A^*A is an elliptic operator of order $2m$ in $\Gamma_Y \times \Gamma_X$. Therefore there exist parametrices R_1 and R_2:

$$R_1(A^*A) \sim I, \qquad (AA^*)R_2 \sim I.$$

If we set

$$E_1 = R_1 A^*, \qquad E_2 = A^* R_2,$$

we have $E_1 A \sim I$ in Γ_Y and $AE_2 \sim I$ in Γ_X, where the sign $\sim$ denotes the equivalence modulo smoothing operators. Moreover, an operator is said to be smoothing in Γ if its symbol and all the derivatives of the symbol decay, as $|\theta| \to \infty$, faster than any power $|\theta|^{-N}$, $\theta \in \Gamma$.

Let $Y = X$. The operator A will be called unitary if it is an elliptic operator of order zero and $A^*A = I$, $AA^* = I$.

The following theorem generalises Theorem 2.1 of Chap. 2 (see Trèves [1982]).

Theorem 4.2. *If A is an unitary Fourier integral operator associated with a local canonical graph C, then the principal symbol of the operator A^*PA for any pseudodifferential operator P is locally equal to the pullback of the principal symbol of P with respect to the local symplectomorphism $T^*X \to T^*X$ whose graph is C.*

§ 5. The Image of the Wave Front Under the Action of a Fourier Integral Operator

5.1. The Singularities of Fourier Integrals. Let $\Omega \subset \mathbb{R}^n$. Let $S(x, \theta)$ be the phase function defined in $\Omega \times \mathbb{R}^N$, and let Ω' be the set of those points of Ω where $d_\theta S(x, \theta) \neq 0$ for $\theta \neq 0$.

If $u \in C_0^\infty(\Omega')$, then

$$I_S(au) = \iint a(x, \theta) e^{iS(x,\theta)} u(x)\, dx\, d\theta = \int A(x) u(x)\, dx, \qquad (7)$$

where

$$A(x) = \int a(x, \theta) e^{iS(x,\theta)} \, d\theta, \qquad x \in \Omega'.$$

Clearly, $A \in C^\infty(\Omega')$. Therefore if the distribution A is defined by (7), then

$$\operatorname{sing\,supp} A \subset \{x \in \Omega, \partial S(x, \theta)/\partial\theta = 0 \text{ for some } \theta \neq 0\}.$$

If $a \in S^m_{\rho,\delta}(\Omega \times \mathbb{R}^N)$, where $a = 0$ in a conical neighbourhood of the set

$$C = \{(x, \theta); x \in \Omega, \theta \in \mathbb{R}^N \backslash \{0\}, \partial S(x, \theta)/\partial\theta = 0\},$$

then the distribution $u \mapsto I_S(au)$, defined above, is a C^∞-function.

More precisely, we have

Proposition 5.1. *Let S be a non-degenerate phase function in the cone $\Gamma \subset \Omega \times \mathbb{R}^N$, and let $a \in S^m_{\rho,\delta}(\Omega \times \mathbb{R}^N)$,* cone supp $a \subset \Gamma \backslash (\Omega \times 0)$. *Assume that $\rho > \delta$ and that either $\rho + \delta = 1$ or S is linear in θ. Then the distribution $u \mapsto I_S(au)$ is a C^∞-function if a has a zero of infinite order at points of the set*

$$C = \{(x, \theta) \in \Gamma, \partial S(x, \theta)/\partial\theta = 0\}.$$

If $a = 0$ at points of C, then there exists a symbol $b \in S^{m+\delta-\rho}_{\rho,\delta}(\Omega \times \mathbb{R}^N)$, with cone supp $b \subset \Gamma \backslash (\Omega \times 0)$, *such that $I_S(au) = I_S(bu)$ for $u \in C^\infty(\Omega)$.*

Definition 5.1. A real-valued function S defined in $X \times Y \times \mathbb{R}^N$, which is infinitely smooth for $\theta \neq 0$ and which is positive homogeneous in θ of degree 1, is known as an *operator phase function* if for each fixed x the differential $d_{y,\theta}S$ does not vanish or if for each fixed y one has $d_{x,\theta}S \neq 0$ for $\theta \neq 0$.

Let

$$C_S = \{(x, y) \in X \times Y, \exists \theta \in \mathbb{R}^N \backslash 0, \partial S(x, y, \theta)/\partial\theta = 0\}.$$

It can be deduced from Proposition 5.1 that

$$\operatorname{sing\,supp} Au \subset C_S \circ \operatorname{sing\,supp} u, \qquad u \in \mathscr{E}'(Y),$$

where C_S is a relation between Y and X. For instance, for pseudodifferential operators

$$\operatorname{sing\,supp} Au \subset \operatorname{sing\,supp} u,$$

that is, pseudolocality holds.

5.2. The Wave Front of the Fourier Integral. The fact that microlocalization is possible under the sign of the Fourier integral enables us to make the above results more precise by replacing the singular support by the wave front of the distribution.

Proposition 5.2. *Let $\Omega \subset \mathbb{R}^n$, Γ be an open cone in $\Omega \times (\mathbb{R}^N \backslash 0)$, and let S be the phase function in Γ. If $a \in S^m_{\rho,\delta}(\Omega \times \mathbb{R}^N)$, $\rho > 0$, $\delta < 1$, and $a = 0$ in a neighbourhood of the set $\theta = 0$ and if* cone supp $a \subset \Gamma$, *then*

$$\operatorname{WF}(A) \subset \{(x, \partial S/\partial x); (x, \theta) \in \operatorname{cone\,supp} a, \partial S(x, \theta)/\partial\theta = 0\},$$

where A is a distribution such that $A(u) = I_S(au)$.

In particular, if a proper pseudodifferential operator P is such that $P \sim 0$ in $\mathrm{WF}(u)$ (see § 4.3 above), then $Pu \in C^\infty$ for all u in $\mathscr{D}'(\Omega)$.

For the general case, we have

Theorem 5.1. *Let $K \in I_\rho^m(X, \Lambda)$, $a \in S_\rho^{m+n/4}(\Lambda, \Omega_{1/2} \otimes L)$, and $K(u) = I_S(au)$ for $u \in \mathscr{D}(X)$. Then $\mathrm{WF}(K) \subset \Lambda$ and $a \in S_\rho^{m+n/4+1-2\rho}$ in $\Lambda \setminus \mathrm{WF}(K)$.*

Using Sobolev spaces, we can obtain the following result (see Hörmander [1971]).

Theorem 5.2. *$I^m(X, \Lambda) \subset H_{(s)}(X)$ if and only if $m + n/4 + s < 0$. Furthermore, if $K(u) = I_S(au)$, $K \in I_\rho^m(X, \Lambda)$ and $a \not\equiv 0$ on the set where $d_\theta S = 0$, then $K \notin H_{(s)}(X)$ for $0 \leqslant m + n/4 + s$.*

5.3. The Action of the Fourier Integral Operator on Wave Fronts. Let A be a Fourier integral operator of the form

$$Au(x) = (2\pi)^{-n} \iint a(x, y, \theta) u(y) e^{iS(x, y, \theta)}\, dy\, d\theta. \tag{8}$$

Its Schwartz kernel is

$$K(x, y) = (2\pi)^{-n} \int a(x, y, \theta) e^{iS(x, y, \theta)}\, d\theta.$$

To be able to apply Theorem 3.5, Chap. 1, we must know the wave front of this kernel.

Lemma 5.1.

$$\mathrm{WF}(K) \subset \left\{(x, y, \xi, \eta);\ \xi = \frac{\partial S(x, y, \theta)}{\partial x}, \eta = \frac{\partial S(x, y, \theta)}{\partial y}, \frac{\partial S(x, y, \theta)}{\partial \theta} = 0\right\}.$$

This lemma can be easily established by finding the Fourier transform of the distribution $K(x, y)$ and then applying the method of stationary phase.

This lemma enables us to find $\mathrm{WF}(Au)$ (Hörmander [1971]).

Theorem 5.3. *Let $u \in \mathscr{E}'(\Omega)$ and let A be an operator of the form* (8). *Assume that the following two conditions hold:*

1. *If* $\dfrac{\partial S(x, y, \theta)}{\partial x} = 0, \dfrac{\partial S(x, y, \theta)}{\partial \theta} = 0$, *then* $\left(y, -\dfrac{\partial S(x, y, \theta)}{\partial y}\right) \notin \mathrm{WF}(u)$;
2. *If* $\dfrac{\partial S(x, y, \theta)}{\partial \theta} = 0, \theta \neq 0$, *then* $\dfrac{\partial S(x, y, \theta)}{\partial y} \neq 0$.

Then the distribution $Au \in \mathscr{D}'(X)$ is defined, and

$$\mathrm{WF}(Au) \subset \left\{(x, \xi);\ \xi = \frac{\partial S(x, y, \theta)}{\partial x}, \eta = -\frac{\partial S(x, y, \theta)}{\partial y}, \frac{\partial S(x, y, \theta)}{\partial \theta} = 0,\right.$$
$$\left.(y, \eta) \in \mathrm{WF}(u)\right\}.$$

This theorem follows easily from Theorem 3.5, Chap. 1. In the particular case $n = N$ and $S(x, y, \theta) = S(x, \theta) - S(y, \theta)$, condition 2 is fulfilled automatically, and the following result holds (Hörmander [1971]).

Theorem 5.4. *Let $u \in \mathscr{E}'(\Omega)$ and let A be an operator of the form*

$$Au(x) = (2\pi)^{-n} \int a(x, \xi)\tilde{u}(\xi)e^{iS(x,\xi)}\, d\xi.$$

Assume that $\partial S(x, \xi)/\partial x = 0$ implies $(\partial S(x, \xi)/\partial \xi, \xi) \notin \mathrm{WF}(u)$. Then the distribution $Au \in \mathscr{D}'(\Omega)$ is defined and

$$\mathrm{WF}(Au) \subset \left\{\left(x, \frac{\partial S(x, \eta)}{\partial x}\right); \left(\frac{\partial S(x, \eta)}{\partial \eta}, \eta\right) \in \mathrm{WF}(u)\right\}.$$

Thus $\mathrm{WF}(Au)$ is contained in the image set of $\mathrm{WF}(u)$ under the canonical transformation $(y, \eta) \to (x, \xi)$ generated by the function $S(x, \eta)$. Since a canonical transformation is invertible, it follows that the inclusion sign in Theorem 5.4 can be replaced by the equality sign in the case of an elliptic operator A.

By way of example, we note that Theorem 5.4 enables us to determine completely the wave front of the solution of the Cauchy problem for strictly hyperbolic equations by means of the wave front of the initial data.

§ 6. Fourier Integral Operators with Complex Phase Functions

6.1. The Complex Phase. In examining operators with complex-valued symbols there arise in a natural manner Fourier integral operators with *complex phase*. These operators are represented microlocally in the form

$$Au(x) = (2\pi)^{-n} \int\int e^{iS(x,y,\theta)} a(x, y, \theta)u(y)\, dy\, d\theta.$$

Such an integral makes sense if $\operatorname{Im} S \geqslant 0$. Besides this, it is usually assumed that

$$S(x, y, t\theta) = tS(x, y, \theta) \qquad \text{for} \quad t > 0;$$

$$dS(x, y, \theta) \neq 0 \qquad \text{if} \quad \operatorname{Im}(x, y, \theta) = 0.$$

The complex phase function S is said to be non-degenerate if the differentials $d(\partial S/\partial \theta_j)$, $j = 1, \ldots, N$, are linearly independent over the field $\mathbb{C}$ at every point of the set C_S where $\partial S/\partial \theta = 0$.

6.2. Almost Analytic Continuation. Let Ω be an open subset in $\mathbb{R}^n$ and $\rho(x)$ a non-negative function in Ω that satisfies a Lipschtiz condition.

Definition 6.1. A function f is said to be *ρ-flat* on Ω if for any compact subset $K \subset \Omega$ and for any integer $N \geqslant 0$ there exists a constant $C = C(K, N) > 0$ such that $|f(x)| \leqslant C|\rho(x)|^N$, $x \in K$.

Two functions f and g are said to be *ρ-equivalent* if the difference $f - g$ is a ρ-flat function. f is said to be flat on the compact set $K \subset \Omega$ if it is ρ-flat with $\rho(x) = \operatorname{dist}(x, K)$.

It is easy to see that if $f \in C^\infty(\Omega)$ and if f is ρ-flat, then any derivative $D^\alpha f$ is also a ρ-flat function. In particular, such a function is flat on K if and only if $D^\alpha f(x) = 0$ for $x \in K$ and for all $|\alpha|$.

Definition 6.2. Let O be an open subset in $\mathbb{C}^n$ and K a closed subset in O. A function $f \in C^\infty(O)$ is said to be *almost analytic* on K if for $j = 1, \ldots, n$ the functions $\bar\partial_j f$ are flat on K.

The set of functions that belong to $C^\infty(O)$ and are almost analytic on K will be denoted by $A(O, K)$. We put $O_{\mathbb{R}} = O \cap \mathbb{R}^n$.

If $f \in A(O, O_{\mathbb{R}})$, $f|_{O_{\mathbb{R}}} = 0$, then f is flat on $O_{\mathbb{R}}$.

Let $z = x + iy$ be coordinates in $\mathbb{C}^n$, and let Ω be an open subset in $\mathbb{R}^n$. We denote by $\tilde\Omega$ the subset $\Omega + i\mathbb{R}^n$ in $\mathbb{C}^n$. We shall identify Ω with $\tilde\Omega \cap \{y = 0\}$.

Proposition 6.1. *Let O be an open subset in $\mathbb{C}^n$ that is contained in $\tilde O_{\mathbb{R}}$. Let $N(O, O_{\mathbb{R}})$ be the space of C^∞-functions on O that are flat on $O_{\mathbb{R}}$. The restriction on $O_{\mathbb{R}}$ defines an isomorphism of the space $A(O, O_{\mathbb{R}})/N(O, O_{\mathbb{R}})$ onto $C^\infty(O_{\mathbb{R}})$.*

Thus a function f in $C^\infty(O_{\mathbb{R}})$ may be regarded as an equivalence class (modulo functions that are flat on $O_{\mathbb{R}}$) of almost analytic functions on $O_{\mathbb{R}}$. A representative of this equivalence class is known as an almost analytic continuation of f in O.

This almost analytic continuation can be constructed in the following manner. Let $h \in C^\infty(\mathbb{R}^1)$, where $h(t) = 1$ for $|t| < 1$ and $h(t) = 0$ for $|t| > 2$. Let $f \in C^\infty(O_{\mathbb{R}})$. We set

$$\tilde f(z) = \sum_{\alpha \in \mathbb{Z}_+^n} \frac{1}{\alpha!}(iy)^\alpha f^{(\alpha)}(x) h\left(\frac{|y|}{\varepsilon_{|\alpha|}}\right),$$

where $\varepsilon_j > 0$ are chosen for $j = 0, 1, \ldots$ in such a way that this series converges in $C^\infty(O_{\mathbb{R}})$.

Definition 6.3. Let O be an open subset in $\mathbb{C}^n$, M a smooth submanifold in O of codimension $2k$, and let K be a closed subset in O. M is said to be *almost analytic* on K if any point $z_0 \in K$ has an open neighbourhood U in O where there exist k smooth complex functions $f_1, \ldots, f_k$ thar are almost analytic on $K \cap U$ and are such that M is defined on U by the equations $f_1(z) = 0, \ldots, f_k(z) = 0$ and $df_1, \ldots, df_k$ are linearly independent. Furthermore, M is locally equivalent to M_1, $M \sim M_1$, if M_1 is defined in U by means of the functions $g_1, \ldots, g_k$ such that $f_j \sim g_j$ for $j = 1, \ldots, k$, that is, $f_j - g_j$ are flat functions on $K \cap U$.

6.3. The Formula for Stationary Complex Phase. Let us consider the integral

$$I(\lambda) = \int e^{i\lambda f(x)} g(x)\, dx,$$

assuming that Im $f \geqslant 0$ in supp g. Suppose that in supp g f has a unique critical point $x = 0$ which is non-degenerate. This last condition implies that the matrix $\|\partial^2 f(x)/\partial x_i \partial x_j\|$ is non-degenerate. Suppose that f depends also on the parameters $t = (t_1, \ldots, t_l)$ that vary in a neighbourhood V of the point O in $\mathbb{R}^l$. Let $\tilde{f}$ be an almost periodic continuation of f in the neighbourhood of the point $x = 0$. Let $x = \varphi(t)$ be the unique solution of the equation $d_x \tilde{f}(\varphi(t), t) = 0$ and let $\tilde{H}(t) = \|\partial^2 \tilde{f}(\varphi(t), t)/\partial x_i \partial x_j\|$. By the hypothesis, $\tilde{H}(O)$ is invertible and Im $\tilde{H}(O)$ is non-negative definite. Hence no eigenvalue of the matrix $\tilde{H}(O)/i$ is real and negative. If the neighbourhood V is sufficiently small, then the matrix $[\tilde{H}(t)/i]^{-1/2}$ makes sense; here $t \in V$ and the branch of the function $\sqrt{z}$ is chosen so that $\sqrt{1} = 1$. As in § 1.3, we can show that

$$I(\lambda) \sim \left(\frac{\pi}{\lambda}\right)^{n/2} \det[\tilde{H}(t)/i]^{-1/2} e^{i\lambda \tilde{f}(\varphi(t), t)} \sum_{j=0}^{\infty} \lambda^{-j} Q_j(D_x) \tilde{g}(\varphi(t)),$$

where Q_j are differential operators, with constant coefficients, of order $2j$, and $Q_0(D_x) = I$. These operators depend on the choice of almost periodic continuations of f and g.

6.4. The Lagrange Manifold. Let M be a real symplectic manifold of dimension $2n$, and let $\tilde{M}$ be its almost analytic continuation in $\mathbb{C}^{2n}$. Let $\Lambda \subset \tilde{M}$ be an almost analytic submanifold containing the real point $\rho_0 \in M$, and let (x, ξ) be real symplectic coordinates in a neighbourhood $W \subset \mathbb{R}^{2n}$ of the point ρ_0. Let $\tilde{W}$ be an open subset in $\mathbb{C}^{2n}$ for which $\tilde{W} \cap \mathbb{R}^{2n} = W$. Assume that $(\tilde{x}, \tilde{\xi})$ are almost analytic continuation of the coordinates in $\tilde{W}$ so that $(\tilde{x}, \tilde{\xi})$ map $\tilde{W}$ diffeomorphically onto an open subset in $\mathbb{C}^{2n}$. Suppose that Λ is defined in a neighbourhood of ρ_0 by the equations $\tilde{\xi} = \partial g(x)/\partial \tilde{x}$, $\tilde{x} \in \mathbb{C}^n$, where g is an almost analytic function such that Im $\tilde{g} \geqslant 0$ in $\mathbb{R}^n$.

Theorem 6.1. *If $(\tilde{y}, \tilde{\eta})$ is another almost analytic continuation of coordinates in $\tilde{W}$ and Λ is defined by the equation $\tilde{\eta} = H(\tilde{y})$ in a neighbourhood of the point ρ_0, then Λ is locally equivalent to the manifold $\tilde{\eta} = \partial h(\tilde{y})/\partial \tilde{y}$, $\tilde{y} \in \mathbb{C}^n$, where h is an almost analytic function and* Im $h \geqslant 0$ *in* $\mathbb{R}^n$.

Definition 6.4. An almost analytic manifold Λ satisfying the hypotheses of Theorem 6.1 at every real point in some real symplectic coordinate system (x, ξ) is known as a *positive Lagrange manifold.*

Definition 6.5. An almost analytic manifold $\Lambda \subset \tilde{M}$ is known as a *strictly positive Lagrange manifold* if dim $\Lambda_{\mathbb{R}} = 2n$ and $\Lambda_{\mathbb{R}}$ is a submanifold in M, $\sigma_\alpha|_{\Lambda_\alpha} \sim 0$ for all local representatives Λ_α of Λ and for all local almost analytic continuations σ_α of the symplectic form σ on $T_\rho(\tilde{M})$, and, finally, if $i^{-1}\sigma(v, \bar{v}) > 0$ for all $v \in T_\rho(\Lambda) \setminus \widetilde{T_\rho(\Lambda_{\mathbb{R}})}$, $\rho \in \Lambda_{\mathbb{R}}$, where $\widetilde{T_\rho(\Lambda_{\mathbb{R}})}$ is the complexification of $T_\rho(\Lambda_{\mathbb{R}})$.

As in the real case, for a conical almost analytic manifold $\Lambda \subset \widetilde{T^*X \setminus O}$, on which $\sigma_\alpha|_{\Lambda_\alpha} \sim 0$ for all local representatives Λ_α and for all almost analytic continuations σ_α of the form σ, the local coordinates x in a neighbourhood of each point $\rho_0 \in \Lambda_{\mathbb{R}}$ may be chosen so that Λ can be represented in the form $\tilde{x} =$

$\partial g(\bar{\xi})/\partial\bar{\xi}$. Here ξ are dual coordinates to x and $(\tilde{x}, \tilde{\xi})$ are almost analytic continuation of the coordinates (x, ξ). The function g here is almost analytic and positive homogeneous of degree one.

Definition 6.6. A C^∞-function $S(x, \theta)$ defined in an open conical set $\Gamma \subset \mathbb{R}^n \times (\mathbb{R}^n \backslash O)$ is known as a *regular phase function* of positive type if

1. It has no critical points;
2. $S(x, t\theta) = tS(x, \theta)$ for $t > 0$;
3. The differentials $d(\partial S/\partial\theta_1), \ldots, d(\partial S/\partial\theta_N)$ are linearly independent over $\mathbb{C}$ on the set $C_{S_\mathbb{R}} = \{(x, \theta) \in \Gamma, \partial S/\partial\theta = 0\}$;
4. $\operatorname{Im} S(x, \theta) \geqslant 0$.

Theorem 6.2. *Let S be a regular phase function of the positive type that is defined in a conical neighbourhood. Let $\tilde{S}(\tilde{x}, \tilde{\theta})$ be an almost analytic homogeneous continuation of S in a conical neighbourhood in $\mathbb{C}^n \times (\mathbb{C}^N \backslash O)$. Let*

$$C_{\tilde{S}} = \{(\tilde{x}, \tilde{\theta}) \in \mathbb{C}^n \times (\mathbb{C}^N \backslash O); \partial\tilde{S}(\tilde{x}, \tilde{\theta})/\partial\tilde{\theta} = 0\}.$$

Then the image $\Lambda_{\tilde{S}}$ of the set $C_{\tilde{S}}$ under the map

$$C_{\tilde{S}} \ni (\tilde{x}, \tilde{\theta}) \mapsto \left(x, \frac{\partial\tilde{S}(\tilde{x}, \tilde{\theta})}{\partial\tilde{x}}\right) \in \mathbb{C}^n \times (\mathbb{C}^n \backslash O)$$

is a local conical positive Lagrange manifold. Furthermore, $\Lambda_{S_{\tilde{\mathbb{R}}}}$ is the imge of $C_{S_{\tilde{\mathbb{R}}}}$, and when $\tilde{S}$ is replaced by an equivalent almost analytic continuation, the manifold $\Lambda_{\tilde{S}}$ is replaced by an equivalent conical positive Lagrange manifold.

6.5. The Equivalence of Phase Functions. Let $S \in C^\infty(\Gamma)$ be a positive type regular phase function. Suppose that $a \in S^m_\rho(\mathbb{R}^n \times \mathbb{R}^N)$ and $\rho > \frac{1}{2}$. Let Γ be an open conical set and let supp a be contained in a closed conical subset in Γ. Consider the integral

$$I_S(au) = \iint e^{iS(x,\theta)} a(x, \theta) u(x)\, dx\, d\theta, \qquad u \in C_0^\infty(\mathbb{R}^n).$$

As in the real case, we can show that

$$\mathrm{WF}(A) \subset \{(x, \partial S(x, \theta)/\partial x), (x, \theta) \in \text{cone supp } a \cap C_{S_\mathbb{R}}\},$$

where A is the distribution defined by $A(u) = I_S(au)$.

Suppose now that $S_1(x, \omega)$ is another positive type regular phase function defined in an open conical set $\Gamma \subset \mathbb{R}^n x(\mathbb{R}^M \backslash O)$. Let $\lambda_0 = (x_0, \xi_0)$, with

$$\xi_0 = \partial S(x_0, \theta_0)/\partial x = \partial S_1(x_0, \omega_0)/\partial x$$

and

$$\partial S(x_0, \theta_0)/\partial\theta = 0, \qquad \partial S_1(x_0, \omega_0)/\partial\omega = 0.$$

Two distributions A and $B \in \mathscr{D}'(\mathbb{R}^n)$ are said to be microlocally equivalent at a point λ_0 if $\lambda_0 \notin \mathrm{WF}(A - B)$. Let $\mathscr{D}'_{\lambda_0}$ be the class of distributions from $\mathscr{D}'(\mathbb{R}^n)$ that are locally equivalent at λ_0.

The phase function S defines a map from $S_\rho^m(\mathbb{R}^n \times \mathbb{R}^N)$ into $\mathscr{D}'_{\lambda_0}$ according to the following rule. If $a \in S_\rho^m(\mathbb{R}^n \times \mathbb{R}^N)$ and $a = a_0$ for large θ in a small conical neighbourhood of the ray $\{(x_0, t\theta_0), t > 0\}$, with supp a_0 lying in this neighbourhood, then the distributions A and A_0 defined by the equations $A(u) = I_S(au)$, $A_0(u) = I_S(a_0 u)$ are equivalent at λ_0.

Definition 6.7. The phase functions S and S_1 are equivalent at a point λ_0 for symbols belonging to S_ρ if the images $S_\rho^\infty(\mathbb{R}^n \times \mathbb{R}^N)$ and $S_\rho^\infty(\mathbb{R}^n \times \mathbb{R}^M)$ under the maps defined by S and S_1 coincide in $\mathscr{D}'_{\lambda_0}$.

Theorem 6.3. *Let S and S_1 be positive type regular phase functions defined in conical neighbourhoods of the points $(x_0, \theta_0) \in \mathbb{R}^n \times (\mathbb{R}^N \backslash O)$ and $(x_0, \omega_0) \in \mathbb{R}^n \times (\mathbb{R}^M \backslash O)$, respectively. Suppose that Λ_S and Λ_{S_1} are equivalent in a neighbourhood of the point $\xi_0 = \partial S(x_0, \theta_0)/\partial x = \partial S_1(x_0, \omega_0)/\partial x$. Then S and S_1 are equivalent at (x_0, ξ_0) for symbols belonging to S_ρ, $\rho > \frac{1}{2}$.*

Definition 6.8. $I_\rho^m(X, \Lambda)$, $\rho > \frac{1}{2}$, is a subspace of distributions A belonging to $\mathscr{D}'(X, \Omega_{1/2})$ for which

1. $\mathrm{WF}(A) \subset \Lambda_{\mathbb{R}}$;
2. For each point $\lambda_0 \in \Lambda_{\mathbb{R}}$ and for any local coordinates $x_1, \ldots, x_n$ in a neighbourhood of the point $\pi(\lambda_0) \in X$ A is of the form $A(u) = I_S(au)$, where S is a positive regular phase function that generates Λ near λ_0 and $a \in S_\rho^{m+(n-2N)/4}(\mathbb{R}^n \times \mathbb{R}^N)$ has support in a small conical neighbourhood of the point $(x_0, \theta_0) \in C_{S_{\mathbb{R}}}$ corresponding to λ_0.

6.6. The Principal Symbol. Let us define the principal symbol for elements belonging to $I_\rho^m(X, \Lambda)$. In the case of a real function S, the principal symbol is a section in the tensor product of the linear bundle of density $\frac{1}{2}$ on Λ and the Maslov linear bundle that gives the transition function of the form i^ν, $\nu \in \mathbb{Z}$.

In the complex case, it is not possible to define almost analytic densities of order $\frac{1}{2}$ on complex manifolds in such a way that transition functions are continuous relative to small perturbations.

Let us start by examining the linearized situation. Let M be a real symplectic vector space of dimension $2n$ with the symplectic bilinear form σ and let $\tilde{M}$ be its complexification. The Lagrange plane $\Lambda \subset \tilde{M}$ is said to be positive (positive definite) if $\operatorname{Im} \sigma(u, \bar{u}) \geqslant 0$ (> 0) for all $u \in \Lambda$, $u \neq 0$. Similarly, one defines negative and negative definite planes. We denote by $\mathscr{L}^-$ the set of all negative definite Lagrange planes.

It is clear that if $L \in \mathscr{L}^-$, then L is transversal to all positive Lagrange planes and is of the form $\tilde{\xi} = B\tilde{x}$, where (x, ξ) are real linear symplectic coordinates in M and B is a symmetric matrix.

Let F be a fixed real Lagrange plane in M and let $\tilde{F}$ be its comlexification.

Definition 6.9. Let $\Lambda \subset \tilde{M}$ be a positive Lagrange plane. A basis $e = (e_1, \ldots, e_n)$ in Λ is said to be *admissible* if there exist a basis $f = (f_1, \ldots, f_n)$ in F and a plane $L \in \mathscr{L}^-$ such that, for all j, e_j is the projection of f_j along L.

The set of all admissible basis will be denoted by $B(\Lambda)$. In particular, $B(\tilde{F})$ is the set of real basis in F. Note that e as defined by f and L is not unique.

Proposition 6.2. *$B(\Lambda)$ is a union of two disjoint connected subsets so that two basis e and e', corresponding to given f and L, belong to the same component if and only if their orientations coincide. Furthermore, there exists a unique function $s = s_\Lambda: B(\Lambda) \times B(\Lambda) \to \mathbb{C}\backslash O$ such that for each compact set $K \subset B(\tilde{F}) \times \mathcal{L}^-$, $s_\Lambda(e, e')$ depends continuously on e, e' and if $e, e' \in E_\Lambda(K)'$. What is more, if we put $e/e' = e_1 \wedge \cdots \wedge e_n/e'_1 \wedge \cdots \wedge e'_n$, then $s^2(e, e') = \pm e/e'$, where the plus sign corresponds to the case when the basis have the same orientation, and $s(e, e')s(e', e'') = s(e, e'')$ if $e, e', e'' \in B(\Lambda)$.*

Suppose now that X is a paracompact smooth n-dimensional manifold and $\Lambda \subset \widetilde{T^*(X)\backslash O}$ is a positive closed conical Lagrange manifold. If $\rho \in \Lambda_{\mathbb{R}}$, then we can define $B(T_\rho(\Lambda))$ by setting, as above, $M = T_\rho(T^*X)$; for F we take here the tangent space to the fibre. For any choice of local coordinates in X there arise natural linear symplectic coordinates in $T_\rho(T^*X)$ so that the spaces $T_\rho(T^*X)$ and $T_\mu(T^*X)$ can be identified provided that $\rho, \mu \in \Lambda_{\mathbb{R}}$ and are close enough. Therefore if we consider the section

$$\Lambda_{\mathbb{R}} \ni \rho \to e(\rho) \in B(T_\rho(\Lambda)),$$

then it makes sense to say that $e(\rho)$ locally (with respect to ρ) belongs to $B_K(T_\rho(\Lambda))$ for some compact set $K \subset B(\tilde{F}) \times \mathcal{L}^-$.

Definition 6.10. Let $\lambda_1, \ldots, \lambda_n$ be almost analytic functions on Λ defined in some complex neighbourhood U^λ, $\lambda = (\lambda_1, \ldots, \lambda_n)$. These functions are referred to as *admissible coordinates* on Λ if

1. $d\lambda_1, \ldots, d\lambda_n$ are linearly independent over $\mathbb{C}$ at real points.
2. Let $(\delta\lambda_1, \ldots, \delta\lambda_n)$ be the dual basis to $(d\lambda_1, \ldots, d\lambda_n)$ in $T_\rho(\Lambda)^*$.

Then $(\delta\lambda_1, \ldots, \delta\lambda_n)$ belongs to the set $B_K(T_\rho(\Lambda))$ locally with respect to $\rho \in U^\lambda \cap \Lambda_{\mathbb{R}}$ for some compact set $K \subset B(\tilde{F}) \times \mathcal{L}^-$. If U^μ, $\mu = (\mu_1, \ldots, \mu_n)$, is another system of local coordinates, then $s(\delta\lambda, \delta\mu)$ is a continuous function in $U^\lambda \cap U^\mu \cap \Lambda_{\mathbb{R}}$ and

$$s^2(\delta\lambda, \delta\mu) = \pm\frac{d(\mu)}{d(\lambda)} = \pm\det\left(\frac{\partial\mu_j}{\partial\lambda_k}\right).$$

Moreover, the derivatives $\partial/\partial\lambda_k$ are defined by the equation $d\mu_j = \sum(\partial\mu_j/\partial\lambda_k)\, d\lambda_k + \sum(\partial\mu_j/\partial\bar{\lambda}_k)\, d\bar{\lambda}_k$. It is clear that $s(\delta\lambda, \delta\mu)$ has a unique (to within equivalence) almost analytic continuation in Λ in some complex neighbourhood of the set $U^\lambda \cap U^\mu \cap \Lambda_{\mathbb{R}}$ and this continuation satisfies the condition

$$(s_{\lambda,\mu})^2 \sim \pm\frac{d(\mu)}{d(\lambda)}.$$

What is more, $s_{\lambda,\lambda} \sim 1$, $s_{\lambda,\mu}s_{\mu,\omega} \sim s_{\lambda,\omega}$ and $s_{\lambda,\mu}$ are continuous with respect to small perturbations of λ, μ under which $\delta\lambda$, $\delta\mu$ remain locally in $B_K(T_\rho(\Lambda))$ for some compact set K from $B(\tilde{F}) \times \mathcal{L}$.

We now define an almost analytic linear Maslov bundle $\mathscr{L}$ on Λ as a family of admissible coordinate systems U^λ on Λ with transition functions $s_{\lambda,\mu}$. The section $f \in \Gamma(\Lambda; \mathscr{L})$ is then defined by an almost analytic function f_λ on U^λ for which $f_\lambda \sim f_{\lambda,\mu} f_\mu$ for all λ and μ.

Theorem 6.4. *Let Λ be a closed conical positive Lagrange manifold in $\widetilde{T^*X\backslash O}$, where X is an n-dimensional paracompact smooth manifold. Then there exists a "natural" bijective linear map*

$$P: \Gamma^{m+n/4}(\Lambda; \mathscr{L}) \to I_\rho^m(X, \Lambda)/I_\rho^{m-1}(X, \Lambda).$$

Here $\Gamma^k(\Lambda; \mathscr{L})$ denotes the space of equivalence classes of homogeneous sections of degree k.

Let us demonstrate how to construct P locally. Let $s \in \Gamma^{m+n/4}(\Lambda; \mathscr{L})$ and let supp s lie in a small neighbourhood of the point ρ_0. Choose local coordinates $x_1, \ldots, x_n$ and a positive type regular phase function $S \in C^\infty(\mathbb{R}^n \times \mathbb{R}^N)$ which generates Λ, in these coordinates, in the neighbourhood of ρ_0. Then there exists an almost analytic function a on Λ which is positive homogeneous of degree $m + n/4 - N/2$, and unique to within equivalence, and which is such that

$$s \sim a\sqrt{dS}.$$

Consider a as a function on $C_{\tilde{S}}$ and its almost analytic homogeneous continuation a on $\mathbb{C}^n \times \mathbb{C}^N$ with support in a small conical neighbourhood of the point (x_0, θ_0) corresponding to ρ_0. The element $P(s)$ is now defined as the distribution

$$(2\pi)^{-(n+2N)/4} \int e^{iS(x,\theta)}\tilde{a}(x, \theta)\, d\theta$$

belonging to the class $I_\rho^m(X, \Lambda)$. That this procedure is well defined has been established by Melin and Sjöstrand [1976].

6.7. Fourier Integral Operators with Complex Phase Function. Let X and Y be smooth paracompact manifolds. If C is an arbitrary manifold in $T^*(X) \times T^*(Y) = T^*(X \times Y)$, then C' is defined to be $\{(x, \xi, y, -\eta); (x, \xi, y, \eta) \in C\}$.

Definition 6.11. A manifold $C \subset T^*(XxY)\backslash O$ is said to be a positive canonical relation if $C' \subset \widetilde{T^*(X \times Y)}\backslash O$ is a closed conical positive Lagrange manifold and $C_R \subset (T^*X\backslash O) \times (T^*Y\backslash O)$.

Let $A \in I_\rho^m(X \times Y, \Lambda)$, where $\rho > \frac{1}{2}$. Let $\Lambda \subset \widetilde{T^*(X \times Y)}\backslash O$ be a closed conical positive Lagrange manifold. Then A is the kernel of a continuous operator $A: C_0^\infty(Y, \Omega_{1/2}) \to \mathscr{D}'(X, \Omega_{1/2})$. We note that $\mathrm{WF}(A) \subset \Lambda_{\mathbb{R}}$. If $C = \Lambda'$ is a canonical relation, then $A: C_0^\infty(Y, \Omega_{1/2}) \to C^\infty(X, \Omega_{1/2})$, and A can be extended to a continuous operator from $\mathscr{E}'(Y, \Omega_{1/2})$ to $\mathscr{D}'(X, \Omega_{1/2})$.

Let X, Y, Z be smooth paracompact manifolds of dimension n_X, n_Y, n_Z, respectively. Suppose that

$$A_1 \in I_\rho^{m_1}(X \times Y, \Lambda_1), \qquad A_2 \in I_\rho^{m_2}(Y \times Z, \Lambda_2) (\rho > \tfrac{1}{2})$$

are proper operators, and $C_j = \Lambda_j'$ are positive canonical relations. Then the composition $A_1 \circ A_2$ is well defined, and

$$\mathrm{WF}'(A_1 \circ A_2) \subset \mathrm{WF}'(A_1) \cdot \mathrm{WF}'(A_2) \subset C_{1\mathbb{R}} \cdot C_{2\mathbb{R}},$$

where $C_{1\mathbb{R}}$, $C_{2\mathbb{R}}$ are relations generated by the intersection of C_1 and C_2 with $\mathbb{R}^n$. This composition is a Fourier integral operator if the following condition is fulfilled.

Let $\Delta = T^*(X) \times \mathrm{diag}\ T^*(Y) \times T^*(Z)$ and let $\tilde{\Delta}$ be its almost analytic complexification. Then $C_1 \times C_2$ intersects $\tilde{\Delta}$ transversally at points lying in $(C_{1\mathbb{R}} \times C_{2\mathbb{R}}) \cap \Delta$, and the natural projection

$$C_{1\mathbb{R}} \times C_{2\mathbb{R}} \to (T^*(X)\backslash O) \times (T^*(Z)\backslash O)$$

is an injective and proper map.

6.8. Some Applications. 1. Let A be a classical pseudodifferential operator of order 1 on a compact manifold X, and let the principal symbol a satisfy the condition

$$\operatorname{Im} a(x, \xi) \geqslant 0, \qquad \forall (x, \xi) \in T^*X\backslash O.$$

It can be shown by the methods of the semi-group theory that there exists a unique smooth functon $U(t)$, with values in the space of bounded linear operators on $\mathscr{D}'(X, \Omega_{1/2})$, defined for $t \geqslant 0$ and satisfying the equation $D_t U = AU$ for $t > 0$, $U(O) = I$.

We can show that $U(t)$ is a Fourier integral operator with complex phase function associated with a positive canonical relation from X into X. In a certain sense, this relation turns out to be the graph of the Hamiltonian flow of the function $-a$ (Trèves [1982]).

2. Let P be a pseudodifferential operator of the principal type (see § 3.1, Chap. 2). Suppose that the point $(x_0, \xi_0) \in T^*X\backslash O$ has an open conical neighbourhood whose image under the map $(x, \xi) \to p(x, \xi)$ does not intersect a certain ray $\{z = \rho e^{i\alpha}, 0 < \rho < \infty\}$ in the complex plane. Then there exists an open conical neighbourhood Γ of the point (x_0, ξ_0) and a linear continuous operator E on $\mathscr{D}'(X, \Omega_{1/2})$ such that $PE \sim I$ in Γ. What is more, it turns out that E can be represented in the form of a Fourier integral operator with a complex phase function (Trèves [1982]).

3. In $\mathbb{R}^2$, consider the operator $P = D_t + it|D_x|$. By means of the Fourier transformation with respect to x, we can easily show that every solution belonging to $S'(R^2)$ of the homogeneous equaiton $Pu = 0$ is of the form $u(t, x) = U(t)\varphi$, where $\varphi \in S'(R)$, and

$$U(t)\varphi = (2\pi)^{-1} \int e^{ix\xi - t^2|\xi|/2} \tilde{\varphi}(\xi)\, d\xi.$$

Although this operator is formally a Fourier integral operator with a complex-valued phase function, it may be regarded as a pseudodifferential operator with amplitude $e^{-t^2|\xi|/2}$ belonging to the class $S^0_{1/2,1/2}$.

Operators with complex phase functions have also been studied by Kucherenko [1977], Mishchenko and others [1978], Nazajkinskij and others [1981], Sjöstrand [1979], Taylor [1981], Trèves [1982], and others.

Chapter 4
The Propagation of Singularities

§ 1. The Regularity of the Solution at Non-characteristic Points

1.1. The Microlocal Smoothness. The results of § 2 of Chap. 1 can be made more precise by calibrating the wave front with the aid of Sobolev spaces.

Definition. A distribution $u \in \mathscr{D}'(\Omega)$ belongs to the space $H_s(x_0, \xi_0)$, where $s \in \mathbb{R}$ and $(x_0, \xi_0) \in T^*\Omega \backslash 0$, if $u = u_1 + u_2$ where $u_1 \in H_s(\Omega)$ and $(x_0, \xi_0) \notin \mathrm{WF}(u_2)$.

The set of those points $(x, \xi) \in T^*\Omega \backslash 0$ where u does not lie in H_s will be denoted by $\mathrm{WF}_s(u)$.

Of course, one can use other spaces such as C^k, $C^{k,\alpha}$ etc. instead of H_s.

Proposition 1.1. *Let A be a pseudodifferential operator of order m, and let the distribution $u \in \mathscr{E}'(\Omega)$ be such that $u \in H_s(x_0, \xi_0)$, where $(x_0, \xi_0) \in T^*\Omega \backslash 0$. Then $Au \in H_{s-m}(x_0, \xi_0)$.*

Using the decomposition of unity, one can easily establish the following result (Hörmander [1983, 1985]).

Proposition 1.2. *If $u \in H_s(x_0, \xi_0)$ for all the points $(x_0, \xi_0) \in T^*\Omega \backslash 0$, then $u \in H_{s,\mathrm{loc}}(\Omega)$.*

On the other hand, since $\mathscr{E}' = \bigcup (H_s \cap \mathscr{E}')$, we have

Proposition 1.3. *If $(x_0, \xi_0) \in \mathrm{WF}(u)$, then there exists an s such that $(x_0, \xi_0) \in \mathrm{WF}_{s+\varepsilon}(u)$ but $(x_0, \xi_0) \notin \mathrm{WF}_{s-\varepsilon}(u)$ for any $\varepsilon > 0$.*

A majority of the results of § 2, Chap. 1, can be reestablished by replacing $\mathrm{WF}(u)$ by $\mathrm{WF}_s(u)$. It is clear from the definition that $\mathrm{WF}_s(u)$ is a conical set. The set $\mathrm{WF}_s(u)$ can also be defined as $\bigcap \operatorname{char} A$, where the intersection is taken over the set of all proper classical pseudodifferential operators A of order zero for which $Au \in H_s$; char A denotes the set of characteristic points of A.

1.2. The Smoothness of Solution at a Non-characteristic Point. Let $u \in \mathscr{E}'(\Omega)$, $P(x, D)u = f$, where P is a pseudodifferential operator of order m.

Theorem 1.1. *If $(x_0, \xi_0) \in T^*\Omega \backslash 0$, $p_0(x_0, \xi_0) \neq 0$ and if $f \in H_s(x_0, \xi_0)$, then $u \in H_{s+m}(x_0, \xi_0)$.*

Proof. To prove the result, it is enough to construct a microlocal left parametrix Q for P. Its symbol $q(x, \xi)$ is constructed in a conical neighbourhood of the point (x_0, ξ_0) such that $q \sim \sum_0^\infty q_j$, where $q_0 = p_0^{-1}$ and where q_1 is determined from the equation

$$q_1 p_0 + q_0 p_1 + \frac{1}{i} \sum_{j=1}^{n} \frac{\partial q_0}{\partial \xi_j} \frac{\partial p_0}{\partial x_j} = 0.$$

The succeeding summands q_j for $j \geqslant 2$, which are homogeneous of degree $-j - m$, are found from the equation

$$\sum_{|\alpha|+k+l=j} \frac{1}{i^{|\alpha|}} \partial_\xi^\alpha q_k \partial_x^\alpha p_l = 0.$$

The operator Q is of order $-m$, and moreover

$$u + Tu = Qf,$$

where T is a smoothing operator so that $Tu \in C^\infty$. This fact and Proposition 1.1 together establish our assertion. □

Corollary 1.1. *If* $Pu = f$, *then*

$$\mathrm{WF}_s(u) \subset \mathrm{Char}\, P \cup \mathrm{WF}_{s-m}(f),$$

where, as before, $\mathrm{Char}\, P = \{(x, \xi) \in \Omega \times (\mathbb{R}^n \backslash 0), p_0(x, \xi) = 0\}$.

Corollary 1.2. *If P is an elliptic operator, then*

$$\mathrm{WF}_s(u) = \mathrm{WF}_{s-m}(Pu).$$

§ 2. Theorems on Removable Singularities

2.1. Removable Singularities in the Right-Hand Sides of Equations. A well-known theorem of Riemann states that a function which is holomorphic in $\Omega \backslash 0$ is holomorphic in Ω if $|f(z)| = 0(|z|^{-1})$.

It is also known that for functions that are harmonic in $\Omega \backslash 0$, Ω is a domain in $\mathbb{R}^n$, a singularity at $x = 0$ is *removable* if $|u(x)| = 0(|x|^{2-n})$ when $n > 2$ and if $|u(x)| = 0\left(\ln \frac{1}{|x|}\right)$ when $n = 2$.

What is less known is that a singularity at the point $x = 0$ is removable for an arbitrary linear differential operator $P(x, D)$ of order m if $|u(x)| = 0(|x|^{m-n})$ for $n \geqslant m$ (Harvey and Polking [1970]). Of course, in this case the solution may have a singularity at $x = 0$ but if $P(x, D)u = 0$ in $\Omega \backslash 0$ then $P(x, D)u = 0$ in Ω.

When $P(x, D)u = 0$ on the set $\Omega \backslash A$, where A is a closed measurable set, a significant role is played by its Hausdorff measure. We recall that the Hausdorff measure of order d is the quantity

$$H_d(A) = \varlimsup_{\varepsilon \to 0} \inf \sum r_j^d,$$

where the lower bound is taken over all the covers of A by the balls $\bigcup B_j$ in which the radius r_j of B_j does not exceed ε.

Theorem 2.1 (Harvey and Polking [1970]). *A singularity on the set A is removable, that is, $Pu = 0$ on $\Omega \backslash A$ implies $Pu = 0$ in Ω if at least one of the following conditions is fulfilled*:

1^0. $u \in W^p_{k,\mathrm{loc}}(\Omega)$, $1 < p < \infty$, $H_{n-(m-k)q}(K) < \infty$, $K \subset\subset A, q = p(q-1)$;

2^0. $u \in W^\infty_{k,\mathrm{loc}}(\Omega)$, $H_{n-m+k}(A) = 0$;

3^0. $u \in C^k(\Omega)$, $H_{n-m+k}(K) < \infty$ for $K \subset\subset A$;

4^0. $u \in C^{k,\alpha}(\Omega)$, $0 < \alpha < 1$, $H_{n-m+k+\alpha}(A) = 0$.

2.2. Removable Singularities in Boundary Conditions. Let Ω be a bounded domain in $\mathbb{R}^n$. Let $P(x, D)$ be a linear differential operator, of order m, with coefficients smooth in Ω. Let Γ, a part of the boundary $\partial\Omega$, be an $(n-1)$-dimensional smooth manifold. Suppose that on Γ there are prescribed linear differential operators $B_1(x, D), \ldots, B_k(x, D)$, with smooth coefficients, of order $m_j \leqslant m - 1$; $1 \leqslant k \leqslant m$, $1 \leqslant j \leqslant k$.

Let A be a compact subset in Γ. Suppose that the function $u(x)$ satisfies the equation $P(x, D)u = 0$ in Ω, and

$$B_j(x, D)u = 0 \qquad \text{on} \quad \Gamma\backslash A, j = 1, \ldots, k.$$

Let us pose the question: when do $B_j u = 0$ on Γ? Suppose that the direction of the normal to Γ is non-characteristic for P. Suppose that the Green's formula

$$\int_\Omega (Pu \cdot \bar{v} - u\overline{P^* v})\, dx = \sum_{j=1}^{t} \int_\Gamma B_j(x, D)u \cdot S_j(x, D)v \, d\Gamma \tag{1}$$

holds for smooth functions v that vanish in a neighbourhood of the set $\partial\Omega\backslash\Gamma$. Here $k \leqslant t \leqslant m$, the operator P^* is formally adjoint to P, the differential operators B_j and S_j have smooth coefficients, the orders μ_j of S_j are distinct, $0 \leqslant \mu_j \leqslant m - 1$, and, finally, in S_j the coefficient of the μ_jth order derivative along the normal does not vanish.

In particular, all these conditions are fulfilled if $m = 2k$ and the operators P, $B_1, \ldots, B_k$ define a regular elliptic problem on Γ (Lions and Magenes [1963]).

Theorem 2.2 (Egorov [1985]). *Suppose that $Pu = 0$ in Ω, $B_j u = 0$ on $\Gamma\backslash A$, $j = 1, \ldots, k$, and suppose that the Green's formula* (1) *holds. Then $B_j u = 0$ on Γ for $j = 1, \ldots, k$ provided that at least one of the following conditions is satisfied*:

1^0. $u \in W^p_{l,\mathrm{loc}}(\Omega)$, $1 < p < \infty$, $-\infty < l < m$, $H_{n-(m-l)q}(K) < \infty$;

2^0. $u \in W^\infty_{l,\mathrm{loc}}(\Omega)$, $-\infty < l < m$, $H_{n-m+l+\alpha}(A) = 0$;

3^0. $u \in C^l(\Omega \cup \Gamma)$, $-\infty < l < m$, $H_{n-m+l}(K) < \infty$;

4^0. $u \in C^{l,\alpha}(\Omega \cup \Gamma)$, $0 < \alpha < 1$, $0 \leqslant l < m$, $H_{n-m+l+\alpha}(A) = 0$.

§ 3. The Propagation of Singularities for Solutions of Equations of Real Principal Type

3.1. The Definition and an Example. Let us recall that an operator $P(x, D)$ of order m is of real principal type if its characteristic form (principal symbol) P_0 is real valued and the form $dP_0(x, \xi)$ is not proportional to the form $\xi\, dx$.

The simplest non-elliptic operator of real principal type is D_1. Consider the equation $D_1 u = f(x)$, where $f \in C^\infty(\Omega)$.

Theorem 1.1 implies that $\xi_1 = 0$ if $(x, \xi) \in \mathrm{WF}(u)$. Therefore for each C the restriction of u to the plane $x_1 = C$ is defined. The bicharacteristics of D_1 are straight lines, parallel to the x_1-axis, in $\mathbb{R}^{2n}$. By Theorem 3.3, Chap. 1, $(x, \xi') \subset \mathrm{WF}(u)_{x=C}$ if $(x, \xi) \in \mathrm{WF}(u)$. But $\psi(D')\varphi(x')u(x) = \psi(D')\varphi(x')u(C, x') + \int \psi(D')\varphi(x')f(t, x')\, dt$. This shows that if $(x^0, \xi^0) \notin \mathrm{WF}(u)$ and $x_1^0 = C$, then no point of the line segment that is parallel to the x_1-axis and lies in $T^*\Omega\backslash 0$ and passes through (x^0, ξ^0) belongs to $\mathrm{WF}(u)$. Hence every segment, lying in $T^*\Omega$, of the bicharacteristic either does not intersect $\mathrm{WF}(u)$ or lies completely in $\mathrm{WF}(u)$. The same is true for $\mathrm{WF}_s(u)$ if $f \in H_s(\Omega)$.

3.2. A Theorem of Hörmander. By means of a canonical transformation a real principal type operator can be reduced to the form $Q(x, D)D_1$, where Q is an elliptic operator. This enables us to examine the general equation $Pu = f$, where P is an mth order pseudodifferential operator of real principal type. Let $f \in C^\infty(\Omega)$. By *bicharacteristics* of P we mean integral curves of the system of equations

$$\frac{dx_j}{dt} = \frac{\partial P_0(x(t), \xi(t))}{\partial \xi_j}, \qquad \frac{d\xi_j}{dt} = -\frac{\partial P_0(x(t), \xi(t))}{\partial x_j}, \qquad j = 1, \dots, n, \tag{2}$$

if $p_0(x(t), \xi(t)) \equiv 0$.

Theorem 3.1 (Duistermaat and Hörmander [1972]). *If $u \in \mathscr{D}'(\Omega)$ and I is a connected piece of the bicharacteristic and if $I \cap \mathrm{WF}(Pu) = \varnothing$, then either $I \subset \mathrm{WF}(u)$ or $I \cap \mathrm{WF}(u) = \varnothing$.*

Proof. We may assume that $m = 1$, because $\mathrm{WF}(Pu)$ and the zero bicharacteristics of P are preserved when P is multiplied by an elliptic operator.

As we saw in § 3.1, there exists a Fourier integral operator A, of order zero, such that $P = A^*D_1A + T$ and $AA^* = I + T_1$, where T, T_1 are smoothing operators. Therefore if we set $Au = v$, then $D_1 v = Af + T_2 v$. Thus the proof reduces to that of the case considered above. Since a transformation involving A associates bicharacteristics of P with straight lines parallel to the x_1-axis, it follows that the theorem is true in the general case too. □

Thus on the complement to $\mathrm{WF}(Pu)$, the set $\mathrm{WF}(u)$ is invariant under translations along trajectories of the Hamiltonian system (2).

In exactly the same way, we can prove

Theorem 3.2. *If $u \in \mathscr{D}'(\Omega)$, I is a connected piece of the bicharacteristic and $I \cap \mathrm{WF}_s(Pu) = \varnothing$, where P is a real principal type pseudodifferential operator of order m, then either $I \subset \mathrm{WF}_{s+m-1}(u)$ or $I \cap \mathrm{WF}_{s+m-1}(u) = \varnothing$.*

3.3. Local Solvability. Suppose that P is a principal type pseudodifferential operator of order m in a domain $\Omega \subset \mathbb{R}^n$ with a real-valued principal symbol. The bicharacteristics of P are the integral curves of (2) along which $p_0(x(t), \xi(t)) = 0$.

Theorem 3.3. *Suppose that P is a pseudodifferential operator with a real principal symbol p_0 and suppose that $d_\xi p_0(x, \xi) \neq 0$ in $T^*\Omega \backslash 0$. Then for each point $x_0 \in \Omega$ we can find a neighbourhood U such that, if $f \in H_s(\Omega)$, there exists a function u in $H_{s+m-1}(\Omega) \cap \mathscr{E}'(\Omega)$ satisfying the equation $Pu = f$ in U. Moreover, $\|u\|_{s+m-1} \leqslant c\|f\|_s$, where the constant c depends only on s.*

Proof. Let U be a neighbourhood of the point x_0 such that every zero bicharacteristic of the function p_0 that passes through the point (x, ξ), $x \in U$ and $\xi \neq 0$, also passes through a certain point (x_1, ξ_1), where $x_1 \in \partial U$ and $\xi_1 \neq 0$. Such a neighbourhood always exists, because P is a principal type operator.

We note that P^* is also a principal type operator and that $\varphi \in H_s(U)$ if $P^*\varphi \in H_{s-m+1}(U)$ and $\varphi \in \mathscr{E}'(U)$. Indeed, by Theorem 3.2, the set of points where $\varphi \in H_s$ is invariant under the Hamiltonian flux H_{p_0}, and because $\varphi = 0$ in a neighbourhood of ∂U, it follows that every zero bicharacteristic contains points at which $\varphi \in C^\infty$. Hence $\varphi \in H_s$ at every point of T^*U. Thus $\varphi \in H_s(U)$, by Proposition 1.2.

By the Banach open mapping theorem, there exists a constant C, depending only on s, such that

$$\|\varphi\|_s \leqslant C\|P^*\varphi\|_{s-m+1}, \qquad \varphi \in C_0^\infty(U).$$

Therefore the integral $\int f\overline{\varphi}\,dx$, where $f \in H_s(\Omega)$, is a continuous linear functional of $\overline{P^*\varphi} \in H_{-s-m+1}$. This implies, by the Hahn-Banach theorem, the existence of a function $u \in H_{s+m-1}$ for which

$$\int f\overline{\varphi}\,dx = \int u \cdot \overline{P^*\varphi}\,dx, \qquad \varphi \in C_0^\infty(U),$$

and $\|u\|_{s+m-1} \leqslant C\|f\|_s$. By definition, u is a generalized solution of the equation $Pu = f$. □

3.4. Semiglobal Solvability. By semiglobal solvability we mean solvability in every compact subset. That the semiglobal solvability is not a consequence of the local solvability becomes clear from the following example.

Example 3.1. Let Ω be the ring

$$\Omega = \{(x, y), 1 \leqslant x^2 + y^2 \leqslant 4\}$$

in the plane. The equation $\partial u/\partial\varphi = f$ is always locally solvable, but for it to be solvable in Ω it is necessary that

$$\int_0^{2\pi} f(r, \varphi)\, d\varphi = 0$$

for $1 \leqslant r \leqslant 2$.

This example demonstrates that the following condition is a natural one.

Condition A. Let $K \ll \Omega$. Each zero bicharacteristic passing through the point (x, ξ), $x \in K$, $\xi \neq 0$, contains points the projection of which onto Ω lies outside K.

Theorem 3.4 (Duistermaat and Hörmander [1972]). *Let P be a principal type pseudodifferential operator of order m, with principal symbol p_0, in a domain $\Omega \subset \mathbb{R}^n$. Let $K \ll \Omega$. Assume that condition A holds. Then the set*

$$N(K) = \{\varphi;\ \varphi \in C_0^\infty(K),\ P^*\varphi = 0\}$$

has a finite dimension. If $f \in H_s(\Omega)$ and $\int f\overline{\varphi}\, dx = 0$ for all $\varphi \in N(K)$, then there exists a function $u \in H_{s+m-1}(\Omega)$ satisfying the equation $Pu = f$ in K which is such that

$$\|u\|_{s+m-1} \leqslant C\|f\|_s,$$

where the constant C does not depend on f.

This theorem can be proved by means of Theorem 3.2 and the closed graph theorem. In doing so, one has to use the fact that each solution of the equation $P^*\varphi = 0$ in $\mathscr{E}'(K)$ belongs to $C_0^\infty(\Omega)$, in view of condition A and Theorem 3.2.

§4. The Propagation of Singularities for Principal Type Equations with a Complex Symbol

4.1. An Example (Hörmander [1979]). Consider the equation

$$(D_1 + ix_1^k D_2)u = 0,$$

where k is a positive odd number. Let $\eta_0 = (0, -1, \eta'')$, $\eta'' \in \mathbb{R}^{n-2}$. Assume that $v(x') \in C^1$, $x' = (x_2, \dots, x_n)$, is a function with compact support for which

$$\mathrm{WF}(v) = \{(0, t\eta_0'),\ t > 0\}$$

(see Example 2.2, Chap. 1). Then $\tilde{v}$ decays rapidly outside each conical neighbourhood Γ of the point η_0'. We set

$$u(x) = (2\pi)^{1-n} \int_{\Gamma_0} e^{\xi_2 x_1^{k+1}/(k+1) + i(x', \xi')}\tilde{v}(\xi')\, d\xi',$$

where Γ_0 is a neighbourhood of η_0' such that $\xi_2 < -C|\xi'|$ in Γ. Then $u \in C^1$, and, moreover, $u \in C^\infty$ for $x_1 \neq 0$. It can be shown that

$$\mathrm{WF}(u) = \{(0, t\eta_0),\ t > 0\};$$

thus the solution has a singularity at $x = 0$ that is not propagated.

4.2. The Fixed Singularity. The phenomenon noted in the above example is a feature of a large class of operators. Suppose that P is a proper pseudodifferential operator of order m in a domain Ω. Let $(x_0, \xi_0) \in T^*\Omega\backslash 0$ be a characteristic point at which $H_{p_0} \neq 0$. By multiplying P by i, if necessary, one can always achieve that $H_{\mathrm{Re}\, p_0} \neq 0$. Then the equation $\mathrm{Re}\, p_0 = 0$ defines a smooth surface passing through the point (x_0, ξ_0) and consisting of bicharacteristics of the function $\mathrm{Re}\, p_0$.

Assume that $\mathrm{Im}\, p_0$ has a zero of finite order $k \neq 0$ on each bicharacteristic of $\mathrm{Re}\, p_0$ that is close to the one passing through (x_0, ξ_0).

This condition signifies, in particular, that $H_{\mathrm{Re}\, p_0}$ cannot have a radial direction because otherwise the function $\mathrm{Im}\, p_0$ would vanish identically on the bicharacteristic passing through (x_0, ξ_0).

Theorem 4.1 (Sato, Kawai and Kashiwara [1973]). *Under the above assumptions, there exists a Fourier integral operator A such that*

$$A^*A = \Lambda^{1-m} + T,$$

$$A^*PA = D_1 + ix_1^k D_2 + T_1,$$

where T and T_1 are operators having symbols of order $-\infty$ in a neighbourhood of the point $(0, \eta_0)$.

Using the example cited above, we have

Corollary 4.1. *If k is odd and $(H_{\mathrm{Re}\, p_0})^k\, \mathrm{Im}\, p_0(x_0, \xi_0) < 0$, then there exists a distribution u for which*

$$Pu \in C^\infty \text{ and } \mathrm{WF}(u) = \{(x_0, t\xi_0), t > 0\}.$$

4.3. A Special Case. Let P be a principal type pseudodifferential operator of order m with principal symbol $p_0 = a + ib$. Assume that the following conditions hold:

1. The Poisson bracket $\{a, b\}$ vanishes at each characteristic point lying in $T^*\Omega\backslash 0$.

2. The fields H_a, H_b and $\sum \xi_j \dfrac{\partial}{\partial \xi_j}$ are linearly independent at each characteristic point belonging to $T^*\Omega\backslash 0$.

Under these conditions, there exists a homogeneous canonical transformation χ of the neighbourhood of the point $(0, \eta^0) \in T^*\mathbb{R}^n\backslash 0$ onto a neighbourhood of the point $(x^0, \xi^0) \in T^*\Omega\backslash 0$ and there exist Fourier integral operators F and F_1, corresponding to χ and χ^{-1}, such that the operator $F_1PF - (D_1 + iD_2)$ has a symbol of order $-\infty$ in a conical neighbourhood of $(0, \eta^0)$.

Condition 1 implies that $\{a, b\} = \alpha a + \beta b$ and $[H_a, H_b] = \alpha H_a + \beta H_b$. By Frobenius theorem, the manifold $p_0 = 0$ is fibered into two-dimensional manifolds the tangent planes to which are generated by H_a and H_b. These manifolds are referred to as bicharacteristics. The fields H_a and H_b define on bicharacteristics an analytic lattice in which the solutions of the equation

$(H_a + iH_b)v = 0$ are analytic functions. For this lattice, we can also define classes of harmonic and superharmonic functions. If $u \in \mathscr{D}'(\Omega)$, then the smoothness of u at a point $(x, \xi) \in T^*\Omega\backslash 0$ can be measured with the aid of the function $s_u^*(x, \xi) = \operatorname{supp}\{s; u \in H_s(x, \xi)\}$. This function is lower semicontinuous and positive homogeneous of degree zero.

Theorem 4.2 (Duistermaat and Hörmander [1972]). *Assume that conditions 1 and 2 are satisfied. Let $u \in \mathscr{D}'(\Omega)$ and $Pu = f$ in Ω. Suppose that B is a domain on a two-dimensional bicharacteristic and s is a superharmonic function in B such that $s \leqslant s_f^*$. Then the function* $\min(s_u^*, s + m - 1)$ *is superharmonic in B.*

4.4. The Propagation of Singularities in the Case of a Complex Symbol of the General Form. The most general results in this direction are due to Hörmander, Shapira, and Dencker. A majority of these results and the corresponding references can be found in Hörmander [1983, 1985]. Here we shall cite only one result of Hörmander.

Theorem 4.3. *Let $u \in \mathscr{D}'(\Omega)$ and $Pu = f$ in Ω. Let $I = \{t \in \mathbb{R}, t_1 \leqslant t \leqslant t_2\}$ and $\gamma(I)$ be a segment of the zero bicharacteristic of the function* $\operatorname{Re} p_0$. *Let* $\operatorname{Im} p_0 \geqslant 0$ *in a neighbourhood of $\gamma(I)$. If $\gamma(I) \cap \mathrm{WF}(f) = \varnothing$ and $\gamma(t_2) \notin \mathrm{WF}(u)$, then $\gamma(I) \cap \mathrm{WF}(u) = \varnothing$. More precisely, if $f \in H_s(\gamma(I))$ and $u \in H_{s+m-1}(\gamma(t_2))$, then $u \in H_{s+m-1}$ at points of $\gamma(I)$.*

If $\operatorname{Im} p_0 \leqslant 0$ in a neighbourhood of $\gamma(I)$, the theorem remains valid, only one has to take the left end instead of the right end of the segment.

§ 5. Multiple Characteristics

5.1. Non-involutive Double Characteristics. Let $P = P_1 P_2 + Q$, where P_1 and P_2 are pseudodifferential operators of order m_1 and m_2, respectively, and the operator Q is of order $m + m_2 - 1$. Assume that the principal symbols p_1 and p_2 are real valued.

We take a point $z_0 \in T^*\Omega\backslash 0$ at which

$$p_1(z_0) = 0, \qquad p_2(z_0) = 0, \qquad \{p_1, p_2\}(z_0) \neq 0. \tag{2}$$

This condition implies that the fields H_{p_1}, H_{p_2} and $\xi\partial/\partial\xi$ are linearly independent at z_0. In particular, an open interval $I \subset \mathbb{R}$, $0 \in I$, can be found such that the bicharacteristics γ_j of p_j passing through z_0 define an injective map

$$\gamma_j\colon I \to \Sigma_j = \{z \in T^*\backslash 0, p_j(z) = 0\}, \qquad \gamma_j(0) = z_0.$$

We note that at those points of the neighbourhood U_{z_0} that lie outside $\Sigma_1 \cap \Sigma_2$, P is a real principal type operator, and for such points the propagation of singularities has been investigated in § 3.

If $I_k = \{t \in I, (-1)^k t > 0\}$, we can define 4 semibicharacteristics at the point 2 as the curves $\gamma_{jk}\colon I_k \to \Sigma_j$ for which $\gamma_{jk} = \gamma_j|_{I_k}$ for $j, k = 1, 2$.

Theorem 5.1. *Let $u \in \mathscr{D}'(\Omega)$, $z_0 \notin \mathrm{WF}(Pu)$, and let (2) hold. Suppose that for each $j \in \{1, 2\}$ there exists a $k \in \{1, 2\}$ such that $\gamma_{jk}(I_k) \cap \mathrm{WF}(u) = \varnothing$. Then $z_0 \notin \mathrm{WF}(u)$.*

Let S_p denote the subprincipal symbol of P.

Theorem 5.2. *Suppose that (2) holds and*

$$iS_P(z_0)/\{p_1, p_2\}(z_0) - \tfrac{1}{2} \notin \{0, 1, 2, \dots\}.$$

Let $u \in \mathscr{D}'(\Omega)$, $z_0 \notin \mathrm{WF}(Pu)$ and $(\gamma_1(I)\backslash z_0) \cap \mathrm{WF}(u) = \phi$. Then $z_0 \notin \mathrm{WF}(u)$. Similarly, if

$$iS_P(z_0)/\{p_1, p_2\}(z_0) + \tfrac{1}{2} \notin \{0, -1, -2, \dots\}$$

and $(\gamma_2(I)\backslash z_0) \cap \mathrm{WF}(u) = \varnothing$, then $z_0 \notin \mathrm{WF}(u)$.

Corollary 5.1. *Assume that (2) holds and that*

$$iS_P(z_0)/\{p_1, p_2\}(z_0) - \tfrac{1}{2} \notin \mathbb{Z}.$$

If $z_0 \notin \mathrm{WF}(Pu)$ and u is a smooth function on certain two semibicharacteristics through z_0, then $z_0 \in \mathrm{WF}(u)$. (These and more general results can be found in Ivrii [1979, 1981], Bove, Lewis and Parenti [1983], Hanges [1979], and Melrose and Sjöstrand [1978, 1982].)

5.2. The Levi Condition. When the operator P has multiple characteristics, its properties may significantly depend on the lower terms. In this case, an important role is played by the *Levi condition*, introduced first in connection with the investigation of the Cauchy problem. This condition signifies that the transport equations are differential equations along bicharacteristics and that their order equals the multiplicity of the corresponding characteristic.

Definition 5.1. An operator $P \in \mathscr{L}^m(\Omega)$ has characteristics of constant multiplicity if its principal symbol p is of the form

$$p = q_1^{r_1} \dots q_s^{r_s}, \qquad \text{where } r_j \in \mathbb{N},$$

and q_j are symbols of principal type operators, and the sets $q_j^{-1}(0)$ do not intersect in $T^*\Omega\backslash 0$.

Definition 5.2. Suppose that $P \in \mathscr{L}^m(X)$, its principal symbol is real valued, and suppose that P has characteristics of constant multiplicity. P is said to satisfy the *Levi condition* $\mathscr{L}(x_0, \xi_0)$ at a point (x_0, ξ_0) in $p^{-1}(0) \subset T^*\Omega\backslash 0$ if

$$e^{-it\varphi}P(ae^{it\varphi}) = 0(t^{m-r_j}), \qquad t \to +\infty$$

for each function $\varphi(x)$ that satisfies the equation

$$q_j(x, d\varphi(x)) = 0 \qquad (\text{if } q_j(x_0, \xi_0) = 0)$$

in a neighbourhood of x_0 and which is such that $d\varphi(x_0) = \xi_0$ and for each function $a \in C_0^\infty(\Omega)$ that has support in a neighbourhood of x_0 where $d\varphi \neq 0$. The operator P is said to satisfy the *Levi condition* ($\mathscr{L}$) if the condition $\mathscr{L}(x_0, \xi_0)$ is satisfied for all the points of $p^{-1}(0)$.

It is clear that ($\mathscr{L}$) is a condition imposed on terms having order $\geqslant m - \max r_j$. This condition was introduced by Flaschka and Strang [1971].

5.3. Operators Having Characteristics of Constant Multiplicity. Let us consider an operator $P \in \mathscr{L}^m(\Omega)$ having characteristics of constant multiplicity and satisfying the Levi condition ($\mathscr{L}$). In this case, in a neighbourhood of each characteristic point (x^0, ξ^0) the operator P can be expressed in the form $\sum_{k=0}^{r_j} B_k Q_j^k + T_0$ for some j, $1 \leqslant j \leqslant s$. Here T_0 is an operator of order zero, Q_j is an operator with real principal symbol $q_j(x, \xi) \in S^{s_j}(\Omega)$, $B_k \in \mathscr{L}^{m-ks_j}(\Omega)$. This representation enables us to obtain the following result which is a particular case of a theorem of Tulovskij [1979].

Theorem 5.3. *Suppose that the operator $P \in \mathscr{L}^m(\Omega)$ has characteristics of constant multiplicity and that it satisfies the Levi condition ($\mathscr{L}$). Let $u \in \mathscr{D}'(\Omega)$. Then the set $\mathrm{WF}(u) \setminus \mathrm{WF}(Pu)$ is contained in $p^{-1}(0)$ and is invariant under the translations along bicharacteristics.*

5.4. Operators with Involutive Multiple Characteristics. We assume now that $\Sigma \subset T^*X \setminus 0$ is a closed conical manifold of codimension d defined by the equations $q_1(x, \xi) = 0, \ldots, q_d(x, \xi) = 0$. Here the functions q_j are smooth, real valued, and are such that:

1. The fields $H_{q_1}, \ldots, H_{q_d}$ and ξD_ξ are linearly independent at every point of Σ;
2. $\{q_j, q_k\} = 0$ on Σ; $j, k = 1, \ldots, d$.

We may assume that q_j are homogeneous in ξ of degree 1. Consider now a proper classical pseudodifferential operator $P \in \mathscr{L}^{m+k}(X)$, where $m \in \mathbb{N}$, $k \in \mathbb{R}$. Let its principal symbol p_{m+k} be such that $p_{m+k} = 0$ on Σ and $p_{m+k} \neq 0$ outside Σ; p_{m+k} has a zero of order m at each point of Σ. Take a point ρ on Σ, and let $a_\rho(t)$ denote a homogeneous polynomial in $t \in T_\rho(T^*(X))/T_\rho(\Sigma) = F_\rho$ of degree m that arises in the Taylor expansion of p_{m+k} at the point ρ. Suppose that

3. $a_\rho(t) \neq 0$ for all $\rho \in \Sigma$ when $0 \neq t \in F_\rho$.

The following condition imposed on the lower terms is a natural generalization of the Levi condition.

4. If $Q_j = q_j(x, D)$, then there exist classical pseudodifferential operators A_α of order k such that $P = \sum_{|\alpha| \leqslant m} A_\alpha Q_1^{\alpha_1} \ldots Q_d^{\alpha_d}$ microlocally.

When $m = 2$, for instance, this condition implies that the subprincipal symbol of P vanishes on Σ.

Theorem 5.4 (Sjöstrand [1976]). *Assume that the conditions 1–4 hold. Let $F = T_\Sigma(T^*X)/T(\Sigma)$, and let $F \times_\Sigma F$ be the product of bundles on Σ. Let $z_1(s, t), \ldots, z_m(s, t)$ be the roots of the equation $a_\rho(s + zt) = 0$; $s, t \in F_\rho$. Assume that for $(s, t) \in F \times_\Sigma F$ and for linearly independent s and t the multiplicity of z_j is a constant that does not exceed 2. If $u \in \mathscr{D}'(X)$, $Pu \in C^\infty(X)$, and $\rho \in \mathrm{WF}(u)$, then $\Gamma_\rho \subset \mathrm{WF}(u)$, where Γ_ρ is an integral manifold, passing through ρ, of the fields $H_{q_1}, \ldots, H_{q_d}$.*

5.5. The Schrödinger Operator. Let Ω be a domain in $\mathbb{R}^n$. Let $P(x, D)$ be a pseudodifferential operator in Ω of order m with real principal symbol $p_0(x, \xi)$. Suppose that P has double characteristics so that the set $\Sigma = \{(x, \xi) \in T^*\Omega \backslash 0, p_0(x, \xi) = 0\}$ is defined by the equations $u_1(x, \xi) = 0, \ldots, u_k(x, \xi) = 0$, where $u_j(x, t\xi) = u_j(x, \xi)$, $t > 0$, and the forms $du_1, \ldots, du_k, \xi\, dx$ are linearly independent.

Assume that Σ is an involutive manifold, that is, $\{u_i, u_j\} = 0$ on Σ. Then Σ is fibered into one-dimensional smooth curves that are integral manifolds for the fields H_{u_j} for each $j = 1, \ldots, k$. We also assume that in a neighbourhood of every point of Σ, the principal symbol can be expressed in the form $\sum a_{ij}u_iu_j$, $i, j = 1, \ldots, k$, and that the matrix $\|a_{ij}\|$ is positive definite. This matrix defines a Riemannian metric g_A on Σ.

We assume further that $p_{m-1}(x, \xi) - \dfrac{1}{2i}\sum \dfrac{\partial^2 p_0(x, \xi)}{\partial x_j \partial \xi_j} < 0$ on Σ.

Theorem 5.5 (Boutet de Monvel [1975]). *Suppose that all the conditions listed above are satisfied. If* $u \in \mathscr{E}'(\Omega)$ and $Pu \in C^\infty(\Omega)$, *then* WF(u) *is a union of geodesics, lying on* Σ, *in the metric* g_A. *Conversely, if* $(x, \xi) \in \Sigma$ *and if* γ, *a geodesic arc in the metric* g_A, *passes through* (x, ξ), *then there exists a distribution* $u \in \mathscr{E}'(\Omega)$ *such that* WF$(u) \cap U = \gamma \cap U$ *and* $Pu \in C^\infty(U)$, *where* U *is a sufficiently small conical neighbourhood of the point* (x, ξ).

As an example of an operator satisfying the hypotheses of Theorem 5.5, one can cite the Schrödinger operator $P = \dfrac{1}{i}\dfrac{\partial}{\partial t} - a^2\Delta$. If $Pu \in C^\infty$, then in this case WF(u) consists of straight lines that are parallel to the subspace $t = 0, \tau = 0$ and lie in the cone $\xi_1 = \cdots = \xi_n = 0, \tau < 0$ (see Boutet de Monvel [1975]).

We remark that several other important results concerning the propagation of singularities for operators with multiple characteristics have been obtained in the works of Bony, Grigis, Ivrii, B. Lascar, R. Lascar, Shapira, Sjöstrand, and others. In particular, see Ivrii [1974, 1980a, 1981], Bony and Schapira [1973], Chazarain [1974], Grigis [1976, 1979], B. Lascar [1981], B. Lascar and Sjöstrand [1982], Menikoff [1979], Sjöstrand [1976], and Uhlmann [1977].

Chapter 5
Solvability of (Pseudo)Differential Equations

§ 1. Examples

1.1. Lewy's Example. Consider the unit sphere Ω in $\mathbb{C}^2 = \{(z_1, z_2): |z_1|^2 + |z_2|^2 = 1\}$. If a change of variables is performed by the formula $z_1' = (z_1 - 1)^{-1}$, $z_2' = z_2(z_1 - 1)^{-1}$, the surface Ω is defined by the equation

$$1 + z_1 + \bar{z}_1 + |z_2|^2 = 0.$$

The operator

$$\mathscr{L} = 2\frac{\partial}{\partial \bar{z}_2} - z_2\left(\frac{\partial}{\partial z_1} + \frac{\partial}{\partial \bar{z}_1}\right)$$

is tangent to Ω. The operator $\mathscr{L}$ is the one that is referred to as Hans Lewy's operator. In 1956, Lewy constructed a function $f(x_1, x_2, x_3) \in C^\infty(\mathbb{R}^3)$ such that the equation

$$\mathscr{L}u \equiv \frac{\partial u}{\partial x_1} + i\frac{\partial u}{\partial x_2} - (x_1 + ix_2)\frac{\partial u}{\partial x_3} = f(x_1, x_2, x_3)$$

has no solution in the class of continuous functions in any subdomain $\omega \subset \mathbb{R}^3$.

1.2. Mizohata's Equation. Consider the equation

$$\frac{\partial u}{\partial x} + ix^k\frac{\partial u}{\partial y} = f(x, y). \tag{1}$$

If k is even, the substitution $t = x^{k+1}/(k+1)$ transforms this equation into the equation

$$\frac{\partial u}{\partial t} + i\frac{\partial u}{\partial y} = g(t, y), \qquad g(t, y) = f(x, y)x^{-k},$$

which is solvable for any distribution $g \in \mathscr{D}'$.

If k is odd, then (1) is, in general, not solvable in any neighbourhood of the origin (see Grushin [1971]).

Indeed, let $\{\Omega_j\}$ be a sequence of disjoint discs converging to the origin, each disc lying in the domain $x > 0$. Let $f = 0$ outside $\bigcup \Omega_j$, $f \in C^\infty$, $f(x, y) = f(-x, y)$, and let $f > 0$ in Ω_j.

If we now set

$$v(x, y) = \frac{u(x, y) + u(-x, y)}{2}, \qquad w(x, y) = \frac{u(x, y) - u(-x, y)}{2},$$

then (1) implies

$$\frac{\partial v}{dx} + ix^k\frac{\partial v}{\partial y} = 0, \qquad \frac{\partial w}{\partial x} + ix^k\frac{\partial w}{\partial y} = f(x, y).$$

Clearly, $v = F(x^{k+1}/(k+1) + iy)$, where F is an analytic function. As to the function w, it satisfies $w(0, y) = 0$ and $w = G(x^{k+1}/(k+1) + iy)$, where G is an analytic function at the points of the complement set to $\bigcup \Omega_j$. It implies that $w = 0$ in $\mathbb{R}^2 \setminus \bigcup \Omega_j$ so that, in particular, $w = 0$ on $\partial\Omega_j$ for $j = 1, 2, \ldots$. But then

$$\iint_{\Omega_j} f(x, y)\, dx\, dy = \int_{\partial\Omega_j} w(\cos\alpha + i\cos\beta)\, ds = 0,$$

contradicting the definition of f.

1.3. Other Examples. Simple considerations show that the set of smooth functions f for which the equation (1) has a solution (even if generalized) constitutes a set of the first category.

We note that for $L = \frac{\partial}{\partial x} - ix\frac{\partial}{\partial y}$ the operator $L\bar{L}$ is of the form $\frac{\partial^2}{\partial x^2} + x^2\frac{\partial^2}{\partial y^2} + i\frac{\partial}{\partial y}$, while the operator

$$P = L\bar{L}\bar{L}L = \left(\frac{\partial^2}{\partial x^2} + x^2\frac{\partial^2}{\partial y^2}\right)^2 + \frac{\partial^2}{\partial y^2}$$

has real coefficients, and the equation $Pu = f$ has no solution at the origin for the set of functions f of second category in $C_0^\infty(\mathbb{R}^2)$ (Trèves).

There exists a function $f \in C^\infty(\mathbb{R}^2)$ such that the equation

$$\mathscr{L}u - fu = 0$$

has no solution, apart from $u = 0$, in the rectangle $|x| < r$, $|y| < r'$.

Kannai [1971] has shown that the equation $Au = f$, where

$$A = D_1 + ix_1 D_2^2,$$

has no solution at the origin, and that A is a hypoelliptic operator.

Hörmander [1963] has established the insolvability of a linear self-adjoint differential operator of second order, with real coefficients, defined by the formula

$$Pu = (x_2^2 - x_3^2)D_1^2 u + (1 + x_1^2)(D_2^2 u - D_3^2 u) - x_1 x_2 D_1 D_2 u - D_1 D_2(x_1 x_2 u)$$
$$+ x_1 x_3 D_1 D_3 u + D_1 D_3(x_1 x_3 u).$$

Like the above cases, for every neighbourhood Ω of the origin there exists a function $f \in C_0^\infty(\Omega)$ such that the equation $Pu = f$ has no solution u in $\mathscr{D}'(\Omega)$. This implies that there is a function $f_0 \in S(\mathbb{R}^3)$ for which the equation $Pu = f_0$ has no solution u in $\mathscr{D}'(\Omega)$ in any neighbourhood of the origin.

Definition 1.1. A pseudodifferential operator P is *solvable* at a point x_0 if there are two neighbourhoods U and V of x_0 such that $U \subset V$ and for each function f in $C_0^\infty(U)$ there exists a distribution u, with support in V, that satisfies the equation $Pu = f$ in U.

In the case of a differential operator, one can take $U = V$.

§ 2. Necessary Conditions for Local Solvability

2.1. Hörmander's Theorem. Hörmander succeeded in explaining the behaviour of the above examples. He [1963] obtained the following result.

Theorem 2.1. *If* $(x_0, \xi_0) \in T^*\Omega\backslash 0$, $p_0(x_0, \xi_0) = 0$ *and* $c_1(x_0, \xi_0) < 0$, *where*

$$c_1(x, \xi) = 2 \operatorname{Im} \sum_{j=1}^{n} \overline{\frac{\partial p_0(x, \xi)}{\partial \xi_j}} \frac{\partial p_0(x, \xi)}{\partial x_j},$$

then the pseudodifferential operator, with principal symbol p_0, *is insolvable at* x_0.

Corollary 2.1. *If* $P(x, D)$ *is a differential operator, and if* $p_0(x_0, \xi_0) = 0$ *and* $c_1(x_0, \xi_0) \neq 0$, *then* P *is insolvable at* x_0.

To prove Corollary 2.1, it is enough to observe that the function $c_1(x, \xi)$ is homogeneous in ξ of degree $2m - 1$, and is, hence, odd.

The proof of Theorem 2.1 is based on the fact that the solvability of P implies

$$\|\varphi\|_{s_1} \leqslant c\|P^*\varphi\|_{s_2}, \qquad \varphi \in C_0^\infty(\omega), \tag{2}$$

with certain real C, s_1 and s_2. For instance, for the Lewy operator $P = \dfrac{\partial}{\partial z} + iz\dfrac{\partial}{\partial t}$ the adjoint operator is of the form $P^* = -\dfrac{\partial}{\partial \bar{z}} + i\bar{z}\dfrac{\partial}{\partial t}$. And the function

$$\varphi_0 = e^{-\lambda(2|z|^2 + t_2 - |z|^4 - 2it + 2it|z|^2)}$$

satisfies the equation $P^*\varphi_0 = 0$. Let $\omega = \{(x, y, t); x^2 + y^2 + t^2 \leqslant 1\}$ and let $h \in C_0^\infty(\omega)$ be a function such that $h = 1$ for $\rho^2 \equiv x^2 + y^2 + t^2 \leqslant 1/4$. Then the function $f = P^*(h\varphi_0)$ vanishes for $\rho \geqslant 1$ and $\rho \leqslant \frac{1}{2}$. It is easy to see that $\|f\|_{s_2} \leqslant C\lambda^{s_2}e^{-\lambda/4}$. On the other hand, $\|h\varphi_0\|_{s_1} \geqslant C_0\lambda^{s_1 - 3/4}$, where the constant $C_0 > 0$ depends on s_1 but not on λ. Substituting the function $\varphi = h\varphi_0$ into (2), we obtain an inequality that cannot hold for $\lambda \geqslant \lambda_0$ if λ_0 is sufficiently large. This establishes the insolvability of P in ω.

In the general case, when P is a pseudodifferential operator of an arbitrary order the proof is carried out along the same lines. The hypotheses of the theorem enable us to construct a function w for which

$$p_0(x, \operatorname{grad} w(x)) = 0, \qquad w(0) = 0, \qquad \operatorname{grad} w(0) = \xi_0$$

and $\operatorname{Im} w(x) > 0$ when $x \neq 0$. We now take for φ a function of the form

$$\varphi(x) = h(x)e^{i\lambda w(x)} \sum_{j=0}^{N} \lambda^{-j}a_j(x), \qquad a_j \in C^\infty(\omega).$$

2.2. The Zero of Finite Order. Let $p_0(x, \xi)$ be the symbol of a principal type operator, $a = \operatorname{Re} p_0$ and $b = \operatorname{Im} p_0$. Consider a bicharacteristic of b, that is, an integral curve of the system of equations

$$\frac{dx_k}{dt} = \frac{\partial b(x(t), \xi(t))}{\partial \xi_k}, \qquad \frac{\partial \xi_k}{\partial t} = -\frac{\partial b(x(t), \xi(t))}{\partial x_k}, \qquad k = 1, \ldots, n,$$

passing through the characteristic point (x_0, ξ_0) when $t = 0$. Clearly, $b(x(t), \xi(t)) \equiv 0$ in this case. Consider the function $\gamma(t) = a(x(t), \xi(t))$. We note that

$$c_1(x(t), \xi(t)) = \gamma'(t)$$

so that the hypotheses of Theorem 2.1 imply that $\gamma(t)$ has a first-order zero at $t = 0$. This theorem can be generalized by replacing its hypothesis by the condition

$$\gamma'(0) = \cdots = \gamma^{(k-1)}(0) = 0, \qquad \gamma^{(k)}(0) > 0, \qquad k \text{ odd} \tag{3}$$

(see Egorov [1971], Nirenberg and Trèves [1970]).

2.3. The Zero of Infinite Order. The following condition (Ψ) is a necessary condition for the solvability of P in the general case (Hörmander [1983, 1985]).

Condition (Ψ). As t increases, the function $\gamma(t)$ does not change its sign from minus to plus at $t = 0$. (Clearly, this condition generalizes condition 3.)

2.4. Multiple Characteristics. Less attention has been paid to the case of multiple characteristics. The necessary conditions known to us for solvability in the case of a characteristic point (x_0, ξ_0) at which $dp_0(x_0, \xi_0) = 0$ lead to the fact that in the neighbourhood of this point there cannot be simple characteristic points where condition (Ψ) is violated. We cite a few results of a different nature.

Theorem 2.2 (Egorov and Rangelov [1977]). *Let* $p_0(x_0, \xi_0) = 0$, $d_{x,\xi}p_0(x_0, \xi_0) = 0$ *and* $(x_0, \xi_0) \in T^*\Omega\backslash 0$. *Suppose that the inequality* $c_1(x, \xi) \geqslant m_0(|x - x_0|^2 + |\xi - \xi_0|^2)$, $m_0 > 0$, *holds in some neighbourhood of* (x_0, ξ_0). *Then the pseudodifferential operator* P, *with principal symbol* p_0, *is insolvable at* x_0.

We now consider the case $P = Q^2 + R$, where the operators Q and R are of order m and $2m - 1$, respectively.

Theorem 2.3 (Popivanov [1975]). *Let Q be an operator of real principal type. If* $q_0(x_0, \xi_0) = 0$, $\operatorname{Re} r_0(x_0, \xi_0) < 0$ *and the function* $\operatorname{Im} r_0$ *has a zero of finite odd order along the bicharacteristic of* q_0 *passing through the point* (x_0, ξ_0), *then* P *is insolvable at* x_0.

Theorem 2.4 (Popivanov [1975]). *Let* $\operatorname{Re} q_0(x, \xi) = a(x, \xi)$ *and* $\operatorname{Im} q_0(x, \xi) = b(x, \xi)$. *Let* $q_0(x_0, \xi_0) = 0$, $\xi_0 \neq 0$, *and let* $k \in \mathbb{N}$ *be odd. Assume that*

$$H_a^j b(x_0, \xi_0) = 0 \qquad \text{for} \quad j < k \qquad \text{and} \quad H_a^k b(x_0, \xi_0) > 0.$$

Then P *is insolvable at* x_0.

§ 3. Sufficient Conditions for Local Solvability

3.1. Operators of Real Principal Type. Theorem 3.2, Chap. 4, which concerns the propagation of singularities, immediately yields theorems on local and semiglobal solvability of the equations $Pu = f$ of real principal type if $d_\xi p_0(x, \xi) \neq 0$ in $T^*\Omega\backslash 0$.

In fact, P^*, the formal adjoint operator to P, is also a real principal type operator. For $x_0 \in \Omega$ a neighbourhood $\omega \ni x_0$ can be found such that every bicharacteristic of P^* that passes through (x_0, ξ), with $|\xi| = 1$, comes out on the boundary $\partial\omega$. Therefore, by Theorem 3.2, Chap. 4,

$$\varphi \in \mathscr{E}'(\omega), \qquad P^*\varphi \in C^\infty(\omega) \Rightarrow \varphi \in C_0^\infty(\omega),$$

$$\varphi \in \mathscr{E}'(\omega), \qquad P^*\varphi \in H_s(\omega) \Rightarrow \varphi \in H_{s+m-1}(\omega),$$

where m denotes the order of P. This implies, in view of the closed graph theorem, that there exists a constant C such that

$$\|\varphi\|_{s+m-1} \leqslant C(\|P^*\varphi\|_s + \|\varphi\|_{s+m-2}), \qquad \varphi \in C_0^\infty(\omega),$$

and if ω is sufficiently small and $s + m \geqslant 1 + n/2$, then

$$\|\varphi\|_{s+m-1} \leqslant C\|P^*\varphi\|_s, \qquad \varphi \in C_0^\infty(\omega).$$

It follows then that $\int f\bar{\varphi}\,dx$, with $f \in H_{-s-m+1}$, is a bounded linear functional of $\overline{P^*\varphi}$ in H_{-s}, that is, there exists an element $u \in H_s$ such that

$$\|u\|_{-s} \leqslant C\|f\|_{-s-m+1}, \qquad \int f\bar{\varphi}\,dx = \int u\overline{P^*\varphi}\,dx, \qquad \varphi \in C_0^\infty(\omega).$$

Hence $Pu = f$ in ω.

If one considers an arbitrary compact set, instead of a small neighbourhood, then it becomes necessary to impose some additional conditions on the behaviour of the bicharacteristics in ω (see § 3.4, Chap. 4).

3.2. Operators of Principal Type. In his works, Hörmander has established that the principal type equation $Pu = f$ is locally solvable if the condition $C_1(x_0, \xi_0) > 0$ holds at each characteristic point (x_0, ξ_0). This result was further generalized by Nirenberg, Trèves, Beals, Fefferman, and the present author (see Egorov [1984]).

The problem of obtaining sufficient conditions was resolved completely in the case of differential operators by Beals and Fefferman [1973]. In this case, the sufficient condition coincides with the following condition, introduced originally by Nirenberg and Trèves:

Condition $(\mathscr{P})$. The function $\gamma(t) = \operatorname{Re} zp_0(x(t), \xi(t))$, where $(x(t), \xi(t))$ is the zero bicharacteristic of the function $\operatorname{Im} zp_0(x, \xi)$, does not change its sign with a change in t; $z \in \mathbb{C}\backslash 0$.

Theorem 3.1. *If condition* $(\mathscr{P})$ *holds and* $d_\xi p_0(x, \xi) \neq 0$, *then the differential operator* P *is locally solvable.*

Nirenberg and Trèves [1970] first established this theorem for operators P having analytic coefficients. For differential operators conditions $(\mathscr{P})$ and (Ψ) are equivalent.

It is not known till now whether or not condition (Ψ) is sufficient for the local solvability of the principal type pseudodifferential equation with a complex-valued symbol.

The most general theorem can be obtained for subelliptic operators. These operators can be defined as follows. Let $a_1 = \operatorname{Re} p_0$, $a_2 = \operatorname{Im} p_0$, and

$$H_i = \sum_{j=1}^{n} \left(\frac{\partial a_i(x,\xi)}{\partial x_j} \frac{\partial}{\partial \xi_j} - \frac{\partial a_i(x,\xi)}{\partial \xi_j} \frac{\partial}{\partial x_j} \right), \qquad i = 1, 2.$$

For $\alpha = (\alpha_1, \ldots, \alpha_r)$, $\beta = (\beta_1, \ldots, \beta_r)$, we denote by $H_1^\alpha H_2^\beta$ the operator $H_1^{\alpha_1} H_2^{\beta_1} \ldots H_1^{\alpha_r} H_2^{\beta_r}$. Let $k(x, \xi)$ be the smallest integer for which

$$H_1^\alpha H_2^\beta a_1(x,\xi) \neq 0, \qquad |\alpha + \beta| = k.$$

If $a_1(x, \xi) \neq 0$, we set $k(x, \xi) = 0$.

Theorem 3.2 (Egorov [1971]). *If condition (Ψ) holds for the operator P and* $\sup k(x, \xi) \leqslant k$ *for* $(x, \xi) \in T^*K \setminus 0$, *then the equation $Pu = f$ is solvable in K when f satisfies a finite number of conditions. More precisely, in this case there are functions $g_1, \ldots, g_N$ in $C_0^\infty(K)$ such that, if*

$$\int f \bar{g}_j \, dx = 0, \qquad j = 1, \ldots, N; \qquad f \in H_s(K),$$

there exists a function $u \in H_{s+m-k(k+1)^{-1}}(\Omega) \cap \mathscr{E}'(\Omega)$ satisfying the equation $Pu = f$ in K. If the diameter of the compact set K is sufficiently small, then the orthogonality condition becomes redundant.

We list a few more conditions that are sufficient for the local solvability.

Theorem 3.3 (Egorov [1984]). *The operator P is locally solvable at a point x_0 if at each characteristic point (x, ξ) condition (Ψ) and at least one of the following conditions hold:*

1. *The forms d* Re *$p_0(x, \xi)$, d* Im *$p_0(x, \xi)$ and $\xi\, dx$ are linearly independent.*
2. *There exist a constant $N > 0$ and a function $\alpha(x, \xi)$, smooth and positive homogeneous of degree $m - 1$, such that the inequality*

$$C_1^0(x,\xi) + \operatorname{Re} \alpha(x,\xi) p_0(x,\xi) + N |p_0(x,\xi)|^2 |\xi|^{-1} \geqslant 0$$

holds in some neighbourhood of the point in question.

3. *Let $a =$* Re *p_0, $b =$* Im *p_0, where a is the symbol of the principal type operator. Let $V'b$ be the derivative of b in the direction of the vector field that is tangent to the surface $a(x, \xi) = 0$ and orthogonal to the field H_a.*

There exists a constant K such that the inequality

$$|V'b(x,\xi)|^2 \leqslant K |b(x,\xi)|$$

holds in some neighbourhood of the point (x_0, ξ_0) for all the points (x, ξ) lying on the surface $a(x, \xi) = 0$, $|\xi| = 1$.

In this case, a neighbourhood $\omega \subset \Omega$ can be found for each real s, $s \leqslant m - 1 + n/2$, such that the equation $Pu = f$, with $f \in H_{s-m+1}(\Omega)$, has a solution $u \in H_s(\Omega) \cap \mathscr{E}'(\Omega)$ in ω. Moreover, $\|u\|_s \leqslant C\varepsilon \|f\|_{s-m+1}$, $\varepsilon =$ diam *ω, where the constant C is independent of f or ε.*

3.3. Operators with Multiple Characteristics. We examine only the case $P = Q^2 + R$, where the orders of Q and R are, respectively, m and $2m - 1$.

Theorem 3.4 (Popivanov [1975]). *Let Q be a principal type operator and $d_\xi q_0(x, \xi) \neq 0$ for $\xi \neq 0$. Assume that condition (Ψ) holds. If $r_0(x, \xi) \equiv 0 \pmod{q_0}$, then for $f \in \mathscr{E}'(\Omega) \cap H_{s-2m+2}(\Omega)$ there exists a function $u \in \mathscr{E}'(\Omega) \cap H_s(\Omega)$ satisfying the equation $Pu = f$ in ω; ω is a sufficiently small neighbourhood of x_0. Moreover,*

$$\|u\|_s \leqslant C(\omega)\|f\|_{s-2m+2} + C_1\|u\|_{s-1}.$$

If $s \geqslant \max(-n/2, -1 - n/2 + m)$, we may take $C_1 = 0$. Here $C(\omega) \to 0$ as diam $\omega \to 0$.

When q_0 is a real-valued function, we can obtain more precise results.

Theorem 3.5 (Popivanov [1975]). *Let Q be a real principal type operator and $d_\xi q_0(x, \xi) \neq 0$. The operator P is locally solvable if at least one of the following conditions is satisfied:*

1. Re $r_0(x, \xi) \geqslant 0$ *if* $q_0(x, \xi) = 0$, $\xi \neq 0$. *In this case the following estimate holds:* $\|\varphi\|_{m-1} \leqslant C(\omega)\|P\varphi\|_{1-m}$, $\varphi \in C_0^\infty(\omega)$.
2. Re $r_0(x, \xi) > 0$ *if* $q_0(x, \xi) = 0$, $\xi \neq 0$. *In this case,*

$$\|\varphi\|_{m-(1/2)} \leqslant C\|P\varphi\|_{(1/2)-m}, \qquad \varphi \in C_0^\infty(\omega).$$

Chapter 6
Smoothness of Solutions of Differential Equations

§ 1. Hypoelliptic Operators

1.1. Definition and Examples. As we know, every classical or generalized solution of the Laplace equation is actually an infinitely differentiable, and even analytic, function. Petrovskij has shown that all the solutions of the equation $P(D)u = 0$, where $P(D)$ is a differential operator with constant coefficients, are analytic functions if and only if $P(D)$ is an elliptic operator. A complete characterization of the differential operators $P(D)$ for which every solution of the equation $P(D)u = 0$ is an infinitely differentiable function was given by Hörmander.

Definition 1.1. A pseudodifferential operator $P(x, D)$ is said to be *hypoelliptic* in a domain $\Omega \subset \mathbb{R}^n$ if the conditions $u \in \mathscr{E}'(\Omega)$ and $P(x, D)u \in C^\infty(\omega)$ imply that $u \in C^\infty(\omega)$ for every subdomain $\omega \subset \Omega$.

Definition 1.2. A pseudodifferential operator $P(x, D)$ is said to be *microlocally hypoelliptic* in a domain $\Omega \subset T^*\mathbb{R}^n$ if the conditions $u \in \mathscr{E}'(\mathbb{R}^n)$ and $\mathrm{WF}(Pu) \cap \omega \neq \varnothing$ imply that $\mathrm{WF}(u) \cap \omega = \varnothing$ for each subdomain $\omega \subset T^*\Omega$.

1.2. Hypoelliptic Differential Operators with Constant Coefficients. A basic feature of such operators is the behaviour of the roots ζ of the equation $P(\zeta) = 0$ as $\zeta \to \infty$. Namely, the condition

$$|\operatorname{Im} \zeta| \to \infty \quad \text{as} \quad |\zeta| \to \infty \text{ on the surface } P(\zeta) = 0 \tag{1}$$

is necessary and sufficient for hypoellipticity (Hörmander [1963]).

By means of the Seidenberg – Tarski theorem, one can show that this condition is equivalent to the following condition:

$$\exists \gamma > 0, \quad C > 0\colon |\zeta|^{\gamma} \leqslant C|\operatorname{Im} \zeta|, \qquad \text{if} \quad P(\zeta) = 0, \quad |\zeta| > C. \tag{2}$$

We can cite equivalent conditions in terms of the behaviour of $P(\xi)$ for real ξ. Namely, the conditions (1) and (2) are equivalent to each of the following three conditions:

$$P^{(\alpha)}(\xi)/P(\xi) \to 0 \qquad \text{as} \quad |\xi| \to \infty, \quad \xi \in \mathbb{R}^n, \quad \forall \alpha \neq 0; \tag{3}$$

$$\exists \gamma > 0, \qquad C > 0\colon |P^{(\alpha)}(\xi)| \leqslant C|P(\xi)||\xi|^{-|\alpha|\gamma}; \quad \forall \alpha; \tag{4}$$

$$\frac{P(\xi + \eta)}{P(\xi)} \to 1 \qquad \text{as} \quad |\xi| \to \infty, \quad \xi \in \mathbb{R}^n, \quad \forall \eta \in \mathbb{R}^n. \tag{5}$$

We note that elliptic operators are characterized by the fact that for them one can take $\gamma = 1$ in (2) and (4). For other hypoelliptic operators, this quantity is less than 1.

The class of semi-elliptic operators, close to the class of elliptic operators, is defined in the following manner. Let $m_1, \ldots, m_n$ be positive integers, and let $\sum \alpha_k/m_k \leqslant 1$ for all those $\alpha = (\alpha_1, \ldots, \alpha_n)$ for which $a_\alpha \neq 0$ in the definition of the operator $P(D) = \sum a_\alpha D^\alpha$.

Let $P_0(D) = \sum' a_\alpha D^\alpha$, where the summation is performed over those α for which $\sum \alpha_k/m_k = 1$. If $p_0(\xi) \neq 0$ for $\xi \neq 0$, then $P(D)$ is called a semi-elliptic operator. In particular, if $m_1 = \cdots = m_k = m$, then such an operator becomes elliptic. For instance, p-parabolic equations, introduced by Petrovskij, are semi-elliptic.

We also note the following theorem (Hörmander [1963]).

Theorem 1.1. *Let $Q(\xi)$ be a polynomial, with real coefficients, of degree m. Let $k \geqslant 2$ be an integer. Suppose that $R(\xi)$ is a positive definite homogeneous polynomial of degree $2km - 2(k - 1)$. Then the operator $P(D) = Q(D)^{2k} + R(D)$ is hypoelliptic.*

1.3. The Gevrey Classes. In proving sufficiency of the algebraic conditions, cited above, one usually constructs a fundamental solution of the operator P. This construction depends significantly on condition (2). If we know the exact value of γ in condition (2), we can obtain estimates for the derivatives of the solution of the equation.

Definition 1.3. By a *Gevrey class* $G_\beta = G_\beta(\Omega)$ ($\beta \geqslant 0$) we mean the space of functions $u(x)$ that are defined in Ω, have derivatives of all orders and satisfy the inequalities

$$|D^\alpha u(x)| \leqslant CB^{|\alpha|}\alpha^{\beta\alpha}$$

for some constants C ad B and for all α.

Here, as usual, $\alpha^{\beta\alpha} = \alpha_1^{\beta\alpha_1} \ldots \alpha_n^{\beta\alpha_n}$. Clearly, these inequalities are equivalent to the inequalities

$$|D^\alpha u(x)| \leqslant CB^{|\alpha|}(\alpha!)^\beta.$$

The class G_1 consists of analytic functions (in each of the variables $x_1, \ldots, x_n$). The projective Gevrey classes $G'_\beta = \bigcap_{\gamma > \beta} G_\gamma(\Omega)$ are also used.

Theorem 1.2. *If the condition* (2) *holds for the operator $P(D)$, then every solution $u(x)$ of the equation $P(D)u = f$, where $f \in G_\beta$ and $\beta\gamma = 1$, belongs to G_β. In particular, if $P(D)$ is an elliptic operator, then every solution u of the equation $P(D)u = f$, with analytic f, is analytic.*

The quantity β can be found in terms of the polynomial $P(\xi)$ by the formula

$$\beta = -\overline{\lim_{\xi \to \infty}} \ln \frac{|\operatorname{grad} P(\xi)|}{|P(\xi)|} \bigg/ \ln |\xi|$$

(Hörmander [1983, 1985]).

1.4. Partially Hypoelliptic Operators. Let us divide the variables $x = (x_1, \ldots, x_n)$ into two groups: $x' = (x_1, \ldots, x_k)$ and $x'' = (x_{k+1}, \ldots, x_n)$. Similarly, $\alpha = (\alpha', \alpha'')$, etc.

Definition 1.4. The operator $P(D)$ is said to be *partially hypoelliptic* in x'' if

$$P(D)u \in C^\infty(\Omega), \qquad D^\alpha u \in \mathscr{L}_2(\Omega), \qquad \forall \alpha = (\alpha', 0) \Rightarrow u \in C^\infty(\Omega)$$

for any subdomain $\Omega \subset \mathbb{R}^n$.

Example 1.1. A differential operator $P(D)$ is partially hypoelliptic in x_n if and only if

$$P(D) = CD_n^p + \sum_{k=0}^{p-1} Q_k(D')D_n^k, \qquad C \neq 0, \qquad D' = (D_1, \ldots, D_{n-1}).$$

In particular, this condition is satisfied if the plane $x_n = 0$ is non-characteristic.

Theorem 1.3 (Hörmander [1983, 1985]). *The operator $P(D)$ is partially hypoelliptic in x'' if one of the following equivalent conditions holds*:

I. $|\operatorname{Im} \zeta| + |\operatorname{Re} \zeta'| \to \infty$ *if* $\zeta \to \infty$ *along the surface* $P(\zeta) = 0$.

II. *There are constants C and $\gamma > 0$ such that*

$$(1 + |\xi|)^\gamma \leqslant C(1 + |\operatorname{Im} \xi''|)(1 + |\xi'|), \qquad P(\xi) = 0, \qquad \xi \in \mathbb{R}^n.$$

III. $\lim P^{(\alpha)}(\xi)/P(\xi) = 0$ *if* $\alpha \neq 0, |\xi''| \to \infty$.

IV. $P(\xi) = \sum_{\alpha''=0} p_\alpha(\xi'')\xi'_\alpha$, *where $p_0(\xi'')$ is a hypoelliptic polynomial in x'' and $p_\alpha(\xi'')/p_0(\xi'') \to 0$ as $|\xi''| \to 0, \alpha \neq 0$.*

We denote by $H_{r,s}(\mathbb{R}_n^+)$ the space of functions with the norm

$$\|u\|_{r,s}^2 = (2\pi)^{-n} \int (1 + |\xi|^2)^r (1 + |\xi'|^2)^s |\tilde{u}(\xi)|^2 \, d\xi,$$

where

$$\xi' = (\xi_1, \ldots, \xi_{n-1}), \qquad \mathbb{R}_n^+ = \{x \in \mathbb{R}^n, x_n \geqslant 0\}.$$

Theorem 1.4 (Hörmander [1983, 1985]). *Let $P(D) = D_n^m + \sum_{\alpha_n < m} a_\alpha D^\alpha$ be a differential operator of order m, with constant coefficients. Let Ω be an open subset in $\mathbb{R}_n^+$. If*

$$P(D)u \in H^{\text{loc}}_{\sigma_1 - m, \tau_1}(\Omega), \qquad u \in H^{\text{loc}}_{\sigma_2, \tau_2}(\Omega),$$

then $u \in H^{\text{loc}}_{\sigma,\tau}(\Omega)$, where $\sigma \leqslant \sigma_1$, $\sigma + \tau \leqslant \sigma_j + \tau_j$, $j = 1, 2$.

Remark. Theorem 1.4 remains valid when a_α is a function of x and belongs to C^∞.

Corollary. *Let the hypotheses of Theorem 1.4 hold, and let $Pu \in C^\infty(\Omega)$. Let there be two sequences $\{\sigma_j\}$ and $\{\tau_j\}$ such that $\sigma_j + \tau_j \to \infty$ and $u \in H^{\text{loc}}_{\sigma_j, \tau_j}(\Omega)$ for all j. Then $u \in C^\infty(\Omega)$.*

1.5. Hypoelliptic Equations in Convolutions. A distribution $\mu \in \mathscr{E}'$ will be said to be invertible if

$$u \in \mathscr{E}', \qquad \mu * u \in C^\infty \Rightarrow u \in C^\infty.$$

Theorem 1.5 (Hörmander [1983, 1985]). *If each distribution u in $\mathscr{D}'(\mathbb{R}^n)$ satisfying the equation $\mu * u = 0$ lies in $C^\infty(\mathbb{R}^n)$, then*

$$|\operatorname{Im} \zeta| / \ln |\zeta| \to \infty \qquad \text{as} \quad \zeta \to \infty \text{ along the surface } \tilde{\mu}(\zeta) = 0. \tag{6}$$

Definition 1.5. A distribution μ is said to be hypoelliptic if μ is invertible and (6) holds.

Theorem 1.6 (Hörmander [1983, 1985]). *The following conditions regarding $\mu \in \mathscr{E}'(R^n)$ are equivalent:*

1. *μ is hypoelliptic;*
2. *$u \in \mathscr{D}'(\mathbb{R}^n)$, $\mu * u \in C^\infty(\mathbb{R}^n) \Rightarrow u \in C^\infty(\mathbb{R}^n)$.*
3. *There exists a distribution (parametrix) F such that $\mu * F - \delta \in C^\infty$ and*

$$ch \text{ sing supp } F \subset ch \text{ sing supp } \mu.$$

Here ch K denotes the convex hull of the set K.

1.6. Hypoelliptic Operators of Constant Strength (Hörmander [1963]). Let $P(x, D)$ be an mth order differential operator with constant coefficients, and let

$$\widetilde{P}(x, \xi) = \left(\sum_\alpha |D_\xi^\alpha P(x, \xi)|^2 \right)^{1/2}.$$

Definition 1.6. An operator $P(x, D)$ is said to be of constant strength in $\Omega \subset \mathbb{R}^n$ if for any point $(x_0, y_0) \in \Omega$ the differential operators $P(x_0, D)$ and $P(y_0, D)$ are such that

$$\tilde{P}(x_0, \xi) \leqslant C_{x_0, y_0} \tilde{P}(y_0, \xi).$$

Theorem 1.7. *Let $P(x, D)$ be a differential operator, with coefficients in $C^\infty(\Omega)$, $\Omega \subset \mathbb{R}^n$, of constant strength in Ω. Suppose that $P(x_0, D)$, where x_0 is some point of Ω, is a hypoelliptic operator. Then the operator $P(x, D)$ is hypoelliptic in Ω. What is more,* $\mathrm{WF}(u) = \mathrm{WF}(Pu)$ *for all $u \in \mathscr{D}'(\Omega)$, that is, P is microlocally hypoelliptic.*

1.7. Hypoelliptic Differential Operators with Variable Coefficients. Let $P(x, D)$ be a differential operator of order m, with smooth coefficients, in a domain $\Omega \subset \mathbb{R}^n$. Let $M_j(x, \xi)$ be functions in $\Omega \times \mathbb{R}^n$ such that

$$1 \leqslant M_j(x, \xi) \leqslant C_x(1 + |\xi|)^{1-d}, \qquad d > 0, \qquad j = 1, \ldots, n,$$

in which the constant C_x is bounded when x varies on compact subsets in Ω. For $\alpha = (\alpha_1, \ldots, \alpha_n) \in \mathbb{Z}_+^n$, $\beta = (\beta_1, \ldots, \beta_n) \in \mathbb{Z}_+^n$, we set

$$M^{\beta - \alpha} = M_1^{\beta_1} \ldots M_n^{\beta_n} M_1^{-\alpha_1} \ldots M_n^{-\alpha_n}.$$

Definition 1.7. An operator $P(x, D)$ is said to satisfy the condition (HE) in Ω if $P(x, D)$ does not vanish identically in any of the components of Ω and

$$|D_\xi^\alpha D_x^\beta P(x, \xi)| \leqslant C_{\alpha, \beta, x}(1 + |\xi|)^{-d|\alpha|} M^{\beta - \alpha}(x, \xi) P(x, \xi)$$

for all α and β, $x \in \Omega$, $\xi \in \mathbb{R}^n$; the constants $C_{\alpha, \beta, x}$ are bounded when x varies on bounded subsets of Ω.

Theorem 1.8 (Hörmander [1961]). *If a differential operator $P(x, D)$ satisfies condition (HE) in Ω, then it is hypoelliptic in Ω.*

The condition (HE) is a natural generalization of condition (4) to operators with variable coefficients.

Example 1.2 (Hörmander [1961]). The operator

$$P(x, D) = I + |x|^{2\nu}(-\Delta)^\mu$$

satisfies condition (HE), and is, hence, hypoelliptic if $\nu > \mu$. In this case, we may take

$$M_j(x, \xi) = M(x, \xi) = (1 + |\xi|^{2\mu})^{1/2\nu}(1 + |x|^{2\nu}|\xi|^{2\mu})^{-1/2\nu}.$$

Example 1.3 (Hörmander [1961]). The operator

$$P(x, D) = D_1^{2m} + D_2^{2k} + ic(x) D_1^a D_2^b + 1$$

satisfies condition (HE), and is, hence, hypoelliptic if $c \in C^\infty$, $c(x) \in \mathbb{R}$ and if either

$$\frac{a-1}{2m} + \frac{b}{2k} < 1, \qquad \frac{a}{2m} + \frac{b-1}{2k} < 1,$$

or

$$\frac{a-1}{2m} + \frac{b}{2k} < 1, \qquad 0 < a < 2m.$$

1.8. Pseudodifferential Hypoelliptic Operators. The preceding theorem can be generalized in a natural manner to the case of pseudodifferential operators.

Theorem 1.9 (Hörmander [1983, 1985]). *Let P be a matrix pseudodifferential operator, $P \in \mathscr{L}^m_{\rho,\delta}(\Omega; \mathbb{C}^k, \mathbb{C}^k)$, $0 \leqslant \delta < \rho \leqslant 1$. Suppose that there are $(k' \times k)$-matrices $A(x, \xi)$ and $B(x, \xi)$ such that the matrix $p(x, \xi)$ (the symbol of P) satisfies the inequality*

$$|A(x, \xi) D^\alpha_\xi D^\beta_x p(x, \xi) \cdot B(x, \xi)| \leqslant C_{\alpha,\beta,K}(1 + |\xi|)^{-\rho|\alpha| + \delta|\beta|}, \qquad x \in K,$$

where K is an arbitrary compact subset in Ω, and α, β are any multi-indices. Assume also that

$$|B(x, \xi)^{-1} p(x, \xi)^{-1} A(x, \xi)^{-1}| \leqslant C_K,$$

$$|A(x, \xi)| + |B(x, \xi)| \leqslant C_K |\xi|^{m'}, \qquad x \in K, \qquad |\xi| \geqslant C_K.$$

Then P is hypoelliptic.

A still more general theorem was obtained by Beals [1977]. Let $\langle \xi \rangle = (1 + |\xi|^2)^{1/2}$. Suppose that φ and Φ are weight functions, that is,

(I) $c \leqslant \Phi(x, \xi) \leqslant C\langle \xi \rangle$, $\quad c\langle \xi \rangle^{\varepsilon - 1} \leqslant \varphi(x, \xi) \leqslant C$;

(II) $\Phi(x, \xi)\varphi(x, \xi) \geqslant c$;

(III) $\Phi(x, \xi)/\varphi(x, \xi) \sim \Phi(y, \eta)/\varphi(y, \eta)$, if $\langle \xi \rangle \sim \langle \eta \rangle$,

(IV) $\Phi(y, \eta) \sim \Phi(x, \xi)$, $\quad \varphi(y, n) \sim \varphi(x, \xi)$, if $|y - x| \leqslant c\varphi(x, \xi)$ and $|\eta - \xi| \leqslant c\Phi(x, \xi)$;

(V) $\Phi(x, \xi) \geqslant c < \langle \xi \rangle^\delta$, $\quad c > 0$, $\quad \delta > 0$;

(VI) $\Phi(x, \xi)\varphi(x, \xi) \geqslant \varepsilon \langle \xi \rangle^\varepsilon$, $\quad \varepsilon > 0$.

Theorem 1.10 (Beals [1977]). *Let H and H_1 be Hilbert spaces and A a pseudodifferential operator, $A = a(x, D)$, with*

$$a(x, \xi) \colon \Omega \times \mathbb{R}^n \to \mathscr{L}(H, H_1),$$

where $\mathscr{L}(H, H_1)$ is the space of bounded linear operators acting from H into H_1. Assume that

$$\|D^\beta_x D^\alpha_\xi a(x, \xi)\| \leqslant C_{\alpha,\beta} \Phi^{M - |\alpha|} \varphi^{m - |\beta|},$$

where Φ and φ are weight functions satisfying the conditions (I)–(VI) *on each compact subset of Ω (here M and m are real numbers). Let $p(x, \xi) \in \mathscr{L}(H, H_1)$, where a and p have left inverses in each compact set $K \Subset \Omega$ for sufficiently large*

$|\xi|$. *Assume that*

$$\|p(D_x^\beta D_\xi^\alpha a)a^{-1}p^{-1}\| \leqslant C_{\alpha\beta}\Phi^{-|\alpha|}\varphi^{-|\beta|},$$
$$\|p\| + \|a^{-1}p^{-1}\| \leqslant C\langle\xi\rangle^N$$

for some N and sufficiently large $|\xi|$. *Finally, assume that either* $p = I$ *or a is invertible for large* ξ. *Then A is a hypoelliptic operator.*

1.9. Degenerate Elliptic Operators. Let $N = k + n$. Let the coordinates of the point $x \in \mathbb{R}^n$ be (x', y), where $x' \in \mathbb{R}^k$ and $y \in \mathbb{R}^n$. Let $m \in \mathbb{N}$ and $\delta m \in \mathbb{N}$. We set

$$m_0 = \{(\alpha, \beta, \gamma): |\alpha| + |\beta| \leqslant m, m + |\gamma| = |\alpha| + (1 + \delta)|\beta|\}.$$

In $\mathbb{R}^N$ we examine the operator

$$p(y, D) = \sum_{m_0} a_{\alpha\beta\gamma}y^\gamma D_{x'}^\beta D_y^\alpha$$

with constant coefficients.

Theorem 1.11 (Grushin [1972]). *Assume that the following conditions hold.*

1. *The operator* $p(y, D_y)$ *is elliptic when* $y \neq 0$.
2. *The equation* $p(y, \xi', D_y)v(y) = 0$, *where* $\xi' \in \mathbb{R}^k$, $|\xi'| = 1$, *has no non-zero solution in* $S(\mathbb{R}_y^n)$.

Then the operator $p(y, D)$ *is hypoelliptic. Conversely, if condition* 1 *holds and* $p(y, D)$ *is hypoelliptic, then condition* 2 *is satisfied.*

Example 1.4. The operator

$$D_y^2 + y^2D_x^2 + \lambda D_x, \qquad \operatorname{Re}\lambda = 0, \qquad x \in \mathbb{R}, \qquad y \in \mathbb{R},$$

is hypoelliptic if and only if $\lambda \neq 2k + 1$, $k \in \mathbb{Z}$.

Example 1.5. The operator

$$D_y + iay^rD_x, \qquad \operatorname{Re} a \neq 0,$$

is hypoelliptic if and only if $r = 2k$, $k \in \mathbb{N}$.

We set

$$m = \{(\alpha, \beta, \gamma): |\alpha + \beta| \leqslant m, m\delta \geqslant |\gamma| \geqslant |\alpha| + (1 + \delta)/|\beta| - m\}.$$

Let us consider the operator

$$p(x, D) = \sum_m q_{\alpha\beta\gamma}(x, D)y^\gamma D_{x'}^\beta D_y^\alpha,$$

where $q_{\alpha\beta\gamma}(x, D)$ are classical pseudodifferential operators of order zero and $q_{\alpha 00}(x, D) = q_{\alpha 00}(x)$ if $|\alpha| = m$. Let $q_{\alpha\beta\gamma}^0(x, \xi, \eta)$ be the leading part of the symbol of the operator $q_{\alpha\beta\gamma}$. We put

$$\mathscr{L}(y; \xi, D_y) = \sum q_{\alpha\beta\gamma}^0(0; \xi, 0)y^\gamma\xi^\beta D_y^\alpha.$$

Theorem 1.12 (Grushin [1972]). *If the conditions* 1 *and* 2 *of the preceding theorem hold for the pseudodifferential operator* $\mathscr{L}(y; D_{x'}, D_y)$, *then the operator* $p(x, D)$ *is hypoelliptic in some neighbourhood of the origin.*

Example 1.6. The operator

$$D_y + iay^r(D_x^2 + D_y^2)^{1/2}, \qquad \operatorname{Re} a \neq 0,$$

is hypoelliptic if r is even or if r is odd and $\operatorname{Re} a > 0$.

Example 1.7. As we saw in Example 1.4, the operator $D_y^2 + y^2 D_x^2 - D_x$ is non-hypoelliptic. However, the operator

$$D_y^2 + y^2 D_x^2 - D_x + ay^2 D_x$$

is hypoelliptic if $a \neq 0$.

Example 1.8. The operator

$$P = D_{k+1}^2 + x_{k+1}^2(D_1^2 + \cdots + D_k^2) \pm D_k,$$

where $k > 1$, is partially hypoelliptic: if $Pu \in C^\infty(V)$, where V is a neighbourhood of the origin and $u \in C^\infty$ in a neighbourhood of the intersection $\partial V \cap \{x; x_k = x_{k+1} = 0\}$, then $u \in C^\infty(V)$.

1.10. Partial Hypoellipticity of Degenerate Elliptic Operators. Let us consider the operator

$$\mathscr{L}(t, D_t, D_x) = \sum_{\substack{j+|\alpha| \leqslant m \\ \sigma+\delta|\alpha|+j \in N}} a_{j\alpha} t^{\sigma+\delta|\alpha|+j} D_t^j D_x^\alpha,$$

where $t \in \mathbb{R}$, $x \in \mathbb{R}^n$, $a_{j\alpha} \in \mathbb{C}$, $\delta \geqslant 0$, $\sigma \in \mathbb{Z}$, and $\sigma + \delta m \in \mathbb{N}$. We assume that

$$\mathscr{L}(\lambda^{-1} t, \lambda\tau, \lambda^\delta \xi) = \lambda^{-\sigma} \mathscr{L}(t, \tau, \xi)$$

for all $\lambda \in \mathbb{R}_+$, $t \in \mathbb{R}$, $\tau \in \mathbb{R}$ and $\xi \in \mathbb{R}^n$.

Theorem 1.13 (Grushin [1972]). *If the operator* $\mathscr{L}(t, D_t, D_x)$ *is elliptic for* $t \neq 0$, *then it is partially hypoelliptic in* $\mathbb{R} \times \mathbb{R}^n$ *(in* $\overline{\mathbb{R}}_+ \times \mathbb{R}^n$*) if and only if the equation* $\mathscr{L}(t, D_t, \xi)u(t) = 0$ *has only the trivial solution in the class* $S(\mathbb{R})$ *(in* $S(\overline{\mathbb{R}}_+)$*) for all* $\xi \in \mathbb{R}^n \backslash 0$.

Let us recall that in this case partial hypoellipticity means that $u \in C^\infty(I' \times \Omega')$ if $\mathscr{L}u \in C^\infty(I' \times \Omega')$ and $u \in C^\infty(I; \mathscr{D}'(\Omega'))$ for any open sets $I' \subset I$, $\Omega' \subset \Omega$.

Example 1.9. The operators

$$tD_t + iD_x + \lambda,$$

$$t^2 D_t^2 + D_x^2 + \lambda t D_t + \mu D_x + \nu$$

are partially hypoelliptic in $\mathbb{R} \times \mathbb{R}$ and in $\overline{\mathbb{R}}_+ \times \mathbb{R}$ for all λ, μ, ν in $\mathbb{C}$.

We now examine the operator

$$\mathscr{L} = \sum_{j+|\alpha|\leqslant m} a_{j\alpha}(t, x) t^{\sigma+[qj]} D_t^j D_x^\alpha,$$

where $q \in \mathbb{R}$, $q \geqslant 1$, $qm \in \mathbb{N}$, and $[qj]$ denotes the integral part of qj. Let $\mathscr{L}_0$ be the principal part of this operator, that is,

$$\mathscr{L}_0 = \sum_{\substack{j+|\alpha|=m \\ qj \in \mathbb{N}}} a_{j\alpha}(0, x) t^{\sigma+qj} D_t^j D_x^\alpha.$$

Theorem 1.14 (Grushin [1972]). *If the operators $\mathscr{L}$ and $\mathscr{L}_0$ are elliptic for $t \neq 0$ (for $t > 0$), then $\mathscr{L}$ is partially hypoelliptic in $\mathbb{R} \times \mathbb{R}^n$ (in $\overline{\mathbb{R}}_+ \times \mathbb{R}^n$).*

Example 1.10. The operator

$$t^3 D_t^2 + t^2 D_t D_x + i D_x^2 + \lambda t^2 D_t + \mu D_x + \nu$$

is partially hypoelliptic in $\mathbb{R} \times \mathbb{R}$ and in $\overline{R}_+ \times R$ for all λ, μ and ν in $\mathbb{C}$. The same is true for the operators

$$t D_t^2 + t D_x^2 + \lambda D_x + \mu D_t,$$

and

$$t D_t^2 + i D_x^2 + \lambda D_t.$$

1.11. Double Characteristics. Let us consider on an n-dimensional infinitely differentiable manifold Ω a certain $d \times d$ system of pseudodifferential operators of the form

$$P(x, D) = I p(x, D) + M_0(x, D),$$

where I is the identity matrix, $p(x, \xi)$ is a homogeneous symbol of degree m and the matrix operator M_0 is of order $\leqslant m - 1$. Although such an operator P is of a very special form, nevertheless it is of sufficient interest and one comes across this operator in the investigation of, for example, the $\bar{\partial}$-Neumann problem.

We assume that the following conditions hold.

1. The set $\operatorname{Char} P = \{(x, \xi) \in T^*\Omega \backslash 0, p(x, \xi) = 0\}$ is a smooth manifold of even codimension $2k$ and is non-involutive, that is, the restriction of the form $\omega = dx \wedge d\xi$ to every space tangential to $\operatorname{Char} P$ is non-degenerate;
2. The ratio $|p(x, \xi)|/d^2(x, \xi)$, where d is the distance from $\operatorname{Char} P$, is locally bounded;
3. If $k = 1$, that is, if $\operatorname{codim} \operatorname{Char} P = 2$, then the number of rotations of p around $\operatorname{Char} P$ is identically zero.

Theorem 1.15 (Boutet de Monvel and Trèves [1974]). *Assume that conditions 1–3 hold for the operator P. For P to be hypoelliptic with a loss of one derivative, that is, in order that*

$$u \in \mathscr{E}'(\omega), \qquad Pu \in H^s_{\mathrm{loc}}(\omega) \Rightarrow u \in H^{s+m-1}_{\mathrm{loc}}(\omega), \qquad \forall \omega \subset \Omega, \qquad s \in \mathbb{R},$$

it is necessary and sufficient that

$$\chi \neq \sum_{j=1}^{k} \mu_j \lambda_j$$

for any non-negative integers $m_1, \ldots, m_k$ *and any eigenvalue* $\chi(x, \xi)$ *of the matrix* $iM_{m-1}(x, \xi)$, *where* $\lambda_1, \ldots, \lambda_k$ *are the eigenvalues of the automorphism* φ *of the normal space to* Char P *defined by the relation*

$$p^*(\varphi u, v) = \bar{\omega}^*(u, v), \qquad \forall u, \quad v \in \mathbb{N}.$$

Here $\bar{\omega}^*(df, dg) = H_f(g)$ and p^* is a bilinear form associated with the quadratic form $\tilde{p}$ defined in the following manner. If Char P is defined by the $2k$ real equations $h_j(x, \xi) = 0$, $j = 1, \ldots, 2k$, and $\det(\{h_j, h_l\}) \neq 0$ (such functions exist in view of condition 1), then

$$p(x, \xi) = \sum c_{jl}(x, \xi) h_j(x, \xi) h_l(x, \xi)$$

and

$$\tilde{p}(x, \xi; \theta) = \sum c_{jl}(x, \xi) \theta_j \theta_l.$$

A typical example of the operator considered here is the one examined by Grushin:

$$D_t^2 + t^2 D_x^2 + \lambda D_x + \mu.$$

Operators having double characteristics have also been studied by Boutet de Monvel [1974, 1975], Boutet de Monvel, Grigis and Helffer [1976], Boutet de Monvel and Trèves [1974], Hörmander [1975], Menikoff [1977a, 1979], and others.

1.12. Hypoelliptic Operators on the Real Line. A complete characterization of those hypoelliptic differential operators whose leading coefficient has zeros of finite order has been obtained by Kannai [1971]. Let $L(x, D) = \sum_{j=0}^{m} a_j(x) D^j$, $a_j \in C^\infty(\mathbb{R})$. Clearly, L is hypoelliptic at all those points where $a_m(x) \neq 0$.

Suppose now that $a_m(x) = 0$ for $x = 0$ and that for $x = 0$ a_m has a zero of finite order. It is well known (see Wasow [1965]) that in this case there exist m linearly independent solutions $u_1(x), \ldots, u_m(x)$ of the equation $\mathscr{L}u = 0$ for which

$$u_i(x) = e^{Q_i(x)} x^{\rho_i} v_i(x),$$

where Q_i are polynomials in x^{-1/q_i} and

$$v_i(x) = \sum_{j=0}^{m_i} v_{i,j}(x) (\log x)^j, \qquad v_{i,j}(x) \sim \sum_{n=0}^{\infty} v_{i,j,n} x^{n/q_i}$$

for $0 \leqslant j \leqslant m_i$, $1 \leqslant i \leqslant m$. Here m_i, q_i are integers, $m_i \geqslant 0$, $q_i > 0$, and the asymptotic series for $v_{i,j}$ do not, in general, converge even when a_j are analytic functions. Let r be the number of functions Q_i that vanish identically, that is,

$$Q_1(x) \equiv \cdots \equiv Q_r(x) \equiv 0.$$

Theorem 1.16. *The operator* $\mathscr{L}$ *is hypoelliptic in a neighbourhood of the origin if and only if* $a_r(0) \neq 0$ *and* $|\text{Re } Q_i(x)| \to \infty$, *for* $r < i \leqslant m$, *as* $x \to 0$.

In some cases it is possible to obtain a more explicit description of the hypoellipticity conditions. Let

$$p(x, \xi) = a_m(x) \prod_{j=1}^{m} (\xi - \xi_j(x)).$$

The operator $\mathscr{L}$ is said to satisfy condition (S) if for each root ζ_j that increases without bound as $x \to 0$ the following statement is true: $\zeta_i(x)/\zeta_j(x) \to 1$ as $x \to 0$ implies $i = j$.

Theorem 1.17. *Suppose that $\mathscr{L}$ satisfies condition (S). For $\mathscr{L}$ to be hypoelliptic in a neighbourhood U of the origin, it is necessary and sufficient that there exists a constant $C > 0$ such that either $|\zeta| < C$ or $|x||\mathrm{Im}\,\zeta| \to \infty$ as $x \to 0$ when $p(x, \zeta) = 0$ for $x \in U$.*

The following theorem provides another class of hypoelliptic operators.

Theorem 1.18 (Kannai [1971]). *Let there exist a neighbourhood U of the origin and a constant C such that*

$$\text{either} \quad |\zeta| < C \quad \text{or} \quad |x||\mathrm{Im}\,\zeta| \to \infty$$

$$\text{and} \quad C|\mathrm{Im}\,\zeta| > |\mathrm{Re}\,\zeta| \quad \text{as } x \to 0$$

for $x \in U$ provided that $p(x, \zeta) = 0$. Then there exist constants ρ and δ such that $0 \leqslant \delta < \rho \leqslant 1$ and an integer m' such that for any non-negative integers α and β and for any compact subset $K \subset U$ constants $C_1(K)$ and $C(\alpha, \beta, K)$ can be found with the properties

$$\text{(a)} \ |p(x, \xi)| \geqslant C_1(K)|\xi|^{m'};$$

$$\text{(b)} \ |p^{(\alpha)}_{(\beta)}(x, \xi)| \leqslant C(\alpha, \beta, K)(1 + |\xi|)^{-\rho\alpha + \delta\beta}|p(x, \xi)|$$

for $x \in K$, $\xi \in \mathbb{R}$ and $|\xi| \geqslant C_1(K)$. If conditions (a) *and* (b) *hold, then $\mathscr{L}$ is a hypoelliptic operator.*

Example 1.11. Let $\mathscr{L}u = x^3u' + (i + x)u$. By Theorem 1.16, $\mathscr{L}$ is hypoelliptic. However, condition (b) is not satisfied for this operator.

Significant results of the theory of hypoelliptic operators have been obtained by Bronshtejn [1976], Ganzha [1986], Grushin [1972], Popivanov [1976], Beals and Fefferman [1976], Bolley, Camus and Helffer [1976], Bolley [1977], Boutet de Monvel [1974], Folland [1977], Grigis [1976], Grigis and Rothschild [1984], Helffer [1976, 1977], Helffer and Nourrigat [1979], Hörmander [1975], Menikoff [1977a, 1979], Metivier [1976, 1980], Parenti and Rodino [1980], Rothschild and Stein [1976, 1979], Trèves [1961], and by others.

§ 2. Subelliptic Operators

2.1. Definition and Simplest Properties. The subelliptic pseudodifferential operators constitute an important class of hypoelliptic operators. In its properties, this class is very close to the class of elliptic operators (see Egorov [1975, 1984], Hörmander [1979]).

Definition 2.1. A pseudodifferential operator P of order m is said to be *subelliptic* in a domain Ω if there are constants $C > 0$ and $\delta \in [0, 1)$ such that

$$\|u\|_{m-\delta} \leqslant C(\|Pu\|_0 + \|u\|_{m-1}), \qquad u \in C_0^\infty(K), \tag{7}$$

for every compact set $K \subset \Omega$.

Such an estimate is valid for $\delta = 0$ if and only if P is elliptic, that is, if $p_0(x, \xi) \neq 0$ for $\xi \neq 0$.

It follows from (7), as is easy to see, that for any real s a constant $C = C(s, K)$ can be found such that

$$\|u\|_{s+m-\delta} \leqslant C(\|Pu\|_s + \|u\|_{s+m-1}), \qquad u \in C_0^\infty(K).$$

It is easy to show, with the aid of the averaging operator, that the theorem on smoothness holds for subelliptic operators; namely,

$$\text{if } u \in \mathscr{E}'(\Omega) \text{ and } Pu \in H_s(K), \qquad \text{then } u \in H_{s+m-\delta}(K).$$

This theorem has its generalization in the microlocal terms: If $u \in \mathscr{E}'(\Omega)$ and $Pu \in H_s(\omega)$, where ω is a subdomain in $T^*\Omega\backslash 0$, then $u \in H_{s+m-\delta}(\omega)$. In particular, P is microlocally hypoelliptic.

If P is subelliptic,then the operator P^* is solvable in each compact set K. Namely, there exist functions $\psi_1, \ldots, \psi_N \in C_0^\infty(K)$, satisfying the equation $P\psi = 0$ in K, such that if $f \in H_s(K)$, with $\int f(x)\overline{\psi_j(x)}\,dx = 0$, then a function u can be found in the class $\mathscr{E}'(\Omega) \cap H_{s+m-\delta}(K)$ that satisfies the equation $P^*u = f$ in K. Moreover, $\|u\|_{s+m-\delta}^k \leqslant C\|f\|_s^k$.

If P is a subelliptic operator on a closed, smooth and compact manifold Ω, then the set

$$\{u \in \mathscr{D}'(\Omega), Pu = 0\}$$

consists of infinitely smooth functions and is a finite-dimensional linear space. The range of this operator, with the domain of definition $\mathscr{D}_P = H_s(\Omega)$, is a closed subspace of $H_{s-m}(\Omega)$.

Example 2.1. For even k, the operator $D_1 + ix_1^k D_2$ is subelliptic with the exponent $\delta = k/(k + 1)$.

Example 2.2. The oblique derivative problem. We examine in the half-space

$$\mathbb{R}_+^{n+1} = \{x = (x_1, \ldots, x_{n+1}), x_{n+1} > 0\}$$

the equation

$$\Delta u = 0, \qquad u \in H^2(\mathbb{R}_+^{n+1})$$

with the boundary condition

$$\frac{\partial u}{\partial x_1} + a x_1^k \frac{\partial u}{\partial x_{n+1}} = f(x) \qquad \text{for} \quad x_{n+1} = 0, \quad a \not\equiv 0.$$

The Fourier transformation reduces this equation to a pseudodifferential equation

$$\frac{\partial v}{\partial x_1} + ax_1^k \Lambda v = f(x),$$

where

$$\Lambda v = (2\pi)^{-n} \int \tilde{v}(\xi)|\xi| e^{ix\xi}\, d\xi.$$

The corresponding operator $iD_1 + ax_1^k \Lambda$ is subelliptic if k is even or if k is odd but a is negative. In this case

$$\delta = k(k+1)^{-1}.$$

Example 2.3. The $\bar{\partial}$-*Neumann problem.* Let Ω be a domain in $\mathbb{C}^n$. In Ω, let us consider a system of equations

$$\frac{\partial u}{\partial \bar{z}_j} = f_j(z, \bar{z}), \qquad j = 1, \ldots, n, \qquad z_j = x_j + iy_j.$$

Take a point A on the boundary $\partial\Omega$. Let us introduce a coordinate system in a neighbourhood of this point in which $\partial\Omega$ is given by the equation $x_n = 0$ and is situated in $x_n > 0$. When $x_n = 0$, the system takes the form

$$\mathscr{L}_j u = g_j, \qquad j = 1, \ldots, n-1,$$

where

$$\overline{\mathscr{L}}_j = \frac{\partial}{\partial x_j} + a_j(x, y)\frac{\partial}{\partial y_n} + i\left(\frac{\partial}{\partial y_j} + b_j(x, y)\frac{\partial}{\partial y_n}\right).$$

This system is subelliptic, with $\delta = \frac{1}{2}$, if the matrix C defined by the relation

$$[\overline{\mathscr{L}}_j, \mathscr{L}_k] = c_{jk}\frac{\partial}{\partial y_n}$$

is positive definite. The matrix C is referred to as the *Levi matrix* and Ω is called a strictly pseudoconvex domain if $C > 0$.

2.2. Estimates for First-Order Differential Operators with Polynomial Coefficients. Such operators serve as models for general subelliptic pseudodifferential operators and are extensively used in investigating the latter (see Egorov [1975, 1984], Hörmander [1979]).

Theorem 2.1. *Let $P(t)$ be a polynomial of degree k in $t \in \mathbb{R}$. The estimate*

$$\|u\|_0 \leqslant C\|u'(t) + \lambda P(t)u\|_0, \qquad \lambda \geqslant 1, \qquad u \in C_0^\infty(\mathbb{R}),$$

holds if and only if $P(t)$ does not change the sign from minus to plus as t increases. If $P(t)$ satisfies this condition, then the following estimate holds:

$$\|u'(t)\|_0 + \lambda\|P(t)u\|_0 + \lambda^{1/(k+1)}\|u\|_0 \leqslant C_1\|u'(t) + \lambda P(t)u\|_0,$$

$$\lambda > 0, \qquad u \in C_0^\infty(\mathbb{R}).$$

Theorem 2.2. *Let $P(t, x)$ be a polynomial of degree k in two variables $(t, x) \in \mathbb{R}^2$, and let $A(t)$ be a polynomial of degree $s < k/2$ whose leading coefficient is 1. The estimate*

$$\|u\|_0 \leqslant C \left\| \frac{\partial u}{\partial t} + \frac{1}{i} A(t) \frac{\partial u}{\partial x} + \lambda P(t, x) u \right\|_0, \qquad \lambda > 0, \qquad u \in C_0^\infty(\mathbb{R}^2),$$

holds if and only if the following conditions are satisfied:

1. *$A(t)$ does not change its sign.*
2. *$B(t, x) = P(t, x) A(t)^{-1}$ is a smooth function in $\mathbb{R}^2$ and* sgn $A(t) \cdot \partial_t B(t, x) \leqslant 0$.
3. *There is a constant $c_0 > 0$ such that*

$$\sum_{i+j \leqslant k-1} |D_t^i [A(t)^j D_x^j D_t B(t, x)]| \geqslant c_0.$$

If conditions 1–3 are satisfied, then the following estimate holds:

$$\lambda^{1/(k+1)} \|u\|_0 + \|\partial u/\partial t\|_0 + \|A(t) \partial u/\partial x + i\lambda P(t, x) u\|_0$$

$$\leqslant C \left\| \frac{\partial u}{\partial t} + \frac{1}{i} A(t) \frac{\partial u}{\partial x} + \lambda P(t, x) u \right\|_0, \qquad u \in C_0^\infty(\mathbb{R}^2).$$

2.3. Algebraic Conditions. In order to investigate the inequality (7), we can use microlocalization. Without loss of generality, we may take $m = 1$.

Theorem 2.3. *If a first-order pseudodifferential operator P is subelliptic, with $0 < \delta < 1$, in a domain Ω, then for every integer $k > 0$ and every compact subset $K \subset \Omega$ a constant $C = C(k, K)$ can be found such that the inequality*

$$\|\psi(y)\|_0 \leqslant C \Big\{ \lambda^\delta \| T_k^{(x,\xi)} p_0(x + y\lambda^{\delta - 1}, \xi + D\lambda^{-\delta}) \psi(y) \|_0$$

$$+ \lambda^{\delta - (k+1)(1-\delta)} \sum_{|\alpha + \beta| \leqslant k+1} \| y^\beta D^\alpha \psi \|_0 \lambda^{|\alpha|(1-2\delta)} \Big\}$$

holds for all $x \in K$, $\xi \in \mathbb{R}^n$, $|\xi| = 1$, $\lambda \geqslant 1$ and for all ψ in $C_0^\infty(\mathbb{R}^n)$. Here we have used the notation

$$T_k^{(x,\xi)} f(y, \eta) = \sum_{|\alpha + \beta| \leqslant k} \frac{1}{\alpha! \beta!} f_{(\beta)}^{(\alpha)}(x, \xi)(y - x)^\beta (\eta - \xi)^\alpha,$$

$$f_{(\beta)}^{(\alpha)}(x, \xi) = (iD_\xi)^\alpha (iD_x)^\beta f(x, \xi).$$

Theorem 2.4. *Suppose that for every compact subset $K \ll \Omega$ there are constants $C = C(K)$, $N = N(K)$ and $\varepsilon_0 = \varepsilon_0(K)$ such that the inequality*

$$\lambda \|\psi\|_0 + \|\bar{P}_\lambda \psi\|_0 \leqslant C \Big\{ \|P_\lambda \psi\|_0 + \lambda^{-\varepsilon_0} \sum_{|\alpha + \beta| \leqslant N} \lambda^{-|\alpha|(k-1)} (\| y^\beta D^\alpha \psi \|_0$$

$$+ \| y^\beta D^\alpha P_\lambda \psi \|_0 + \| y^\beta D^\alpha \bar{P}_\lambda \psi \|_0) \Big\},$$

where, with $t \geqslant 1, |\xi| \geqslant 1, p_0 \in C^\infty(T^*\Omega)$,

$$P_\lambda(y, D)\psi = \lambda^k T_k^{(x,\xi)} p_0(x + y\lambda^{-1}, \xi + D\lambda^{-k})\psi \qquad \text{and} \quad p_0(x, t\xi) = tp_0(x, \xi),$$

holds for all $x \in K, |\xi| \geqslant \frac{1}{2}, \lambda \geqslant 1$ *and for all* $\psi(y) \in C_0^\infty(\mathbb{R}^n)$. *Then the estimate*

$$\|u\|_{1/(k+1)} + \|P^*u\|_0 \leqslant C_K(\|Pu\|_0 + \|u\|_0), \qquad u \in C_0^\infty(K),$$

holds for the operator P *with principal symbol* p_0.

An analysis of microlocal conditions, similar to the one carried out for conditions of Theorems 2.3 and 2.4, enables us to formulate algebraic conditions that are necessary and sufficient for the operator P to be subelliptic (Egorov [1975]).

The first of these conditions is necessary for the inequality (7) to hold for some δ (not necessarily from $[0, 1)$).

(ψ) Let $a_1 = \operatorname{Re} p_0$ and $a_2 = \operatorname{Im} p_0$. Consider an integral curve of the system

$$\dot{x}(t) = \partial_\xi a_2(x(t), \xi(t)), \qquad \dot{\xi}(t) = -\partial_x a_2(x(t), \xi(t)),$$

which is known as a bicharacteristic of the function a_2. Put $h(t) = a_1(x(t), \xi(t))$. If $a_2(x(t), \xi(t)) = 0$ and $h(t_0) < 0$, then $h(t) \leqslant 0$ for $t \geqslant t_0$. Similar condition holds if p_0 is replaced by ip_0.

In order to formulate the second condition (B), let us consider the operators

$$H_i = \sum_{j=1}^n \left[\partial_j a_i(x, \xi) \frac{\partial}{\partial \xi_j} - \partial_{\xi_j} a_i(x, \xi) \frac{\partial}{\partial x_j} \right], \qquad i = 1, 2.$$

For $\alpha = (\alpha_1, \dots, \alpha_r)$, $\beta = (\beta_1, \dots, \beta_r)$, we set

$$H_1^\alpha H_2^\beta = H_1^{\alpha_1} H_2^{\beta_1} \dots H_1^{\alpha_r} H_2^{\beta_r}.$$

We take $k(x, \xi) = 0$ if $p_0(x, \xi) \neq 0$, and if $p_0(x, \xi) = 0$, then $k(x, \xi)$ is the smallest integer k for which

$$H_1^\alpha H_2^\beta a_1(x, \xi) \neq 0, \qquad |\alpha + \beta| = k.$$

$$(B) \sup k(x, \xi) \leqslant k < \infty, \qquad (x, \xi) \in T^*\Omega, \qquad x \in K, \qquad |\xi| = 1.$$

Theorem 2.5. *The inequality* (7) *holds with* $\delta = k/(k+1)$ *if and only if conditions* (ψ) *and* (B) *are satisfied. If* $p_0(x_0, \xi_0) = 0$, $(x_0, \xi_0) \in T^*\Omega \backslash 0$ *and* $H_1^\alpha H_2^\beta a_1(x_0, \xi_0) = 0$ *for* $|\alpha + \beta| < l$, *then* (7) *cannot hold for* $\delta < l(l+1)^{-1}$, $x_0 \in K$.

Thus to every subelliptic operator P there corresponds a natural number k such that $\delta = k(k+1)^{-1}$. The subelliptic operators arise when elliptic operators degenerate to order k.

Theorem 2.6. *If* P *is a subelliptic operator, then for every compact subset* $K \subset \Omega$ *and every real* s *there exists a constant* $C = C(s, K)$ *such that*

$$\|P^*u\|_s \leqslant C(\|Pu\|_s + \|u\|_{s+m-1}), \qquad u \in C_0^\infty(K).$$

Theorem 2.7. *Let P be a pseudodifferential operator satisfying condition (B). A necessary and sufficient condition for P to be hypoelliptic is that condition (ψ) holds.*

For the case of a differential operator, Theorem 2.5 assumes the following form.

Theorem 2.8. *For a differential operator P to be subelliptic, it is necessary and sufficient that the function $k(x, \xi)$ assumes at all points of $T^*\Omega$ even values not exceeding a certain integer k. In this case, $\delta = k(k+1)^{-1}$. If the function k assumes odd values at a point (x_0, ξ_0), the inequality (7) cannot hold in any compact set K containing x_0 no matter what the value of $\delta \in \mathbb{R}$. In this last case, P is non-hypoelliptic in every neighbourhood of this point.*

We also remark that Theorem 2.7 implies that among operators satisfying condition (B) we do not have any hypoelliptic operators but not subelliptic operators.

§3. Hypoelliptic Differential Operators of Second Order

3.1. The Sum of the Squares. We consider a second-order operator

$$\mathscr{L} = \sum_{j=1}^{r} X_j^2 + X_0 + c, \tag{8}$$

where $X_0, X_1, \ldots, X_r$ are first-order differential operators with real coefficients that are smooth in Ω, $c \in C^\infty(\Omega)$, and where Ω is a bounded domain in $\mathbb{R}^n$. As usual, $[A, B]$ denotes the commutator $AB - BA$.

Theorem 3.1. *Suppose that at every point of Ω there are n linearly independent operators among X_{j_1}, $[X_{j_1}, X_{j_2}]$, $[X_{j_1}, [X_{j_2}, X_{j_3}]]$, $\ldots$, where $j_i = 0, 1, \ldots, r$. Then the operator $\mathscr{L}$ is hypoelliptic.*

If the Lie algebra generated by $X_0, X_1, \ldots, X_r$ is of constant rank, which is less than n, in a neighbourhood of some point $x_0 \in \Omega$, then there exists, by Frobenius's theorem, a local transformation of coordinates the application of which results in $\mathscr{L}$ acting only with respect to the variables $x_1, \ldots, x_{n-1}$. If, moreover, the homogeneous equation $\mathscr{L}u = 0$ possesses even one non-trivial solution u, then $\mathscr{L}$ cannot be hypoelliptic because the same equation is satisfied by a function u_1 that equals u for $x_n < c$ and equals zero for $x_n > c$, where c is some constant.

The first example of the operator $\mathscr{L}$ satisfying the hypotheses of the theorem was considered in 1935 by Kolmogorov. It was the operator

$$\mathscr{L} = \frac{\partial^2}{\partial x^2} + x\frac{\partial}{\partial y} - \frac{\partial}{\partial t}.$$

The proof of Theorem 3.1, given by Hörmander [1983, 1985], uses simple geometrical considerations. Let X be a real smooth vector field in $T(\Omega)$. Let us consider a one-parameter group of transformations in Ω that is generated by X. To do this we determine φ as a solution of the equation

$$\frac{\partial\varphi(t, x)}{\partial t} = X\varphi(t, x)$$

with the initial condition $\varphi(0, x) = x$. Clearly, $\varphi(s, \varphi(t, x)) = \varphi(t + s, x)$. We now set

$$e^{tX}u(x) = u(\varphi(t, x))$$

so that $d(e^{tX}u)/dt = e^{tX}Xu = Xe^{tX}u$. Let

$$|u|_{X,s} = \sup_{0<|t|<\varepsilon} \|e^{tX}u - u\| \cdot |t|^{-s}, \qquad u \in C_0^\infty(\Omega),$$

where $\|\cdot\|$ denotes the norm in $\mathscr{L}_2$. We put $I = (i_1, \ldots, i_k)$ for $0 \leqslant i_j \leqslant r$, and

$$X_I = [X_{i_1}, [X_{i_2}, \ldots, [X_{i_{k-1}}, X_{i_k}], \ldots]].$$

Let $|I| = k$.

Theorem 3.2. *Let $X_j \in T(\Omega)$, $0 < s_j \leqslant 1$, $j = 0, 1, \ldots, r$. Let*

$$m_j = 1/s_j, \qquad m(I) = 1/s(I) \equiv \sum_{j=1}^{k} 1/s_{i_j}, \qquad I = (i_1, \ldots, i_k),$$

and let σ be a positive number. If t is a sufficiently small positive number, then

$$\|e^{t^{m(I)}X_I}u - u\| \leqslant C_1 t \sum_{j=0}^{r} |u|_{X_j, s_j} + C_2 t|u|_\sigma, \qquad u \in C_0^\infty(K),$$

where C_1 and C_2 are constants in which C_1 depends only on r and σ but not on $X_0, \ldots, X_r, s_0, \ldots, s_r$.

In proving this theorem one has to make a significant use of the Campbell-Hausdorff formula.

3.2. A Necessary Condition for Hypoellipticity. We have shown in § 1 that a hypoelliptic differential operator with constant coefficients that is not elliptic must have multiple characteristics. This result can be extended to an operator with variable coefficients whose principal symbol is real (see Hörmander [1963]).

Theorem 3.3. *Let $P(x, D)$ be a differential operator, with C^∞-coefficients, in a domain $\Omega \subset \mathbb{R}^n$. Let its principal symbol $p(x, \xi)$ be real valued. If $(x_0, \xi_0) \in T^*\Omega \backslash 0$ and if*

$$p(x_0, \xi_0) = 0, \qquad \partial p(x_0, \xi_0)/\partial \xi_1 \neq 0,$$

then P cannot be hypoelliptic.

Thus, if $P(x, D)$ is a second-order differential operator with real principal part, then the quadratic form corresponding to this principal part is semi-definite.

We observe that (see Beals and Fefferman [1976], Ganzha [1986]) the sign of this quadratic form may change, all the same, from one point x to another point. For instance, the operator

$$x_1^3 \frac{\partial^2}{\partial x_2^2} + \frac{\partial}{\partial x_1}$$

is hypoelliptic in $\mathbb{R}^3$ (Kannai [1971]).

Theorem 3.4. *Let*

$$\mathscr{L}(x, D) = \sum_{i,j=1}^{n} a_{ij}(x) \frac{\partial^2}{\partial x_i \partial x_j} + \sum_{j=1}^{n} b_j \frac{\partial}{\partial x_j} + C(x) \tag{9}$$

be a differential operator with real C^∞-smooth coefficients. If $x_1 \mathscr{L}_0(x, \xi) > 0$, where $\mathscr{L}_0(x, \xi) = \sum a_{ij}(x)\xi_i\xi_j$, $x_1 \neq 0$, $\xi \neq 0$ and

$$a_{11}(x) = 0, \qquad D_1 a_{11}(x) = 0, \qquad b_1(x) > 0 \qquad \text{when} \quad x_1 = 0,$$

then $\mathscr{L}$ is hypoelliptic. Conversely, if $x_1 \mathscr{L}_0(x, \xi) > 0$ for $x_1 \neq 0$, $\xi \neq 0$, and $\mathscr{L}$ is hypoelliptic, then $a_{11}(x) = 0$, $D_1 a_{11}(x) = 0$ for $x_1 = 0$ and $b_1(x) \geqslant 0$ for $x_1 = 0$.

3.3. Operators with a Non-negative Quadratic Form. Let $\mathscr{L}(x, D)$ be given by (9). Theorem 3.1 can be extended to such operators that are more general than those of form (8).

Theorem 3.5 (Olejnik and Radkevich [1969]). *Let $\mathscr{L}_0(x, \xi) \geqslant 0$ in Ω. Let*

$$A^{(j)}(x, \xi) = \partial \mathscr{L}_0(x, \xi)/\partial \xi_j, \qquad A_{(j)}(x, \xi) = \partial \mathscr{L}_0(x, \xi)/\partial x_j,$$

$$\mathscr{L}_1(x, \xi) = \sum_{j=1}^{n} b_j(x)\xi_j.$$

Consider the Lie algebra generated by $A^{(j)}$, $A_{(j)}$ and $\mathscr{L}_1, j = 1, \ldots, n$, with respect to the operation of computing the Poisson brackets for each pair of functions. Assume that the inequality

$$\sum_{|k| \leqslant m} |A_{j_1 \ldots j_k}(x, \xi)| \geqslant C_0 > 0, \qquad |\xi| = 1, \qquad x \in K,$$

holds for the brackets $A_{j_1, \ldots, j_k} = \{X_{j_1}, \ldots, \{X_{j_{k-1}}, X_{j_k}\}, \ldots\}$, where X_{j_l} are some of the functions $A^{(j)}$, $A_{(j)}$ or $\mathscr{L}_1$, and K is a compact set in Ω. Then $\mathscr{L}$ is hypoelliptic in K and

$$\|\varphi u\|_{s+\varepsilon(K)} \leqslant C(K, s)(\|\varphi_1 \mathscr{L} u\|_s + \|\varphi_1 u\|_\gamma),$$

where $u \in C_0^\infty(K)$, φ, $\varphi_1 \in C_0^\infty(K)$, $\varphi\varphi_1 \equiv \varphi$, $\varepsilon(K) > 0$, and $\gamma = \text{const.} < s + \varepsilon(K)$.

A more precise analysis shows that $\varepsilon = m^{-1}$.

Example 3.1. The operator

$$P(x, D) = a(x)\Delta + \partial/\partial x_1,$$

where $a \in C^\infty(\mathbb{R}^n)$, $a(x) > 0$ for $x \neq 0$ and $a^{(\alpha)}(0) = 0$ for all α, is hypoelliptic in $\mathbb{R}^n$.

Example 3.2. The operator

$$D_1^2 + x_1^{2k} D_2^2 + i x_2^l D_2, \qquad k, l \geqslant 0,$$

is hypoelliptic.

Example 3.3. If ν is not an even integer, then the operator

$$P = |x|^2 \Delta - (\nu n + \nu(\nu - 2))$$

is not hypoelliptic in a domain Ω that contains the origin because $P(|x|^\nu) = 0$.

§ 4. Analytic Hypoellipticity

4.1. Elliptic Operators. We had pointed out in § 1 of Chap. 4 that for differential operators $P(D)$ having constant coefficients all the solutions of the equation $P(D)u = 0$ are analytic if and only if $P(D)$ is an elliptic operator. The matter becomes more complicated when the operators $P(x, D)$ have variable coefficients.

Theorem 4.1 (Olejnik and Radkevich [1972]). *Let $P(x, D)$ be a matrix differential operator whose coefficients are analytic in Ω and which is elliptic in the sense of Douglis and Nirenberg. If the vector function $f(x)$ is analytic in Ω, then every solution $u \in \mathscr{D}'(\Omega)$ of the system of equations $P(x, D)u = f(x)$ is analytic in Ω.*

Example 4.1. All the solutions of the equation $(D_1 + i x_1^2 D_2)u = f(x_1, x_2)$ in $\Omega \subset \mathbb{R}^2$ are analytic functions if f is analytic. But the operator $D_1 + i x_1^2 D_2$ is not elliptic.

Definition 4.1. An operator $P(x, D)$ is said to be *analytically hypoelliptic* in Ω if the facts that $u \in \mathscr{D}'(\Omega)$ and Pu is an analytic function in ω imply that u is analytic in ω for each subdomain $\omega \subset \Omega$.

It follows from what has been said above that the class of analytically hypoelliptic differential operators is wider than that of the elliptic differential operators.

4.2. The Analytic Wave Front. Let $A(\Omega)$ denote the space of functions that are analytic in an open set $\Omega \subset \mathbb{R}^n$. We recall that $f \in A(\Omega)$ if there are constants A and B such that

$$|D^\alpha f(x)| \leqslant A B^{|\alpha|} \alpha!, \qquad x \in K, \qquad \forall \alpha,$$

for every compact set K in Ω.

Proposition 4.1. *For a distribution u in $\mathscr{D}'(\Omega)$ to be an analytic function in Ω, it is necessary and sufficient that for every point $x_0 \in \Omega$ there exist a neighbourhood $U(x_0)$ and a sequence $\{\varphi_N\} \in C_0^\infty(\Omega)$ such that*

$$\varphi_N = 1 \text{ in } U(x_0), \qquad |\widetilde{\varphi_N u}(\xi)| \leqslant C^{N+1} N! (1 + |\xi|)^{-N}$$

where C does not depend on N.

Definition 4.2. By the *support of analytic singularities of* a distribution u (denoted by $\operatorname{sing\,supp}_a u$) we mean the intersection of all closed sets on the complement of each of which u is analytic.

Proposition 4.1 enables us to extend the notion of analyticity to the cotangent bundle space.

Definition 4.3. A distribution $u \in \mathscr{D}'(\Omega)$ is said to be *analytic* at a point $(x_0, \xi_0) \in T^*\Omega\backslash 0$ if there exist an open neighbourhood U of x_0 in Ω, an open cone Γ, in $\mathbb{R}^n$, containing ξ_0, a constant $C > 0$, and a sequence $\{\varphi_N\}$ of functions in $C_0^\infty(\Omega)$ such that $\varphi_N = 1$ in U and $\varphi_N = 0$ outside some compact set $K \subset \Omega$ which is independent of N, and

$$|\widehat{\varphi_N u}(\xi)| \leqslant C^{N+1} N!(1 + |\xi|)^{-N}, \qquad \xi \in \Gamma.$$

The distribution u is said to be analytic in an open conical subset $\Gamma \subset T^*\Omega\backslash 0$ if it is analytic at every point $(x, \xi) \in \Gamma$.

The complement of the set of those points of $T^*\Omega\backslash 0$ where u is analytic is known as the *analytic wave front* of u and is denoted by $WF_a(u)$.

Theorem 4.2 (Hörmander [1983, 1985]). *The projection onto Ω of the analytic wave front $WF_a(u)$ of a distribution $u \in \mathscr{D}'(\Omega)$ coincides with the set* $\operatorname{sing\,supp}_a u$, *the support of analytic singularities.*

Example 4.2. Let $\xi_0 \in \mathbb{R}^n$, $|\xi_0| = 1$, and $0 < \varepsilon < 1$. Then the analytic wave front of the function

$$u(x) = \int e^{ix\xi - |\xi|^{1-\varepsilon} - |x|^2|\xi| - |\xi - |\xi|\xi_0|}\, d\xi$$

consists of a single ray $\{(x, \xi): x = 0, \xi = t\xi_0, t > 0\}$, and $WF(u) = \varnothing$, that is, $u \in C^\infty(\mathbb{R}^n)$. Another definition of the analytic wave front can be found in § 3.2 of Chap. 8.

4.3. Analytic Pseudodifferential Operators. We consider the integral operator

$$Ku(x) = \int K_0(x, y, x - y) u(y)\, dy.$$

Theorem 4.3 (Trèves [1982]). *Let the function $K_0(x, y, z)$ be analytic in $\Omega \times \Omega \times \mathbb{R}^n\backslash 0$. Assume that the following condition holds:*

For every compact subset $\mathscr{M} \subset \Omega$ there exist an integer $m > 0$ and a constant $C > 0$ such that the inequalities

$$\sup_{x \in \mathscr{M}} \left| \int [D_x^\alpha D_y^\beta K_0(x, y, z)]_{z = x - y} f(y)\, dy \right| \leqslant C^{|\alpha + \beta| + 1} \alpha! \beta! \sup_{x \in \mathscr{M}} \sum_{|\gamma| \leqslant m} |D^\gamma f(x)| \quad (10)$$

hold for all $f \in C_0^\infty(\mathscr{M})$ and for all $\alpha, \beta \in \mathbb{Z}_+^n$.

Then the operator K is analytically pseudolocal, that is, for $u \in \mathscr{E}'(\Omega)$ the distribution Ku is analytic on each open subset where u is analytic.

Definition 4.4. A continuous linear operator $\mathscr{E}'(\Omega) \to \mathscr{D}'(\Omega)$ is called *analytically smoothing* if its image is contained in $A(\Omega)$.

Proposition 4.2. *Let the amplitude $a(x, y, \xi) \in C^\infty$, defined on $\Omega^{\mathbb{C}} \times \Omega^{\mathbb{C}} \times \mathbb{R}^n$, be holomorphic in x and y. Suppose that for each compact set $\mathscr{M}^{\mathbb{C}}$ in $\Omega^{\mathbb{C}} \times \Omega^{\mathbb{C}}$ there exists a constant $C > 0$ such that*

$$|a(x, y, \xi)| \leqslant Ce^{-|\xi|/C}, \qquad \forall (x, y) \in \mathscr{M}^{\mathbb{C}}, \qquad \xi \in \mathbb{R}^n.$$

Then the operator A, given by

$$Au(x) = \iint a(x, y, \xi) e^{i(x-y, \xi)} u(y)\, dy\, d\xi,$$

is analytically smoothing.

Here, and in the sequel, $\Omega^{\mathbb{C}}$ denotes an open neighbourhood of $\Omega \subset \mathbb{R}^n$ in $\mathbb{C}^n$.

Theorem 4.4 (Trèves [1982]). *Let $a(x, y, \xi) \in C^\infty(\Omega \times \Omega \times \mathbb{R}^n)$. Suppose that this function can be extended to a C^∞-function of (x, y, ζ) that is holomorphic in x, y on the open set*

$$\{(x, y, \zeta) \colon (x, y) \in \Omega^{\mathbb{C}} \times \Omega^{\mathbb{C}}, \qquad \zeta \in \mathbb{C}^n, \qquad |\operatorname{Im} \zeta| < \delta_0 |\operatorname{Re} \zeta|\}, \tag{11}$$

where $\delta_0 > 0$. Suppose that the following condition holds: For every compact set $\mathscr{M} \subset \Omega \times \Omega$ and every $\varepsilon > 0$ there exist an $r > 0$, a $\delta \in (0, \delta_0)$ and $C, C' > 0$ such that the estimates

$$|a(x, y, \xi)| \leqslant C(1 + |\xi|)^m, \qquad \xi \in \mathbb{R}^n,$$
$$|a(x, y, \zeta)| \leqslant C' e^{\varepsilon\delta|\operatorname{Re} \zeta|},$$
$$|\partial_{\bar{\zeta}} a(x, y, \zeta)| \leqslant Ce^{-|\operatorname{Re} \zeta|/C}$$

hold for (x, y, ζ) from the set (11) that satisfy the condition

$$\operatorname{dist}((x, y), \mathscr{M}) < r, \qquad |\operatorname{Im} \zeta| < \delta |\operatorname{Re} \zeta|.$$

Then the kernel

$$\mathbb{K}_0(x, y, z) = (2\pi)^{-n} \int e^{iz\xi} a(x, y, \xi)\, d\xi$$

satisfies (10) *and the operator A, defined by the formula*

$$Au(x) = \iint a(x, y, \xi) u(y) e^{i(x-y)\xi}\, dy\, d\xi,$$

is analytically pseudolocal.

4.4. Necessary Conditions for Analytic Hypoellipticity. Let Ω be a bounded domain in $\mathbb{R}^{n+1}$, and $x = (x_0, x_1, \ldots, x_n)$.

Definition 4.5. A function u in $C^\infty(\Omega)$ belongs to the class $A(\Omega)$ if for each subdomain $K \Subset \Omega$ there exists a constant $\delta(K) > 0$ such that u can be continued analytically to $V(u)$ as a function of $x + iy$ in the domain $Q_\delta = \{x \in K, -\delta(K) < y_j < \delta(K), j = 0, 1, \ldots, n\}$ and u is bounded in Q_δ.

Definition 4.6. A function u in $C^\infty(\Omega)$ belongs to the class $A_j(\Omega)$ if for each subdomain $K \Subset \Omega$ there exists a constant $\delta_j(K)$ such that u admits analytic continuation to $V_j(u)$ in the variable x_j as the function $u(x_0, \ldots, x_{j-1}, x_j + iy_j, x_{j+1}, \ldots, x_n)$ in a domain

$$Q_{j,\delta} = \{x \in K, -\delta_j(K) < y_j < \delta_j(K)\},$$

and, moreover, u and $\partial u/\partial x_k$ are bounded in $Q_{j,\delta}$ for all $k \neq j$.

If every weak solution u in $\mathscr{D}'(\Omega)^N$ of the system of equation

$$\mathscr{L}u = 0, \tag{12}$$

with analytic coefficients, belongs to $A_j(\Omega)$, then for every compact set $K \subset \Omega$ constants $\delta_j(K)$ and $C = C(K)$ can be found such that

$$\sup_{Q_{j,\delta}} |V_j(u)| \leqslant C \|u\|_{B^N}. \tag{13}$$

Here B^N is a Banach space, containing the solutions of (12), such that the convergence in the norm $\|\cdot\|_{B^N}$ implies the convergence in $\mathscr{D}'(\Omega)^N$. This is a consequence of the Baire-Hausdorff theorem. By choosing those solutions of (12) for which (13) is valid, we obtain a class of systems that are not analytically hypoelliptic.

Theorem 4.5 (Olejnik and Radkevich [1975]). *Let a solution $u_\rho \in \mathscr{D}'(\Omega)$ of (12) of the form $u_\rho(x) = e^{i\rho x_j} v_\rho(x)$, $\rho > 1$, be such that $u_\rho \in A_j(\Omega)$, where $\delta_j(K)$ is independent of ρ and the functions $|V_j(v_\rho)|$ and $|\partial V_j(v_\rho)/\partial x_l|$, $l \neq j$, are bounded in $Q_{\delta,j}(K)$ by a constant $C(K)$ that is independent of ρ. Let $B^N(\Omega)$ be a Banach space contained in $\mathscr{D}'(\Omega)^N$ and endowed with a stronger topology, and, moreover,*

$$\|u_\rho\|_{B^N} \leqslant e^{C_1 \rho^\mu},$$

where $\mu < 1$ and the constants C_1 and μ do not depend on ρ. Suppose that there is a sequence of points $x(\rho) \in K$ for which

$$|V_j(v_\rho(z_{\rho,h}))| > e^{-C_2 \rho^\sigma},$$

where $z_{\rho,h} = (x_0(\rho), \ldots, x_j(\rho) + ih, \ldots, x_n(\rho)) \in Q_{\delta,j}(K)$, $-\delta_1 \leqslant h \leqslant 0$, and the constants $C_2 > 0$ and $\sigma < 1$ do not depend on ρ or h. Then the system (12) is not analytically hypoelliptic in the variable x_j.

Example 4.3. The operator

$$P = D_t^2 + t^2 D_y^2 + D_z^2$$

is not analytically hypoelliptic in t or y, because the equation $Pu = 0$ has a solution

$$u_\rho(x) = e^{i\rho y - \rho t^2/2 + \sqrt{\rho}(z-1)}.$$

In this case, $\|u\|_{B^1} = \sup_\Omega |u|$. However, all the solutions of $Pu = 0$ are analytic in z (Olejnik and Radkevich [1975]).

Example 4.4. The system of differential equations $\mathscr{L}u = f$, with constant coefficients, has for analytic vector function f solutions that are non-analytic in x_j if the equation $\det \mathscr{L}_0(\xi) = 0$ has a real solution $\xi = (\xi_1, \ldots, \xi_n)$ and $\xi_j \neq 0$. The leading part of the symbol $\mathscr{L}_0(\xi)$ may be determined in the "Douglis-Nirenberg sense": there are integers $(s_1, \ldots, s_N)$ and $(t_1, \ldots, t_N)$ such that the order of the operator $\mathscr{L}_{ij}$ does not exceed $s_i + t_j$ and $\mathscr{L}_0(\xi) = \|\mathscr{L}_{ij}^0(\xi)\|$ and the polynomial $\mathscr{L}_{ij}^0$ is homogeneous of degree $s_i + t_j$ (Olejnik and Radkevich [1972]).

Example 4.5. If the function $x_0 \operatorname{Im} a_j(x)$ does not change sign in a neighbourhood of the origin, then the solutions of the equation

$$\frac{\partial u}{\partial x_0} + \sum_{k=1}^n a_k(x) \frac{\partial u}{\partial x_k} + c(x)u = 0$$

are non-analytic in x_j. This is true, for example, for the Mizohata operators $D_t + it^{2k+1}D_x$ (Olejnik and Radkevich [1972]).

Example 4.6. Let $P_1(D_x)$, $P_2(D_y)$ and $P_3(D_z)$ be homogeneous elliptic differential operators, with constant coefficients, of order $2m$.

The operator

$$P_1(D_x) + |x|^{2p}P_2(D_y) + |x|^{2q}P_3(D_z),$$

where $x = (x_1, \ldots, x_k)$, $y = (y_1, \ldots, y_l)$, $z = (z_1, \ldots, z_m)$ and p and q are non-negative integers, is analytically hypoelliptic if and only if $p = q$. It fails to be analytically hypoelliptic in y if $q < p$ and in z if $q > p$ (see Olejnik and Radkevich [1972]).

4.5. Differential Equation of the Second Order. In a domain $\Omega \subset \mathbb{R}^n$, we consider an operator $\mathscr{L}$ defined by the formula

$$\mathscr{L}u = \sum_{i,j=1}^n a_{ij}(x) \frac{\partial^2 u}{\partial x_i \partial x_j} + \sum_{j=1}^n b_j(x) \frac{\partial u}{\partial x_j} + c(x)u,$$

with real analytic coefficients a_{ij}, b_j and c. In this case, Theorem 4.5 assumes a simpler form.

Theorem 4.6 (Olejnik and Radkevich [1975]). *Let*

$$\sum_{j,k=1}^n |a_{k_j}(x)| + \sum_{j=1}^n |b_j(x)| \neq 0, \qquad \sum_{j,k=1}^n a_{k_j}(x)\xi_i\xi_j \geqslant 0$$

everywhere in Ω. $\mathscr{L}$ is hypoelliptic in Ω if and only if the Lie algebra generated by the system of vector fields $(\mathscr{L}_0, \mathscr{L}_1, \ldots, \mathscr{L}_n)$, where

$$\mathscr{L}_0(x) = \left(b_1(x) - \sum_{k=1}^n \frac{\partial a_{1k}(x)}{\partial x_k}, \ldots, b_n(x) - \sum_{k=1}^n \frac{\partial a_{nk}(x)}{\partial x_k} \right),$$

$$\mathscr{L}_j(x) = (a_{j1}(x), \ldots, a_{jn}(x)), \qquad j = 1, \ldots, n,$$

has rank n at every point of Ω.

Necessary conditions for the analytic hypoellipticity are given in the following theorem.

Theorem 4.7 (Popivanov [1976]). *Let* $\sum_{j,k=1}^{n} a_{k_j}(x)\xi_i\xi_j \geqslant 0$ *in* Ω *and* $\sum |a_{k_j}(0)| \neq 0$. *Suppose that there exists an analytic function* $H(x)$ *such that* $H(0) = 0$, grad $H(0) \neq 0$ *and*

$$\sum_{1}^{n} a_{k_j}(x)\frac{\partial H(x)}{\partial x_k}\frac{\partial H(x)}{\partial x_j} = 0$$

at each point $x \in \Omega$ *where* $H(x) = 0$. *Then the operator is analytically hypoelliptic at the origin.*

The next theorem gives sufficient conditions for the analytic hypoellipticity when $n = 2$.

Theorem 4.8. *Let*

$$\mathscr{L}u = \frac{\partial^2 u}{\partial x^2} + x^{2k}\frac{\partial^2 u}{\partial y^2} + a(x, y)\frac{\partial u}{\partial y} + b(x, y)u.$$

Suppose that for some $\delta > 0$ *one of the following conditions holds in* $Q_\delta(\Omega) = \{(x, y + it), (x, y) \in \Omega, |t| < \delta\}$:

1. $|\mathrm{Im}\, b|^2 \leqslant M_1 |\mathrm{Im}\, a|, \quad t\, \mathrm{Im}\, a(x, y + it) \leqslant 0;$
2. $|\mathrm{Im}\, b| \leqslant M_2 |x|^k, \quad t\, \mathrm{Im}\, a(x, y + it) \leqslant 0;$
3. $|\mathrm{Im}\, b| \leqslant M_3 |x|^k, \quad |\mathrm{Im}\, a| \leqslant M_4 |x|^{2k},$

where $M_1, \ldots, M_4$ *are certain constants. If* f *is an analytic function in* Ω, *then every solution of the equation* $\mathscr{L}u = f$ *belonging to* $\mathscr{D}'(\Omega)$ *is analytic in* Ω (*Popivanov* [1976]).

Example 4.7 (Grushin). All the solutions of the equation

$$\frac{\partial^2 u}{\partial x^2} + 2x^l a_{12}(x, y)\frac{\partial^2 u}{\partial x \partial y} + x^{2l} a_{22}(x, y)\frac{\partial^2 u}{\partial y^2} + b_1(x, y)\frac{\partial u}{\partial x} + x^l b_2(x, y)\frac{\partial u}{\partial y} + c(x, y)u = 0$$

are analytic in Ω. Here l is a non-negative integer, and b_1, b_2, c are complex-valued analytic functons, a_{12}, a_{22} are real analytic functions, and $a_{22} - (a_{12})^2 > 0$ in Ω.

Example 4.8. The equation

$$\frac{\partial^2 u}{\partial x^2} + x^{2k}\frac{\partial^2 u}{\partial y^2} + i\lambda x^{k-1}\frac{\partial u}{\partial y} = 0$$

has a solution that is non-analytic in y for some $\lambda \in \mathbb{R}$. This value of λ is determined from the condition that the equation $g'' - x^{2k}g + \lambda x^{k-1}g = 0$ has a non-trivial solution g belonging to the class $S(\mathbb{R})$.

Example 4.9. Let I be a subset of the set $\{1, \ldots, n\}$. We consider the operator

$$A = \Delta_t + \left(\sum_{i \in I} t_i^2\right)^l Q(x, D_x), \qquad \Delta_t = \sum_{i=1}^{n} \frac{\partial^2}{\partial t_i^2}, \qquad l \geqslant 1, \qquad l \in \mathbb{Z},$$

where Q is a second-order elliptic operator and $x = (x_1, \ldots, x_m)$. A is analytically hypoelliptic if and only if $I = \{1, \ldots, n\}$.

4.6. The Gevrey Classes. The properties of the second-order operators of the form

$$\mathscr{L} = \sum_{j=1}^{r} X_j^2 + X_0 + c, \tag{8}$$

where X_j, $j = 0, 1, \ldots, r$, are first-order differential operators with real coefficients, have been investigated in the Gevrey classes.

Let

$$G_\gamma = \left\{u;\, u \in C^\infty(\Omega),\, \sup_K |D^\alpha u(x)| \leqslant M_K^{|\alpha|+1}(\alpha!)^\gamma,\, K \Subset \Omega\right\},$$

and $I = (i_1, \ldots, i_k)$, $0 \leqslant i_j \leqslant r$, $i_j \in \mathbb{Z}$. We put

$$X_I = [X_{i_1}, [\ldots [X_{i_{k-1}}, X_{i_k}] \ldots]]$$

and

$$\frac{1}{\rho(I)} = \sum_{j=1}^{k} \frac{1}{\rho_{i_j}}, \qquad \text{where} \quad \rho_{i_j} = \begin{cases} 1, & \text{if } i_j = 1, 2, \ldots, r, \\ \frac{1}{2}, & \text{if } i_j = 0. \end{cases}$$

Assume that the fields $(X_{I_1}, \ldots, X_{I_n})$ constitute a basis for vector fields in K. Set

$$\rho_K = \inf_{1 \leqslant l \leqslant n} \rho(I_l).$$

Theorem 4.9 (Derridj and Zuilly [1973]). *If the Lie algebra generated by the fields $X_0, X_1, \ldots, X_r$ is of rank n, then*

$$u \in \mathscr{D}'(\Omega), \qquad Pu \in G_\gamma(K) \Rightarrow u \in G_\gamma(K),$$

provided that $\gamma > 2/\rho_K$ and the coefficients of P lie in $G_\gamma(K)$.

If the Lie algebra generated by $X_1, \ldots, X_r$ is of rank n, then the statement is valid for $\gamma > 1/\rho_K$.

If the vector space generated by the fields X_i, $[X_i, X_j]$, where $i, j = 1, \ldots, r$, is of dimension n at each point in Ω, then the same statement is true for all $\gamma \geqslant 2$.

Suppose that the operator $\mathscr{L}$ is of the form

$$\mathscr{L} = P\left(t, x, \frac{\partial}{\partial t}\right) + \left(\sum_{i=1}^{n} t_i^{2k_i}\right)^l Q\left(x, \frac{\partial}{\partial x}\right), \qquad x \in \mathbb{R}^m, \qquad t \in \mathbb{R}^n,$$

where P and Q are second-order elliptic operators, and k_i and l are positive integers. If the coefficients of P and Q belong to $G_\gamma(\Omega)$ and if $\mathscr{L}u \in G_\gamma(\Omega)$ and $u \in \mathscr{D}'(\Omega)$, then $u \in G_\gamma(\Omega)$ for each $\gamma \geqslant 1$ (Derridj and Zuilly [1973]).

4.7. Generalized Analytic Hypoellipticity. In some cases it proves convenient to examine generalized analytic functions, that is, functions $f(z) = u(x, y) + iv(x, y)$ that satisfy, instead of the Cauchy-Rimann equations, the equations

$$\frac{\partial u}{\partial x} = a_1(x, y)\frac{\partial v}{\partial y} + b_1(x, y)u + c_1(x, y)v,$$

$$\frac{\partial v}{\partial x} = -a_2(x, y)\frac{\partial u}{\partial y} + b_2(x, y)u + c_2(x, y)v,$$

where all the coefficients are bounded, and $a_1(x, y) \geqslant a_0$, $a_2(x, y) \geqslant 0$, $a_0 =$ const > 0.

Such a situation arises, for example, in the study of operators that are obtained from the Laplace operator by means of a smooth but non-analytic transformation of independent variables.

Theorem 4.10 (Landis and Olejnik [1974]). *Let $\mathscr{L}$ be a second-order elliptic differential operator in $\mathbb{R}^n$.* Let

$$Pu = \frac{\partial}{\partial t}\left(a(t)\frac{\partial u}{\partial t} + b(t)u\right) + \mathscr{L}_x u,$$

where $a(t) \geqslant a_0 > 0$, and a, b are sufficiently smooth functions. Let $Pu(t, x) = 0$ in the ball $\{(t, x): t^2 + |x|^2 < R\}$. Then the function u can be extended with respect to t to some domain of the complex plane $t +$ is as a generalized analytic function: there exist functions $U(t, s, x)$ and $V(t, s, x)$ such that

$$a(t)\frac{\partial U}{\partial t} + b(t)U = \frac{\partial V}{\partial s}, \qquad \frac{\partial U}{\partial s} = -\frac{\partial V}{\partial t},$$

$$U(t, 0, x) = u(t, x), \qquad V(t, 0, x) = 0.$$

A similar theorem holds for the solutions of the parabolic equation

$$u_t = (a(x_0)u_{x_0} + b(x_0)u)_{x_0} + \mathscr{L}_x u.$$

In this case, u is extended to the complex plane as a generalized functon of x_0.

By means of such extension we can obtain uniqueness theorems, theorems on continuous dependence, Hadamard's theorems, etc. (Landis and Olejnik [1974]).

Chapter 7
Transformation of Boundary-Value Problems

§ 1. The Transmission Property

1.1. Operators in a Half-Space. Let $\Omega = \{x \in \mathbb{R}^n, x_n > 0\}$, $\bar{\Omega}$ the closure of Ω, and let Γ be the boundary of Ω. We denote by D the line $x_1 = x_2 = \cdots = x_{n-1} = 0$ and by D_+ the intersection $D \cap \Omega$. As usual, $H_s(\Omega)$ denotes the space of restrictions, to Ω, of distributions belonging to $H_s(\mathbb{R}^n)$ and $\mathring{H}_s(\Omega)$ denotes the subspace of distributions in $H_s(\Omega)$ whose supports lie in $\bar{\Omega}$.

It can be shown that

$$H_s(\Omega) = \mathscr{L}^2(\Gamma, H_s(D_+)) \cap \mathscr{L}^2(D_+, H_s(\Gamma))$$

if $s \geqslant 0$ and

$$H_s(\Omega) = \mathscr{L}^2(\Gamma, H_s(D_+)) + \mathscr{L}^2(D_+, H_s(\Gamma))$$

if $s \leqslant 0$.

Let $\xi' = (\xi_1, \dots, \xi_{n-1})$ be the variable dual to x', and let

$$F_{x'} f(\xi', x_n) = \int f(x) e^{-ix'\xi'} \, dx.$$

Proposition 1.1. *Let $a(\xi', x_n) \in C^\infty(\bar{\Omega})$. Let*

$$|D_n^j a(\xi', x_n)| \leqslant C_{j,N,K}(1 + |x_n|)^{-N} \tag{1}$$

for any $j \in \mathbb{Z}_+$, $N \in \mathbb{Z}$ and $\xi' \in K$, where K is a compact set in $\mathbb{R}^{n-1}$. Assume that

$$a(\lambda\xi', x_n) = \lambda^r a(\xi', \lambda x_n) \tag{2}$$

for $\lambda \geqslant 1$ and $|\xi'| \geqslant 1$. Then

$$\|a(\xi', x_n)\|_{H_s(D_+)} \leqslant C(1 + |\xi'|)^{s+r-1/2},$$

where the constant C depends on a, s and r.

The quantity r is called the degree of a. Consider the operator $A\colon S'(\Gamma) \to \mathscr{D}'(\Omega)$ for which

$$Af(x) = (2\pi)^{1-n} \int a(\xi', x_n) \tilde{f}(\xi') e^{ix'\xi'} \, d\xi'.$$

Clearly, $Af \in C^\infty$ for $x_n > 0$.

Proposition 1.2. *The operator $A\colon H_{s-(1/2)}(\Gamma) \to H_{s-r}(\Omega)$ is continuous.*

Example 1.1. The Poisson operator, which gives a solution of the Dirichlet problem

$$\Delta(A\varphi) = 0 \quad \text{for} \quad x_n > 0, \qquad A\varphi = \varphi \quad \text{for} \quad x_n = 0,$$

in a half-space, has the symbol

$$a(\xi', x_n) = e^{-x_n|\xi'|}$$

of degree 0.

1.2. The Transmission Property (Boutet de Monvel [1966]). Let $A \in C^\infty(\mathbb{R}^n)$ and $A(t\xi) = t^r A(\xi)$ for $t \geqslant 1$. Let

$$a(\xi', x_n) = \int A(\xi) e^{-i\xi_n x_n} \, d\xi_n.$$

Then the function a satisfies (1) and is of degree $r + 1$.

Definition 1.1. An operator $A(D)$ has the *transmission* property if the function $a(\xi', x_n)$ is infinitely differentiable for $x_n = +0$.

Proposition 1.3. *The operator* $A(D)\colon H_s(\Omega) \to H_{s-r}(\Omega)$ *is continuous for* $s > -\frac{1}{2}$.

Let us cite explicit conditions on $A(\xi)$ under which $A(D)$ has the transmission property.

Proposition 1.4. *Let* $n = 1$. *For the operator* $A(D)$ *to have the transmission property it is necessary and sufficient that* $A(-\xi) = e^{-\pi i r} A(\xi)$ *for* $|\xi| > |\xi_0|$.

Example 1.2. Let $n = 1$. The convolution operators with $\ln x$ or $|x|^\alpha$ for non-integral α or with x^{-k}, where k is a positive integer, do not have the transmission property. This property is possessed by differential operators, by operators with kernels $|x|^k$ or x^k, k is a positive integer, or by operators with the kernel $\theta(-x)|x|^\alpha$, where α is not an integer, and $\theta(x) = 1$ when $x > 0$ and $\theta(x) = 0$ when $x < 0$.

In the general case, the function $A(\xi)$ can be expressed in the form

$$A(\xi) = a_0^\pm |\xi_n|^r + a_1^\pm(\xi')|\xi_n|^{r-1} + \cdots + a_p^\pm(\xi')|\xi|^{r-p} + R_p^\pm(\xi', \xi_n)$$

where $R_p^\pm(\xi', \xi_n)$ is of the order $|\xi_n|^{r-p-1}$ as $\xi_n \to \pm\infty$ and the coefficients $a_k^\pm(\xi')$ are homogeneous polynomials of degree k. These coefficients arise in the Taylor expansion of $A(\xi)$ in a neighbourhood of the half-lines $\xi' = 0$, $\xi_n > 0$ or $\xi_n < 0$.

Proposition 1.5. *The operator* $A(D)$ *has the transmission property if and only if*

$$a_k^-(\xi') = e^{i(r-k)\pi} a_k^+(\xi'), \qquad \xi' \in \mathbb{R}^n.$$

This condition is always fulfilled if, for example, $A(\xi)$ is a rational function when $|\xi| \geqslant |\xi_0|$.

Example 1.3. The inverse of the operator $\dfrac{\partial}{\partial \bar z} = \dfrac{\partial}{\partial x_1} + i\dfrac{\partial}{\partial x_2}$ in the complex plane $z = x_1 + ix_2$ has the symbol $A(\xi) = -i(\xi_1 + i\xi_2)^{-1}$. The corresponding function $a(\xi_1, x_2)$ is of the form

$$a(\xi_1, x_2) = \begin{cases} -ie^{-\xi_1 x_2} & \text{if } \xi_1 > 0, \\ 0 & \text{if } \xi_1 < 0. \end{cases}$$

Thus $A(D)$ has the transmission property.

For operators with variable symbols $a(x, \xi)$ the following definition proves useful.

Definition 1.2. By a *Poisson operator* on $\bar\Omega = \overline{\mathbb{R}^n_+}$, we mean a continuous linear operator $K\colon C_0^\infty(\Gamma) \to C^\infty(\Omega)$ for which there exists a

1. continuous extension $K\colon \mathscr{E}'(\Gamma) \to C^\infty(\Omega)$,
2. sequence $\{s_j\} \in \mathbb{R}$, $s_j \to -\infty$ such that

$$K(f, g, \lambda) = e^{-i\lambda g(x')} K(fe^{i\lambda g})(x', x_n/\lambda) \sim \sum k_j(f, g)\lambda^{s_j}$$

for each $f \in C_0^\infty(\Gamma)$, $g \in C^\infty(\Gamma)$ which are such that g assumes real values and $dg \neq 0$ on supp f. And for each $p \in \mathbb{Z}$ and each α, the function

$$\lambda^{-s_N} x_n^p D_x^\alpha \left[K(f, g, \lambda) - \sum k_j(f, g) \lambda^{s_j} \right]$$

is bounded for $\lambda \geqslant 1$, and uniformly bounded for bounded $|x'|$ and x_n/λ when g runs through a compact set in $C^\infty(\Gamma)$.

It can be shown that this definition is invariant under diffeomorphisms of the half-space $\Omega = \mathbb{R}_+^n$. This enables us to introduce analogous notions for an arbitrary smooth manifold with boundary.

Let V be a smooth manifold. Let $\bar{\Omega}$ be a smooth manifold, with boundary Γ, having the same dimension as that of V and immersed regularly in V.

Definition 1.3. A pseudodifferential operator P, defined on V, acts on $\bar{\Omega}$ if $P(u^0)|_{\bar{\Omega}} \in C^\infty(\bar{\Omega})$ when $u \in C_0^\infty(\bar{\Omega})$; u^0 denotes a continuation of u to $V \backslash \Omega$ by zero.

We denote by P^Ω the operator $u \to P(u^0)|_\Omega$.

Definition 1.4. A *Poisson kernel* on $\bar{\Omega}$ is a linear operator K: $C_0^\infty(\Gamma) \to C^\infty(\Omega)$ such that

1. there exists a continuous extension K: $\mathscr{E}'(\Gamma) \to C^\infty(\Omega)$,
2. for each open submanifold $\Omega'(\bar{\Omega}' \subset \bar{\Omega})$ that is isomorphic to $\bar{\mathbb{R}}_\infty^n$ and has an open boundary $\Gamma' \subset \Gamma$ the operator $K_{\Omega'}$: $C_0^\infty(\Gamma') \to C_0^\infty(\bar{\Omega}')$, defined by the formula $K_{\Omega'}(u) = K(u)|_\Omega$, is an isomorphic image of the Poisson kernel on $\mathbb{R}_+^n$.

Let us again take $\bar{\Omega} = \mathbb{R}_+^n$, $\Gamma = \mathbb{R}^{n-1}$ and $V = \mathbb{R}^n$. Let P be a pseudo-differential operator on V with the symbol $p(x, \xi) \sim \sum p_j(x, \xi)$. We denote by $\tilde{p}_j(x, \xi', t)$ the Fourier transform of the function $p_j(x, \xi)$ with respect to ξ_n. Since p_j is a homogeneous function in ξ of degree s_j, we have

$$\tilde{p}_j\left(x, \lambda\xi', \frac{t}{\lambda}\right) = \lambda^{s_j+1} \tilde{p}_j(x, \xi', t) \qquad \text{for} \quad \xi' \neq 0, \qquad \lambda > 0.$$

Theorem 1.1. *The following statements are equivalent*:

I. *The operator P acts on $\bar{\Omega}$.*

II. *For each integer p the operator K_p: $C_0^\infty(\Gamma) \to \mathscr{D}'(\Omega)$, defined by the formula $K_p(u) = P(u \otimes \delta^{(p)}(x_n))|_\Omega$, is a Poisson kernel.*

III. *For each j the function $\tilde{p}_j(x, \xi', t) \in C^\infty$ for $\xi' \neq 0$, $t \geqslant a|x_n|$ ($a =$ const. > 0).*

IV. *For each j the function $p_j(x, \xi', \xi_n) - e^{-i\pi s_j} p_j(x, \xi', -\xi_n)$ is flat along the manifold $x_n = 0$, $\xi' = 0$, $\xi_n > 0$ (that is, it vanishes together with all its derivatives).*

When these conditions are fulfilled, we say that P has the transmission property along Γ. In this case, the operator P^Ω: $C_0^\infty(\bar{\Omega}) \to C^\infty(\Omega)$ is extended to a continuous operator $H_{\text{comp}}^s(\bar{\Omega}) \to H_{\text{loc}}^{s-s_0}(\bar{\Omega})$. If, for instance, P is an elliptic operator having the transmission property, then there exists a parametrix for P and it also possesses the same property. This enables us to develop a theory of boundary-value problems for elliptic pseudodifferential operators having the transmission property (see Boutet de Monvel [1966], Rempel and Schulze [1982]).

1.3. Application to the Study of Lacunae. Let Ω be a domain in $\mathbb{R}^n$. A *lacuna of a distribution* $f \in \mathscr{D}'(\Omega)$ is an open subset of Ω where $f = 0$.

Definition 1.5. Let U be an open subset of Ω and $x_0 \in \partial U$. A distribution f in $\mathscr{D}'(\Omega)$ is said to be *sharp* at the point x_0 on the side U if there exists a function $g \in C^\infty(\Omega)$ such that $f - g = 0$ in $U \cap V$ for some neighbourhood V of x_0.

Proposition 1.6. *Let $n = 1$. Let $a(\xi) \sim \sum_{k=0}^{\infty} a_{s-k}(\xi)$ be a classical symbol of degree $s \in \mathbb{C}$. The distribution $\tilde{a}(-x)$ is sharp at the point $x = 0$ from the right (left) if*

$$a_{s-k}(\xi) = e^{i\pi s}(-1)^k a_{s-k}(-\xi) \qquad \text{for} \quad \xi > 0 \text{ (for } \xi < 0), \qquad k = 0, 1, \ldots$$

Under this condition, the function $a_{s-k}(\xi)$ has a holomorphic extension to the upper half-plane $\operatorname{Im} \xi > 0$ so that the support of $\tilde{a}(-x)$ lies on the negative semi-axis. Proposition 1.6 enables us to investigate the case of an aribtrary n too. Let

$$P \mapsto \tilde{P} \sim \sum (-1)^{m-k} P_{m-k}(x, -D)$$

be an involution of the algebra of pseudodifferential operators (for integral m). A conical Lagrange manifold $\Lambda \subset T^*\Omega \backslash 0$ is symmetrical if $\Lambda = -\Lambda$, that is, $(x, \xi) \in \Lambda \Leftrightarrow (x, -\xi) \in \Lambda$. Let Λ be symmetrical. A *transmission* is an involution τ of the space I_Λ^t of Fourier integrals of degree t, associated with Λ, under which $\tau(Pf) = \tilde{P}(\tau f)$ (mod C^∞) for each $f \in I_\Lambda^t$. Locally, integrals in I_Λ^t are of the form

$$f(x) = \int e^{iS(x,\theta)} a(x, \theta)\, d\theta,$$

where S is the generating function for Λ and a is a symbol of degree $s = t + n_x/4 - n_\theta/2$. Since Λ consists, locally, of two connected parts Λ_+ and $\Lambda_- = -\Lambda_+$, it follows that every distribution f from I_Λ^t can be written in the form

$$f(x) = \int [e^{iS(x,\theta)} a^+(x, \theta) + e^{-iS(x,\theta)} a^-(x, \theta)]\, d\theta.$$

Proposition 1.7. *The transmission τ is locally of the form*

$$f \mapsto \tau f = \int \left[e^{iS(x\theta)} \frac{\tilde{a}_-(x, \theta)}{\lambda} + e^{-iS(x,\theta)} \lambda \tilde{a}_+(x, \theta) \right] d\theta,$$

where $\lambda \in \mathbb{C}\backslash 0$.

The number λ is unique but depends on the choice of S. If we move over to another representation, λ is replaced by $i^{-\mu}\lambda$, where μ is the signature of the matrix of second-order derivatives that defines the Maslov cocycle. For the existence of a global transmission, it is necessary that $\mu \equiv 0 \pmod 2$.

Let us consider a differential operator $P(t, x, D_t, D_x)$, of degree m, which is strongly hyperbolic in t. Let E denote a fundamental solution of the Cauchy problem:

$$PE = 0, \qquad \left(\frac{\partial}{\partial t}\right)^k E(0, x) = \begin{cases} 0 & \text{if } k < m - 1, \\ \delta(x - x_0) & \text{if } k = m - 1. \end{cases}$$

The distribution E can be represented in the form

$$E(x) = \int a(x, \theta) e^{iS(x,\theta)}\, d\theta,$$

where $S(x, -\theta) = -S(x, \theta)$, and $a_{m-k}(x, -\theta) = (-1)^{m-k} a_{m-k}(x, \theta)$ if $a \sim \sum a_{m-k}(x, \theta)$. Moreover,

$$m = 1 - \deg P, \qquad n_\theta = n_x.$$

For instance, if $P = P(D_t, D_x)$, then

$$E(x) = \sum_1^m \int e^{ix\xi + it\psi_j(\xi)} a_j(\xi)\, d\xi,$$

where ψ_j are the roots of the characteristic equation $P_m(\psi_j(\xi), \xi) = 0$ and a_j is the symbol with degree $1 - m$.

A point on the cone Λ is called regular if at this point the projection $(x, \xi) \to x$ has the maximum rank $n_x - 1$. In a neighbourhood of such a point Λ is a hypersurface, and E, as it comes closer to Λ along the normal, has always a discontinuity of the first kind. In particular, if n_x is odd, then the limit of E will be finite or infinite simultaneously from both sides of Λ. If n_x is even, one of the limits will be finite while the other infinite. Thus diffusion takes place (Gårding [1971]).

§ 2. Distributions on a Manifold with Boundary

2.1. The Distribution Spaces (Melrose [1981]). Let M be a closed smooth manifold with boundary ∂M. Consider the function space

$$\dot{C}^\infty(M) = \{f \in C^\infty(M);\ D^\alpha f = 0 \text{ on } \partial M, \forall \alpha\}.$$

It is a space in which f is a continuous function on $\overline{M}$ for each differential operator P. Let Ω be the linear bundle of densities on M. We set

$$\dot{\mathscr{D}}'(M) = (C^\infty(M, \Omega))'.$$

Consider the space of vector fields

$$V(M) = \{v \in C^\infty(M, TM),\ v\text{'s are tangential to } \partial M\}.$$

Definition 2.1. The space $\dot{A}(M) \subset \dot{\mathscr{D}}'(M)$ consists of all those distributions u for which

$$v_1^{k_1} \dots v_N^{k_N} u \in H^s(M)$$

for all k_j and for all finite collection of fields $v_j \in V$ with a fixed $s \in \mathbb{R}$.

Theorem 2.1. $\dot{A}(M) = \dot{I}(M, N^*(\partial M))$. *Here* $\dot{I}(M, N^*(\partial M))$ *is the subspace of distributions u on M that are smooth functions in* $\mathring{M}$ *and are such that u has the representation*

$$u(x, y) = \frac{1}{2\pi} \int e^{ix\xi} a(y, \xi)\, d\xi, \qquad (x < \varepsilon),$$

with $a \in S^m_{1,0}(\mathbb{R}^n, \mathbb{R})$, *if* $\dim M = n + 1$ *and* (x, y) *are local coordinates on M and* $y \in \mathbb{R}^n$, $x \geqslant 0$. *It is a distribution space associated with the Lagrange manifold* $N^*\partial M$ (*see* § 3, *Chap.* 3).

As a corollary to Proposition 5.1, Chap. 3, we obtain the following result. If $u \in \dot{A}(M)$, then $\mathrm{WF}(u^0) \subset N^*(\partial M)$, where u^0 is the continuation of u beyond M by zero. Thus

$$u \in \dot{A}(M), \qquad \mathrm{WF}(u) = \varnothing \Leftrightarrow \dot{C}^\infty(M).$$

2.2. Contracted Cotangent Bundle (Melrose [1981])

Proposition 2.1. *There exists a natural vector bundle* $\tilde{T}M$ *on M and a bundle map* $\tilde{i}\colon \tilde{T}M \to TM$ *such that the induced map* $\tilde{i}_*\colon C^\infty(M, \tilde{T}M) \to C^\infty(M, TM)$ *is an embedding whose image coincides with V.*

This bundle is obtained from V by means of the following equivalence relation. If $m \in \mathring{M}$, then $v \sim w \Leftrightarrow vf(m) = wf(m)$ for every f in $C_0^\infty(M)$; if $m \in \partial M$, an additional condition has to be imposed, namely, $d(vg)(m) = d(wg)(m)$ for every g in $C^\infty(M)$ that vanishes on ∂M. Thus the image of the map $\tilde{i}$ coincides with $T\mathring{M} \cup T\partial M$. The dual map

$$\tilde{i}^*\colon T^*M \to \tilde{T}^*M$$

is a smooth map between the vector bundles and is, like $\tilde{i}$, of rank n on $\mathring{M}$ and of $n - 1$ on ∂M.

Proposition 2.2. *The image of the map* $\tilde{i}^*$ *can be identified canonically with* $T^*\partial M \cup T^*\mathring{M}$.

Assume that (x, y) are, as before, local coordinates on M so that $x > 0$ on $\mathring{M}$ and x has a zero of exactly order 1 on ∂M; $y \in \mathbb{R}^n$. Then the fields in V are of the form

$$\sum_{j=1}^n \alpha_j(x, y) D_{y_j} + \alpha_0(x, y) x D_x, \qquad \alpha_j \in C^\infty.$$

Every point p in $\tilde{T}^*M$ is of the form

$$p = \sum_{j=1}^n \eta_j(x, y)\, dy_j + \lambda(x, y)\, dx/x; \qquad \eta_j, \lambda \in C^\infty.$$

If (x, y, ξ, η) are the usual coordinates in T^*M, then

$$\tilde{i}^*(x, y, \xi, \eta) = (x, y, x\xi, \eta), \qquad \text{that is, } \lambda = x\xi.$$

The map $\tilde{i}^*$ induces the embedding

$$C^\infty(\tilde{T}^*M) \to C^\infty(T^*M).$$

And the function φ in $C^\infty(T^*M)$ lies in $C^\infty(\tilde{T}^*M)$ if and only if $\partial_x^k \varphi|_{x=0}$ is a polynomial in ξ of degree $\leqslant k$.

Proposition 2.3. *The embedding $\dot{C}^\infty(M) \subset \dot{A}(M)$ is continuous and its image dense in $\dot{A}(M)$. The same is true for the embedding $C^\infty(M) \subset \dot{A}(M)$.*

We denote by $\dot{B}(M)$ the space of those distributions $u \in \dot{\mathscr{D}}'(M)$ for which $u|_{\dot{M}} \in \dot{A}'(M)$.

Since $\dot{A}(M)$ consists of distributions that are smooth in all the directions except along the normal to ∂M, it is natural to expect that the distributions belonging to $\dot{A}'(M)$ must be smooth in the direction of the normal.

Proposition 2.4. *If $u \in \dot{B}(M)$ or $\dot{A}'(M)$, then the restriction $Pu|_{\partial M} \in \mathscr{D}'(\partial M)$ is well defined for each differential operator P. Furthermore,*

$$\dot{B}(M) = \dot{A}(M) \oplus \mathscr{D}'(M, \partial M),$$

where

$$\mathscr{D}'(M, \partial M) = \{u \in \dot{\mathscr{D}}'(M);\ \operatorname{supp} u \subset \partial M\}.$$

Most important for applications is the following corollary.

Proposition 2.5. *Let P be a differential operator on M, and let the map*

$$P\colon \dot{A}(M) \to \dot{A}(M)$$

be surjective mod $\dot{C}^\infty(M)$. *Then*

$$u \in \dot{\mathscr{D}}'(M), \qquad Pu \in \dot{B}(M) \Rightarrow u \in \dot{B}(M).$$

Thus boundary values of any order are well defined for u.

For instance, non-characteristic operators satisfy these conditions.

§ 3. Completely Characteristic Operators

3.1. Pseudodifferential Operators and their Kernels. The space of vector fields $V(M)$ can be viewed as the space of first-order differential operators. Let us define $\mathrm{Diff}_b(M)$ as the space of differential operators generated by the elements of $C^\infty(M)$ and $V(M)$. Near the boundary, these operators in local coordinates are of the form

$$P = \sum_{|\alpha|+k \leqslant m} a_{\alpha,k}(x, y) D_y^\alpha (x D_x)^k, \qquad a_{\alpha,k} \in C^\infty.$$

In microlocalization, it is natural to consider pseudodifferential operators with symbols of the form $\tilde{a}(x, y, \xi, \eta) = a(x, y, x\xi, \eta)$, where $a(x, y, \lambda, \eta) \in S^m_{1,0}(Z \times \mathbb{R}^{n+1}_{(\lambda,\eta)})$ and $Z = \overline{\mathbb{R}}^+ \times \mathbb{R}^n$. These symbols satisfy the estimates

$$|D_y^\alpha D_\eta^\beta D_x^k D_\xi^l \tilde{a}(x, y, \xi, \eta)| \leqslant C|x|^l (1 + |\xi|)^k (1 + |\eta| + |x\xi|)^{m-k-l-|\beta|}.$$

These estimates coincide with the standard ones when $x > 0$. The kernel of a pseudodifferential operator of this form is given by the function

$$k(z, z') = (2\pi)^{-n-1} \int e^{i(x-x')\xi + i(y-y')\cdot\eta} a(x, y, x\xi, \eta)\, d\xi\, d\eta.$$

Here $z = (x, y)$ and $z' = (x', y')$. Replacing x and ξ by $s = x'/x$ and $\lambda = x\xi$, respectively, we have

$$k\left(\frac{x'}{s}, y, x', y'\right) = \frac{1}{x} k_0(s, x, y, y'),$$

where

$$k_0(s, x, y, y') = \frac{1}{2\pi} \int e^{i(1-s)\lambda + i(y-y')\eta} a(x, y, \lambda, \eta)\, d\lambda\, d\eta.$$

Proposition 3.1. $k \in \mathscr{D}'(\mathbb{R}_x^+ \times \mathbb{R}^n \times \mathbb{R}_{z'}^{n+1})$. *The distribution* k_0 *has a singularity only when* $s = 1$ *and* $y = y'$, *and it decays rapidly as* $s \to \infty$.

3.2. The Transmission Property. We note that in the definition of k_0, x occurs only as a parameter. It is clear from the definition of k that this kernel is infinitely smooth when

$$(x', y', y) \in K \Subset \mathbb{R}^+ \times \mathbb{R}^n \times \mathbb{R}^n, \qquad 0 < x < x'/2,$$

and it vanishes at $x = 0$ together with all its derivatives.

In order to study the behaviour of k for $x > 0$ as $x' \to 0$, we must examine the behaviour of k_0 as $s \to +0$.

We assume that the following transmission conditions are fufilled:

$$\int e^{i\lambda} \lambda^k a(x, y, \lambda, \eta)\, d\lambda \equiv 0, \qquad \forall k \in \mathbb{N}. \tag{3}$$

Then we can assert that $k \in C^\infty$ for

$$(x, y, y') \in K \Subset \mathbb{R}^+ \times \mathbb{R}^n \times \mathbb{R}^n, \qquad 0 \leqslant x' \leqslant x/2$$

and that it vanishes at $x' = 0$ together with all its derivatives.

The first impression is that conditions (3) are very stringent. However, we have

Proposition 3.2. *For every* $a' \in S_{1,0}^\infty(Z \times \mathbb{R}^{n+1})$, *there exists a symbol* $b \in S_{1,0}^{-\infty}(Z \times \mathbb{R}^{n+1})$ *such that the symbol* $a = a' - b$ *satisfies* (3).

Proposition 3.3. *If the conditions* (3) *hold, then the operator* $A\colon \mathscr{E}'(Z) \to \dot{\mathscr{D}}'(Z)$ *is continuous and*

$$A(C_0^\infty(Z)) \subset C^\infty(Z), \qquad A^*(\dot{\mathscr{E}}'(Z)) \subset \dot{\mathscr{D}}'(Z).$$

If an operator A *possesses these properties, then its symbol is defined uniquely and satisfies* (3).

3.3. Completely Characteristic Operators

Definition 3.1. An operator A is said to be *completely characteristic* if its kernel $k_A \in \dot{C}^\infty(Z \times Z)$.

We denote by $\mathscr{L}_b^m(Z)$ the set of such operators of order m. These operators are smoothing, $A: \dot{\mathscr{E}}'(Z) \to \dot{C}^\infty(Z)$. The singularities of the kernels of such operators may lie beyond the diagonals also; thus, strictly speaking, these operators are not pseudodifferential operators. However, if $A \in \mathscr{L}_b^m(Z)$, then

$$\mathrm{WF}(k_A) \subset N^*\Delta \cup N^*(\partial Z \times \partial Z),$$

where Δ is the diagonal in $Z \times Z$ and $\partial Z \times \partial Z = \{z, z', x = 0, x' = 0\}$. In this way we find that the singularities of k_A are quite harmless.

The above definitions are invariant under a smooth transformation of variables that preserves the boundary ∂Z. Thus we can define the space $\mathscr{L}_b^m(M)$ for every smooth manifold M with boundary. It can be shown that every operator $A \in \mathscr{L}_b^\infty(M)$ maps each of the spaces $\mathscr{D}'(M)$, $C^\infty(M)$, $\dot{C}^\infty(M)$, $\dot{\mathscr{D}}'(M, \partial M)$, $\dot{\mathscr{D}}'(M)$ and $\mathscr{L}_2(M)$ into itself.

These operators constitute a ring that is closed under the operation of transition to the adjoint operator. Moreover,

$$\sigma_m(A^*) = \overline{\sigma_m(A)}, \qquad \sigma_{m+m'}(A \cdot B) = \sigma_m(A) \cdot \sigma_{m'}(B).$$

For every operator A in $L_b^m(M)$ an operator $A_\partial \in L^m(\partial M)$ can be defined such that

$$Au|_{\partial M} = A_\partial(u|_{\partial M}) \qquad \text{and} \qquad \sigma_m(A_\partial) = \sigma_m(A)|_{T^*\partial M}$$

for all u in $\dot{A}'(M)$ or in $B'(M)$. The space $\mathscr{L}_b^{-\infty}(M) = \bigcap \mathscr{L}_b^m(M)$ consists of those $A \in \mathscr{L}_b^\infty(M)$ that map $\dot{\mathscr{D}}'(M)$ into $\dot{A}(M)$.

3.4. The Boundary Wave Front. Let $u \in \dot{\mathscr{D}}'(M)$. Let $\operatorname{sing\,supp}_b(u) = M \backslash \{m \in M;$ $m \in B$, where B is an open subset in M and there is a $v \in \dot{A}(M)$ such that $v = u$ in $B\}$.

We define the characteristic set $\Sigma_b(A) = \{p \in \tilde{T}^*M \backslash 0;\ \sigma_m(A)$ is not elliptic in any of the cones containing $p\}$.

Definition 3.2. By the *boundary wave front* for $u \in \dot{\mathscr{D}}'(M)$ we mean the set

$$\mathrm{WF}_b(u) = \bigcap_A \{\Sigma_b(A);\ Au \in \dot{A}^*(M)\}.$$

The composition of symbols in local coordinates obeys the usual rules. If A and B are pseudodifferential operators with symbols a and b and if at least one of them is proper, then the operator $C = A \circ B \pmod{\mathscr{L}_b^{-\infty}(Z)}$ has the symbol

$$c \sim \sum_\alpha \sum_{p \leqslant k} \frac{1}{\alpha!p!(k-p)!} \partial_\eta^\alpha \partial_\lambda^k a(x, y, \lambda, \eta) \lambda^p x^{k-p} D_y^\alpha D_x^{k-p} D_\lambda^p b(x, y, \lambda, \eta).$$

Using this formula, we can construct a parametrix at elliptic points. In particular, the formula implies that

$$u \in \dot{\mathscr{D}}'(M), \qquad \mathrm{WF}_b(u) = \varnothing \Leftrightarrow u \in \dot{A}(M);$$

$$u \in A'(M) \Rightarrow \mathrm{WF}(Au|_{\partial M}) \subset \mathrm{WF}_b(u) \cap T^*\partial M.$$

Example 3.1. Let $n = 1$, $z = \mathbb{R}_x^+ \times \mathbb{R}_y$, and $P = xD_x + iD_y$. The operator P has an elliptic symbol $\sigma_1(P) = \lambda + i\eta$, and therefore

$$u \in \dot{\mathscr{D}}'(Z), \qquad Pu \in \dot{A}(Z) \Rightarrow u \in \dot{A}(Z).$$

The substitution $x = e^{-s}$ transforms P into $-D_s + iD_y$. Suppose that the function $f(s + iy)$ is holomorphic in the semistrip $s \geqslant 1/\varepsilon$, $|y| < \varepsilon$, and $|f(s + iy)| \leqslant Ce^{c_1 s}$. Then the function $u(x, y) = f(s + iy)$ satisfies the equation $Pu = 0$ for $0 < x < \delta$, $|y| < \varepsilon$, and has a finite order at $x = 0$. That is, it can be continued to an element u' that belongs to $\mathscr{D}'((0, \delta) \times (-\varepsilon, \varepsilon))$. We may assume that $Pu' = 0$. Hence

$$f(s + iy) = \frac{1}{2\pi} \int e^{i\xi e^{-s}} a(y, \xi)\, d\xi,$$

where $a \in S^m_{1,0}(\mathbb{R} \times \mathbb{R})$ for some m.

§4. Canonical Boundary Transformation

4.1. The Generating Function. Let us examine a canonical transformation of the contracted cotangent bundle $\tilde{T}^*M$. Clearly, at interior points it is the usual canonical transformation. In local canonical coordinates, the symplectic form has the representation

$$\alpha = \xi\, dx + \sum_{j=1}^{n} \eta_j\, dy_j.$$

Since $\tilde{i}^*(x, y, \xi, \eta) = (x, y, \lambda, \eta)$, where $\lambda = x\xi$, it follows that the canonical form on $\tilde{T}^*M$ is represented by the formula

$$\tilde{\alpha} = \lambda d \log x + \sum_{j=1}^{n} \eta_j\, dy_j.$$

Definition 4.1. By a *canonical boundary transformation* from M to M' we mean an infinitely smooth map $\chi\colon \Gamma \to \tilde{T}^*M' \backslash 0$, where Γ is an open cone in $\tilde{T}^*M \backslash 0$ and $\tilde{\chi}^*\tilde{\alpha}' = \tilde{\alpha}$.

Proposition 4.1. *If $\tilde{\chi}$ is a canonical boundary transformation of M into M' and if (x, y, λ, η) and $(x', y', \lambda', \eta')$ are the canonical coordinates in $\tilde{T}^*M$ an $\tilde{T}^*M'$, then*

$$\tilde{\chi}^*(x') = \mu_0(x, y, \lambda, \eta)x,$$

$$\tilde{\chi}^*(\lambda') = \lambda + \mu_1(x, y, \lambda, \eta)x,$$

where $\mu_0 > 0$. In particular, the projection $\tilde{\chi}$ onto T^*M defined a smooth canonical transformation $\tilde{i}^{*-1}(\Gamma) \to T^*M' \backslash 0$.

It is clear that the surface $x = \lambda = 0$ is invariant under the transformation $\tilde{\chi}$, and hence the map

$$\partial\tilde{\chi}: T^*\partial M \cap \Gamma \to T^*\partial M'$$

is defined.

4.2. The Operator of Principal Type. To illustrate the application of canonical boundary transformations, let us examine a problem involving an operator of real principal type.

Suppose that $p \in C^\infty(\tilde{T}^*M\backslash 0)$ assumes real values and that it is a homogeneous function in λ, η of degree 1. Let $p(\bar{\rho}) = 0$, where $\bar{\rho} \in \partial\tilde{T}^*M\backslash 0$. Suppose that on $\partial\tilde{T}^*M$ the differentials dp and $d\lambda$ are linearly independent at $\bar{\rho}$. If $\bar{\rho} \in T^*\partial M$, we impose an additional condition that dp and $\alpha_{\partial M}$ are linearly independent at $\bar{\rho}$.

Proposition 4.2. *Under the above hypotheses, there exists a canonical boundary transformation $\tilde{\chi}: \tilde{T}^*M \supset \Gamma \to \tilde{T}^*Z$, defined in some canonical neighbourhood of $\bar{\rho}$, such that*

$$\tilde{\chi}(\bar{\rho}) = (0, 0, \bar{\lambda}, \bar{\eta}), \qquad \bar{\eta} = (\bar{\eta}_1, 0, \ldots, 0) \qquad \text{and} \quad p = \tilde{\chi}^*(\eta_1).$$

4.3. The Differential Operator of Second Order. Let P be a second-order linear differential operator of real principal type. Suppose that the boundary ∂M does not have any characteristic direction for P. This implies that the forms dp and $\xi\, dx$ are linearly independent at interior points while linearly independent are their restrictions to $\partial T^*M\backslash 0$ at those boundary points where $p = 0$. If ∂M is given in the local coordinates by the equation $x = 0$, then

$$p = g(x, y)[\xi^2 + \xi a(x, y, \eta) + b(x, y, \eta)],$$

where $g \neq 0$ for $x = 0$.

Proposition 4.3 (Melrose [1981]). *If the boundary ∂M does not contain characteristic points for a real principal type operator P, then there exist local coordinates (x, y) such that*

$$p = \pm(\xi^2 + r(x, y, \eta)).$$

In a neighbourhood of every point $(0, y, \xi, \eta)$ at which $r > 0$ ($r < 0$), p can be reduced to the form $\xi^2 + \eta_n^2$ (respectively, $\xi^2 - \eta_n^2$) by means of a canonical boundary transformation.

If $r = 0$ at some point $(0, y, \xi, \eta)$, we assume that

$$q = \{p, \{p, x\}\} \neq 0.$$

When $q > 0$, the corresponding point is known as a point of *diffraction*, and by a homogeneous canonical transformation p can be reduced to the form

$$\xi^2 - x\eta_n^2 + \eta_1\eta_n.$$

When $q < 0$, the point is known as a *gliding* point, and the canonical form for p is the function

$$\xi^2 + x\eta_n^2 + \eta_1\eta_n.$$

When $q = 0$, there exists an uncountable set of non-equivalent symbols even if $\{p, \{p, \{p, x\}\}\} \neq 0$.

§ 5. Fourier Integral Operators

5.1. The Generating Function of the Canonical Boundary Transformation. Let M and N be smooth manifolds with boundary. Let $\tilde{\chi}$: $\tilde{T}^*M \supset \Gamma \to \tilde{T}^*N$ be a boundary canonical transformation. By introducing local coordinates (x, y), we may assume that

$$M = N = Z = \mathbb{R}_+ \times \mathbb{R}^n$$

and

$$\Gamma_1 \subset \overline{\mathbb{R}}_x^+ \times \mathbb{R}_y^n \times \mathbb{R}_{y'}^n \times (\mathbb{R}_{(\mu,\theta)}^{N+1} \backslash 0)$$

is an open cone and $(\mu, \theta) \in \mathbb{R} \times \mathbb{R}^N$. A function $\varphi \in C^\infty(Z \times \mathbb{R}^n \times \mathbb{R}^{N+1} \backslash 0)$ is known as a phase function in Γ_1 if it is real valued, homogeneous in (μ, θ) of degree 1 and

$$(\varphi_y, \mu\varphi_\mu, \varphi_\theta) \neq 0.$$

Because the canonical boundary transformations are of a specical form, we are forced to impose rather stringent non-degeneracy conditions as follows.

If $\dfrac{\partial\varphi}{\partial\theta} = 0$, then $\dfrac{\partial\varphi}{\partial\mu} \neq 0$ and

$$\det\begin{bmatrix} \varphi_{yy'} & \varphi_{y\theta} & \varphi_{y\mu} \\ \varphi_{\theta y'} & \varphi_{\theta\theta} & \varphi_{\theta\mu} \\ \mu\varphi_{\mu y'} & \mu\varphi_{\mu\theta} & (\mu\varphi_\mu)_\mu \end{bmatrix} \neq 0.$$

If φ is such a phase function, then under these conditions the set

$$C_\varphi = \left\{(x, y, y', \mu, \theta) \in \Gamma_1, \frac{\partial\varphi}{\partial\theta} = 0\right\}$$

is a conical submanifold in Γ_1 of codimension N.

The map

$$P_\varphi\colon C_\varphi \ni (x, y, y', \mu, \theta) \mapsto \left(x, y, \frac{\partial\varphi}{\partial x} + \mu\frac{\partial\varphi}{\partial\mu}, \frac{\partial\varphi}{\partial y}; x\frac{\partial\varphi}{\partial\mu}, y', \mu, -\frac{\partial\varphi}{\partial y'}\right) \in \tilde{C}_\varphi$$

is a local diffeomorphism onto a conical submanifold in $\tilde{T}^*Z \times \tilde{T}^*Z$.

Proposition 5.1. *If φ is a non-degenerate phase function, then, near the boundary $x = 0$, the manifold $C_{\tilde{\varphi}}$ is the graph of a canonical boundary transformation. Conversely, the graph of a canonical boundary transformation admits locally such a parametrization.*

5.2. The Fourier Integral Operator. Let φ be a non-degenerate phase function, and let $a \in S^m_{1,0}$. Consider the integral

$$k(z, z') = \int e^{i\varphi(z, y', x\xi, \theta) - ix'\xi} a(z, y', x\xi, \theta)\, d\theta\, d\xi,$$

that defines the Schwartz kernel of the Fourier integral operator associated with $\tilde{\chi}$. Assume that the support of a lies in the interior of the cone Γ_1 where φ is non-degenerate. We denote the set of these a's by $\dot{S}^m(\Gamma_1)$. The integral k makes sense when $x > 0$. In order to define it for $x = 0$ also, we require that the formal series vanish. We set

$$\mathscr{L}_k(\varphi)a = \int_0^\infty e^{i\varphi(0, y, y', \mu, \theta)} \frac{\partial^k}{\partial x^k} a(0, y, y', \mu, \theta) x^k\, d\mu,$$

and formulate

Proposition 5.2. *The map* $\mathscr{L}\colon S^{-\infty}(Z \times \mathbb{R}^n \times \mathbb{R}^{N+1}) \to \prod_{k=0}^\infty S^{-\infty}$ *is surjective.*

Let $\dot{S}^m(\Gamma_1, \varphi)$ denote the subspace of symbols a from $\dot{S}^m(\Gamma)$ for which the transmission conditions are satisfied. Then

$$\dot{S}^m(\Gamma_1, \varphi)/\dot{S}^{-\infty}(\Gamma_1, \varphi) \cong \dot{S}^m(\Gamma_1)/\dot{S}^{-\infty}(\Gamma_1).$$

Proposition 5.3. *If* $a \in \dot{S}^m(\Gamma_1, \varphi)$, *then* $k(z, z')$ *is a Schwartz kernel for the operator* $F\colon \dot{\mathscr{E}}'(Z) \to \mathscr{D}'(Z)$. *Moreover,*

$$F(\dot{C}_0^\infty(Z)) \subset \dot{C}^\infty(Z), \qquad F(C_0^\infty(Z)) \subset C^\infty(Z)$$

and

$$Fu|_{x=0} = F_0(u|_{x=0}) \qquad \text{for} \quad u \in C_0^\infty(z),$$

where F_0 *is the Fourier integral in* $\mathbb{R}^n$ *associated with* $\partial\tilde{\chi}$.

Let $A \in \mathscr{L}_b^m(M)$ be an operator of real principal type. To this operator there corresponds the Hamiltonian vector field

$$H_a = x\frac{\partial a}{\partial \lambda}\partial_x - x\frac{\partial a}{\partial x}\partial_\lambda + \sum_{j=1}^n \left(\frac{\partial a}{\partial \eta_j}\partial_{y_j} - \frac{\partial a}{\partial y_j}\partial \eta_j\right).$$

in $\tilde{T}^*M\backslash 0$. This field is tangential to the manifold

$$\Sigma_b(A) = \{\rho \in \tilde{T}^*M\backslash 0, a(\rho) = 0\}$$

and is tangential even to its invariantly defined components $\Sigma_b(A) \cap \partial\tilde{T}^*M$ and $\Sigma_b(A) \cap T^*\partial M$, because $x = 0$ on the first of these and on the second $x = \lambda = 0$.

Theorem 5.1. *If* $A \in \mathscr{L}_b^m(M)$ *is a real principal type operator, then for* $u \in \dot{\mathscr{D}}'(M)$ *the set*

$$\mathrm{WF}_b(u)\backslash \mathrm{WF}_b(Au) \subset \Sigma_b(A)$$

is the union of bicharacteristics of the field H_a *in* $\Sigma_b(A)$.

Chapter 8
Hyperfunctions

§ 1. Analytic Functionals

1.1. Definition and the Basic Properties. We can introduce analytic functionals in analogy with the distribution spaces (see Martineau [1961], Sato [1970a], Schapira [1971], Hörmander [1983]).

Definition 1.1. Let $K \Subset \mathbb{C}^n$ and A the space of entire analytic functions in $\mathbb{C}^n$. By the space of *analytic functionals* $A'(K)$ we mean the space of those linear forms on A for which the inequality

$$|u(\varphi)| \leqslant C_\omega \sup_\omega |\varphi|, \qquad \varphi \in A,$$

holds no matter what the neighbourhood ω of the compact set K. We define $A'(\mathbb{R}^n) = \bigcup_{K \Subset \mathbb{R}^n} A'(K)$.

Example 1.1. The functional $u(\varphi) = \sum a_\alpha D^\alpha \varphi(0)/\alpha!$ belongs to $A'(0)$ if and only if $|a_\alpha| \leqslant C_\varepsilon \varepsilon^{|\alpha|}$ for all $\varepsilon > 0$. This functional is a distribution only if the sum is finite.

Every smooth curve joining the points 0 and 1 in $\mathbb{C}$ is a support of the functional $u(\varphi) = \int_0^1 \varphi(z)\, dz$, $\varphi \in A(\mathbb{C})$. Therefore, in contrast to distributions, $u \in A'(K_1) \cap A'(K_2)$ does not necessarily imply that $u \in A'(K_1 \cap K_2)$. This is, however, true when K_1 and $K_2 \subset \mathbb{R}^n$.

Theorem 1.1. *If $u \in A'(\mathbb{R}^n)$, then there exists a smallest compact set $K \subset \mathbb{R}^n$ such that $u \in A'(K)$. This set is called the support of u.*

That the space $A'(K)$ is complete follows from the following

Theorem 1.2. *Let K_0 and K be compact sets in $\mathbb{R}^n$, $K_0 \subset K$, and let $u_j \in A'(\mathbb{C}^n)$. Assume that the following conditions hold:*

1. *For each compact neighbourhood V of K in $\mathbb{C}^n$, the elements $u_j \in A'(V)$ for large j.*
2. *For each compact neighbourhood V_0 of K_0 in $\mathbb{C}^n$ the difference $u_j - u_k \in A'(V_0)$ for large j and k.*

Then there exists an element $u \in A'(K)$ such that

$$u - u_j \in A'(V_0) \textit{ for large } \quad j$$

in every compact neighbourhood V_0 of K_0. The last condition defines u uniquely with respect to mod $A'(K_0)$.

The next result replaces the existence of the decomposition of unity.

Theorem 1.3. *If $K_1, \ldots, K_r$ are compact subsets in $\mathbb{R}^n$ and if $u \in A'(K_1 \cup \cdots \cup K_r)$, then there are $u_j \in A'(K_j)$ such that $u = u_1 + \cdots + u_r$.*

1.2. Operations on Analytic Functionals. The operations that are admissible in $\mathscr{D}'$ can be extended to the space $A'(\mathbb{R}^n)$.

1. If $u \in A'(\mathbb{R}^n)$, then $\partial_j u \in A'(\mathbb{R}^n)$, and this functional is defined by the formula $\partial_j u(\varphi) = -u(\partial_j \varphi)$.

2. If $u \in A'(K)$, $K \subset \mathbb{R}^n$, and if the function f is analytic in a neighbourhood ω of K, then the product fu is defined and given by the formula $fu(\varphi) = u(f\varphi)$ provided that the function φ is analytic in ω.

3. If $u \in A'(\mathbb{R}^n)$ and $v \in A'(\mathbb{R}^m)$, then $u \otimes v \in A'(\mathbb{R}^{n+m})$, and $(u \otimes v)(\varphi) = u_x(v_y(\varphi(x, y))) = v_y(u_x(\varphi(x, y)))$, where $\varphi(x, y) \in A(\mathbb{C}^{n+m})$.

4. Every functional $K \subset A'(\mathbb{R}^n \times \mathbb{R}^m)$ defines an operator $\mathscr{K}$ on those functions v that are analytic in a neighbourhood of the projection onto $\mathbb{R}^m$ of the support of K. Moreover, $\mathscr{K} v \in A'(\mathbb{R}^n)$, and $(\mathscr{K} v)(\varphi) = K(\varphi \otimes v)$ for $\varphi \in A(\mathbb{C}^n)$.

5. If $u \in A'(K)$, where K is a compact subset in $\mathbb{R}^n$, and if f is a real analytic bijective map of the open set $\omega \subset \mathbb{R}^n$ onto a neighbourhood of K, with $f^{-1} = h$, then the pullback $f^*u \in A'(f^{-1}K)$ is defined, and

$$(f^*u)(\varphi) = u((\varphi \circ h)|\det h'|),$$

provided that φ is analytic in ω.

6. For $u, v \in A'(\mathbb{R}^n)$, the convolution $u * v$ is defined as the value of the pullback of $u \otimes v$ under the map $(x, y) \to (y, x - y)$ onto the function $\varphi \equiv 1$.

7. If K is a compact set in $\mathbb{R}^n$ and $u \in A'(K)$, then the Fourier transform

$$\tilde{u}(\zeta) = u(e^{-i(\cdot, \zeta)})$$

is an entire analytic function such that

$$|\tilde{u}(\zeta)| \leqslant C_\varepsilon e^{H_K(\operatorname{Im} \zeta) + \varepsilon|\zeta|}, \qquad \zeta \in \mathbb{C}^n,$$

for each $\varepsilon > 0$. Here $H_K(\xi) = \sup_{x \in K} \langle x, \xi \rangle$. Conversely, every entire function satisfying this inequality determines a unique element in $A'(K)$.

§ 2. The Space of Hyperfunctions

2.1. Definition and the Basic Properties. The hyperfunctions in $\mathbb{R}^n$ are defined as local equivalent to analytic functionals having compact supports in $\mathbb{R}^n$.

Definition 2.1. If X is a bounded open set in $\mathbb{R}^n$, then the space $B(X) = A'(\overline{X})/A'(\partial X)$ is called the space of *hyperfunctions* in X.

Just as in the case of distributions, the support of a hyperfunction u is the smallest closed set outside which $u = 0$.

Theorem 2.1. *Let X_j be bounded open sets in $\mathbb{R}^n$ and $X = \bigcup X_j$. If $u_j \in B(X_j)$ and $u_i = u_j$ in $X_i \cap X_j$ for all i and j, then there exists a unique element $u \in B(X)$ such that $u = u_j$ in X_j for each j.*

It is clear that the embedding $\mathscr{D}'(X) \to B(X)$ is injective. The elements of $B(X)$ with compact supports can be identified with those elements of $A'(\mathbb{R}^n)$ that have supports in X. The operations on analytic functionals, described in § 1.2, can be extended to $B(X)$. What is more, the space $B(X)$ can be defined for every real analytic manifold X.

Theorem 2.2. *Let X be a real analytic manifold and Y an open subset of X. Then every hyperfunction $u \in B(Y)$ is a restriction to Y of the hyperfunction v belonging to $B(X)$ and having support in $\overline{Y}$.*

2.2. The Analytic Wave Front of a Hyperfunction. Let $I(\xi) = \int_{|\omega|=1} e^{-(\omega,\xi)}\, d\xi$. $I(\xi) = 2ch\xi$ if $n = 1$. If $n > 1$, this function can be explicitly written in terms of the Bessel function, and

$$I(\xi) = I_0(|\xi|), \qquad I_0(\rho) = (2\pi)^{(n-1)/2} e^{\rho} \rho^{(1-n)/2}[1 + 0(1/\rho)].$$

Let us set

$$K(z) = (2\pi)^{-n} \int e^{i(z,\xi)}\, d\xi / I(\xi).$$

The function $K(z)$ is defined and is analytic for $z \in \mathbb{C}^n$, $|\operatorname{Im} z| < 1$.

Definition 2.2. For $u \in A'(\mathbb{R}^n)$, $\mathrm{WF}_A(u)$ is the set of those points $(x, \xi) \in \mathbb{R}^n \times (\mathbb{R}^n \backslash 0)$ for which the function $K{*}u = u_t(K(z - t))$ is non-analytic at the point $x - i\xi/|\xi|$.

The projection of $\mathrm{WF}_A(u)$ onto $\mathbb{R}^n$ coincides with the set $\operatorname{sing\,supp}_A u$. Since $\mathrm{WF}_A(u)$ is defined at a point x by the local properties of u, Definition 2.2 can be extended to $B(X)$ for any subdomain $X \subset \mathbb{R}^n$. It can be shown (Sjöstrand [1982]) that $\mathrm{WF}_A(u) = \mathrm{WF}_a(u)$ for $u \in \mathscr{D}'(X)$ (see Definition 4.3 of Chap. 6).

2.3. Boundary Values of a Hyperfunction. Let X be an open set in $\mathbb{R}^n$ and Γ an open connected cone in $\mathbb{R}^n \backslash \{0\}$. Let Z be an open set in $\mathbb{C}^n$ such that

$$Z \supset \{z \in \mathbb{C}^n, \operatorname{Re} z \in X_1, \operatorname{Im} z \in \Gamma_1, 0 < |\operatorname{Im} z| < \gamma\}$$

for some $\gamma > 0$ and every open subset $X_1 \Subset X$ and every closed convex cone $\Gamma_1 \subset \Gamma \cup \{0\}$.

Theorem 2.3. *Let f be an analytic function in Z. Then for each open subset $X_1 \Subset X$:*

(1) *There exists an element $f_{X_1} \in A'(\overline{X}_1)$, independent of Γ_1, such that the analytic functional*

$$f_{X_1}(\varphi) - \int_{X_1} f(x + iy)\varphi(x + iy)\, dx, \qquad \varphi \in A,$$

has a support in every small neighbourhood of the set ∂X provided that $y \in \Gamma_1$ and $|y|$ are sufficiently small. This element is determined to within elements of $A'(\partial X_1)$, and, hence, it defines uniquely a hyperfunction in $B(X_1)$.

(2) *If* $X_2 \subset X_1$, X_2 *is an open set in* $\mathbb{R}^n$, *then* $f_{X_1} - f_{X_2} \in A'(\overline{X}_1 \backslash X_2)$; *thus there exists a unique element* $f_X \in B(X)$ *such that the restrictions of* f_X *and* f_{X_1} *to* X_1 *coincide for every* X_1.

(3) *If* $|f(x+iy)| \leqslant C|y|^{-N}$, $x \in X_1$, $y \in \Gamma_1$, $|y| < \gamma$, *then the restriction of* f_X *to* X_1 *equals the limit in the sense of the distribution theory.*

(4) *If* f *admits analytic continuation to a neighbourhood of the set* ∂X_1, *then the functionals* $\int_{X_1} f(x+iy)\varphi(x)\,dx$, $\varphi \in A$, *converge, as* $y \to 0$, $y \in \Gamma_1$, *in* $A'(\overline{X}_1)$ *to the element* f_{X_1} *defined in* (1).

(5) *If* $f_X = 0$ *for some non-empty open set* $X_1 \subset X$ *and if* Z *is connected, then* $f = f_X = 0$.

(6) $\mathrm{WF}_A(f_X) \subset X \times (\Gamma^0 \backslash 0)$, *where the cone* Γ^0 *is dual to* Γ.

Definition 2.3. The hyperfunction $f_X \in B(X)$ is called the boundary value of f belonging to Γ, and is denoted by $b_\Gamma f$.

Theorem 2.4. *If* X *is an open set in* $\mathbb{R}^n$, $u \in B(X)$, *and if*

$$\mathrm{WF}_A(u) \subset X \times (\Gamma^0 \backslash 0),$$

where Γ *is an open convex cone in* $\mathbb{R}^n \backslash 0$, *then there exists a function* f, *analytic in* Γ, *such that* $u = b_\Gamma f$.

As a corollary to Theorem 2.4, we can obtain the classical "edge of the wedge" theorem.

Theorem 2.5. *Let the functions* $f^\pm$ *be analytic in*

$$Z^\pm = \{x + iy; x \in X, \pm y \in \Gamma, |y| < \gamma\},$$

where $\gamma > 0$ *and* Γ *is an open convex cone. If* $f_0 = b_\Gamma f^+ = b_{-\Gamma} f^-$, *then* f^0 *is an analytic function that is the analytic continuation of* f^+ *and* f^-.

The following generalization of this theorem is due to Martineau.

Theorem 2.6. *Let* $\Gamma_1, \ldots, \Gamma_j$ *be closed cones in* $\mathbb{R}^n \backslash 0$ *and* $\bigcup \Gamma_j = \mathbb{R}^n \backslash 0$. *Let* X *be a bounded open set in* $\mathbb{R}^n$ *and* $u \in B(X)$. *Then there exist* $u_j \in B(X)$ *such that* $u = \sum u_j$ *and* $\mathrm{WF}_A(u_j) \subset \mathrm{WF}_A(u) \cap (X \times \Gamma_j)$. *If* $u = \sum u_j'$ *is another decomposition, then* $u_j' = u_j + \sum u_{jk}$, *where* $u_{jk} \in B(X)$, $u_{kj} = -u_{jk}$ *and* $\mathrm{WF}_A(u_{jk}) \subset X \times (\Gamma_j \cap \Gamma_k)$.

§ 3. Solutions of Differential Equations

3.1. The Cauchy Problem. Let $P(x, D)$ be a differential operator, with real analytic coefficients, in an open set $X \subset \mathbb{R}^n$. Clearly,

$$\mathrm{WF}_A(P(x, D)u) \subset \mathrm{WF}_A(u)$$

for $u \in B(X)$. The reverse inclusion also holds. Namely, we have

Theorem 3.1. *If $P(x, D)$ is a differential operator of order m, with real analytic coefficients in X, and if $u \in B'(X)$, then $\mathrm{WF}_A(u) \subset \mathrm{Char}\, P \cup \mathrm{WF}_A(Pu)$. Moreover, if a point (x_0, ξ_0) is non-characteristic for P, then for each hyperfunction $f \in B(X)$ a hyperfunction $u \in B(X)$ can be found such that $(x_0, \xi_0) \notin \mathrm{WF}_A(P(x, D)u - f(x))$.*

For every solution $u \in B(X)$ of the equation $P(x, D)u = 0$, we can determine the Cauchy data on a hypersurface X_0 provided that it is non-characteristic. To simplify the discussion, assume that $u \in B(X_+)$, where $X_\pm = \{x \in X, x_n \gtrless 0\}$ and $X_0 = \{x \in X, x_n = 0\}$. Let $u_0 = u$ in X_+ and $u_0 = 0$ in X_-. Then the hyperfunction $P(x, D)u_0$ has a support in X_0.

Theorem 3.2. *Let $P(x, D)$ have real analytic coefficients. Let X_0 be non-characteristic. Then every hyperfunction $f \in B(X)$, with support in X_0, has a unique representation of the form*

$$f = P(x, D)v + \sum_{j<m} v_j \otimes D_n^j \delta(x_n),$$

where $v \in B(X)$, $\operatorname{supp} v \subset X_0$, $v_j \in B(X'_0)$ for $0 \leqslant j < m$, and $X_0 = X'_0 \times \{0\}$.

As a corollary to this theorem, we have

Theorem 3.3. *If $P(x, D)$ has real analytic coeffients in X and if X_0 is non-characteristic for P, then every hyperfunction $u \in B(X_+)$ satisfying the equation $P(x, D)u = 0$ has a unique extension u_0 to X that vanishes on X_- and is such that*

$$P(x, D)u_0 = \sum_{j<m} v_j \otimes D_n^j \delta(x_n)$$

for some $v_j \in B(X'_0)$ for $0 \leqslant j < m$.

3.2. The Analytic Wave Front. We now present an equivalent definition for the analytic wave front that is due to Bros and Iagolnitzer. For $u \in A'(\mathbb{R}^n)$, let us define an entire function $T_\lambda u$, depending on a positive parameter λ, by the formula

$$T_\lambda u(z) = u_y(e^{-\lambda(z-y)^2/2}), \qquad z \in \mathbb{C}^n.$$

T_λ is closely related to the Fourier transform of u, because

$$e^{i\lambda(x,\xi) - \lambda\xi^2/2} T_\lambda u(z) = u_y(e^{i\lambda(y,\xi) - \lambda(x-y)^2/2})$$

is the Fourier transform of $ue^{-\lambda(x-y)^2/2}$ at the point $-\lambda\xi$. The multiplier $e^{-\lambda(x-y)^2/2}$ replaces the cutoff function used in defining the wave front of a distribution. It is natural to expect that $(x_0, \xi_0) \in \mathrm{WF}_A(u)$ if and only if $u_y(e^{i\lambda(x,\xi) - \lambda(x-y)^2/2})$ is $0(e^{-c\lambda})$, as $\lambda \to \infty$, for some $c > 0$ in a neighbourhood of the point $(x_0, -\xi_0)$. In fact, we have

Theorem 3.4 (Bony [1976–1977]). *Let $u \in A'(\mathbb{R}^n)$ and $(x_0, \xi_0) \in T^*(\mathbb{R}^n)\backslash 0$. Then $(x_0, \xi_0) \notin \mathrm{WF}_A(u)$ if and only if there exists a neighbourhood V of the point $x_0 - i\xi_0$ and positive constants C and c such that*

$$|T_\lambda u(z)| \leqslant Ce^{\lambda(|\xi_0|^2/2 - c)}, \qquad z \in V, \qquad \lambda > 0.$$

Let M be a subset in $\mathbb{R}^n$. If $x_0 \in M$, then $T_{x_0}(M)$ is defined as the set of all the limits of the sequences $t_j(x_j - x_0)$ as $t_j \to +\infty$, $x_j \in M$. This set is known as the tangential cone to M. It is a closed cone and is independent of the choice of local coordinates. Thus it can be defined for $M \subset X$, where X is a C^1-smooth manifold. If F is a closed subset of a C^2-smooth manifold, then $N_e(F) \subset T^*X\backslash 0$ is defined as the set of all the points (x_0, ξ_0) such that $x_0 \in F$ and such that there is a real-valued function $f \in C^2(X)$ for which $df(x_0) = \xi_0 \neq 0$ and $f(x) \leqslant f(x_0)$ for $x \in F$. We set $N_i(F) = \{(x, \xi); (x, -\xi) \in N_e(F)\}$ and $N(F) = N_e(F) \cup N_i(F)$.

Theorem 3.5. *If $u \in B(X)$ and $x_0 \in X$, then*

$$\overline{N}(W_0) \subset \partial W_0 \times T_{x_0}(\operatorname{supp} u),$$

where $W_0 = \{\xi \in T^*_{x_0}(X); (x_0, \xi) \in \mathrm{WF}_A(u)\}$ is regarded as a subset of $T^*_{x_0}(X)\backslash 0$.

The proof of this theorem, due to Kashiwara, is based on Holmgren's theorem.

Let us cite some of the corollaries of this theorem.

Corollary 3.1. *Let $W_1 \subset W_2$ be open convex sets in $T^*_{x_0}(X)\backslash 0$. Assume that*
(1) $W_1 \cap \mathrm{WF}_A(u)_{x_0} = \varnothing$;
(2) *Every hyperplane, with normal lying in $T_{x_0}(\operatorname{supp} u) \cap (-T_{x_0}(\operatorname{supp} u))$, that intersects W_2 also intersects W_1.*
Then $W_2 \cap \mathrm{WF}_A(u)_{x_0} = \varnothing$.

Corollary 3.2. *If $(x_0, \xi_0) \in N_e(\operatorname{supp} u)$, then $(x_0, \xi) \in \mathrm{WF}_A(u) \Rightarrow (x_0, \xi + t\xi_0) \in \mathrm{WF}_A(u)$ for all $t \in \mathbb{R}$.*

Let, once again, $X_\pm = \{x \in X, x_n \gtrless 0\}$ and $X_0 = \{x \in X, x_n = 0\}$. Let $P(x, D)$ be a differential operator, with analytic coefficients, of order m in X, and X_0 a non-characteristic plane.

Theorem 3.6. *If $u \in B(X_+)$ is a solution of the equation $P(x, D)u = 0$ and if $(x_0', \xi_0') \notin \mathrm{WF}_A(D_n^j u|_{x_n=0})$, $j = 0, 1, \ldots, m-1$, then there exists an $\varepsilon > 0$ such that $(x, \xi) \notin \mathrm{WF}_A(u)$ provided that*

$$0 < x_n < \varepsilon, \qquad |x' - x_0'| + |\xi' - \xi_0'| < \varepsilon.$$

Remark 3.1. Analogous statement fails to hold for $\mathrm{WF}(u)$ even if $P(D)$ has constant coefficients. For example, if $P(D) = D_1 D_2 + D_3^2 + D_4^2$, we can construct a solution $u_0 \in C^2$ of the equation $P(D)u_0 = 0$ for which $\operatorname{sing\,supp} u_0 = \{(t, 0, \ldots, 0), t \geqslant 0\}$. Then the function $u = \sum a_j u_0(x_1, x_2, x_3, x_4 - 1/j) \in C^2(\mathbb{R}^4)$ and $D_4 u \in C^\infty(\mathbb{R}^3)$ for $x_4 = 0$ but $\operatorname{sing\,supp} u = \{(x_1, 0, 0, x_4)$, where $x_4 = 0$ or $x_4 = 1/k$, $k = 1, 2, 3, \ldots\}$ if $a_j \to 0$ rapidly enough.

The theorems on the absence of solutions to equations for which the value of the function $c_1(x, \xi)$ is negative at a characteristic point (x_0, ξ_0) (see p. 86) can be extended to the space of hyperfunctions. For instance, the equation $D_1 u + i x_1 D_2 u = f$ may not have a solution in the class of hyperfunctions even if $f \in C^\infty$. What is more, the following result holds.

Theorem 3.7 (Schapira [1971]). *Let P be a first-order differential operator in $\mathbb{R}^n$ with analytic coefficients in a neighbourhood of the origin. Suppose that the principal part P_1 of P does not vanish at the origin. Suppose further that there exists a function w, analytic in a neighbourhood of* 0, *such that $P_1 w = 0$ and*

$$\operatorname{Im} w(x) = 0 \Leftrightarrow x = 0.$$

Then for any neighbourhood Ω, no matter how small, of the origin there exists an $\varepsilon > 0$ such that $PB(\Omega) \supset H_\varepsilon(\Omega)$, where H_ε is the space of functions that are holomorphic for $|\operatorname{Im} z| < \varepsilon$.

§4. Sheaf of Microfunctions

4.1. Traces of Holomorphic Functions. Let us consider in $\mathbb{R}^n$ the sheaves of analytic functions, infinitely smooth functions, distributions and hyperfunctions

$$A \subset C_\infty \subset \mathscr{D}' \subset B.$$

The embedding of these sheaves can be described as follows.

Let Γ be an open convex cone in $\mathbb{R}^n$ and ω an open set in $\mathbb{R}^n$. Let the function f be holomorphic on the intersection of the set $\omega \times i\Gamma$ with a complex neighbourhood of ω in $\mathbb{C}^n$. Then we can define the boundary value $b(f) \in B(\omega)$.

If $|f(x+iy)| \leqslant C|y|^{-N}$ for some integer N, then $b(f) = \lim_{y \to 0, y \in \Gamma} f(x+iy) \in \mathscr{D}'$ and the limit exists in the space $\mathscr{D}'(\omega)$.

If f and all the derivatives $D^\alpha f$ are bounded (or if $|D^\alpha f(z)| \leqslant C_\alpha |y|^{-N}$ for all α and a certain N that is independent of α), then the limit exists in C^∞.

Conversely, let $u \in \mathscr{D}'(\omega)$ and let $\{\Gamma_\alpha\}$ be a finite family of open convex cones in $\mathbb{R}^n$ such that the interiors of the dual cones Γ_α^0 cover S^{n-1}. Then there are $f_\alpha \in A(\omega + i\Gamma_\alpha)$ such that $|f_\alpha(x+iy)| \leqslant C|y|^{-N}$, $u = \sum b(f_\alpha)$. Actually, it is enough that $\omega \times \bigcup(\operatorname{int} \Gamma_\alpha^0)$ covers $\mathrm{WF}_A(u)$.

Similarly, if $u \in C^\infty(\omega)$ and $\bigcup \operatorname{int} \Gamma_\alpha^0$ covers $\mathrm{WF}_A(u)$, then there exist $f_\alpha \in A(\omega + i\Gamma_\alpha)$ such that f_α and their derivatives of any order are bounded, and $u = \sum b(f_\alpha)$. If $u \in \mathscr{D}'(\omega)$ and $\bigcup \operatorname{int} \Gamma_\alpha^0 \supset \mathrm{WF}(u)$, then there exist a function $\varphi \in C^\infty(\omega)$ and functions $f_\alpha \in A(\omega + i\Gamma_\alpha)$, with $|f_\alpha(x+iy)| \leqslant C|y|^{-N}$, such that $u = \varphi + \sum b(f_\alpha)$.

4.2. The Definition of a Sheaf of Microfunctions

Definition 4.1. A sheaf of *microfunctions* $C(\mathbb{R}^n \times S^{n-1})$ is a sheaf constructed by means of the presheaf

$$U \to C(U) = B(\omega)/\{u \in B(\omega), \mathrm{WF}_A(u) \cap U = \varnothing\}$$

for $U \subset \omega \times S^{n-1}$.

To every hyperfunction $u \in B(\omega)$ there corresponds canonically an element in $C(\omega \times S^{n-1})$. This element is zero if and only if $u \in A(\omega)$.

The corresponding definitions can also be given for sheaves $\mathscr{D}'$ and C^∞ on $\mathbb{R}^n \times S^{n-1}$. Namely, to them there correspond presheaves

$$U \to \mathscr{D}'(\mathbb{R}^n)/\{u \in \mathscr{D}'(\mathbb{R}^n),\ \mathrm{WF}_A(u) \cap U = \varnothing\},$$

$$U \to C^\infty(\mathbb{R}^n)/\{u \in C^\infty(\mathbb{R}^n),\ \mathrm{WF}_A(u) \cap U = \varnothing\}.$$

A microfunction $u \in C(U)$ lies in $\mathscr{D}'(u)$ (in $C^\infty(U)$) if for each compact subset $K \subset U$ a distribution v (a function in C^∞) can be found such that $u = w$ in a neighbourhood of K, where w denotes the element in $C^\infty(\omega \times S^{n-1})$ that corresponds canonically to v.

For $u \in C(U)$, we denote by $\mathrm{WF}_A(u)$ the support of u and by $\mathrm{WF}(u)$ the complement to the largest open set V for which $u|_V \in C^\infty(V)$.

4.3. Pseudodifferential Operators. Let Ω be an open conical subset in $T^*\mathbb{R}^n$.

Definition 4.2. A sequence $\{p_j(x, \xi)\}_{j \in \mathbb{Z}}$ of functions, holomorphic in Ω, is said to be a sequence of symbols (of a pseudodifferential operator) if it satisfies the following three conditions:

1. $p_j(x, \xi)$ is homogeneous in ξ of degree j.
2. For each $\varepsilon > 0$ and for each compact subset $K \subset \Omega$ there exists a constant $C = C_{\varepsilon,K}$ such that

$$\sup_K |p_j(x, \xi)| \leqslant C\varepsilon^j/j!, \qquad j \geqslant 0.$$

3. For each compact subset $K \subset \Omega$ there exists a constant $R = R_K$ such that

$$\sup_K |p_j(x, \xi)| \leqslant R^{-j}/(-j)!, \qquad j < 0.$$

The operator $P(x, D_x) = \sum_{j \in \mathbb{Z}} p_j(x, D_x)$ is characterized by the condition

$$P(x, D_x)\delta((x, \eta) - \lambda) = \sum_{j \in \mathbb{Z}} p_j(x, \eta)\delta^{(j)}((x, \eta) - \lambda)$$

and is known as a pseudodifferential operator. If $P(x, D) = \sum_{j \leqslant m} p_j(x, D_x)$, then P is referred to as a pseudodifferential operator of finite order.

P is determined by its kernel $K(x, x - y)$ by the formula

$$P(x, D_x)u = \int K(x, x - y)u(y)\, dy.$$

The kernel $K(x, y) = P(x, D_x)\delta(x - y)$ can be defined by the formula $K(x, y) = \sum \hat{p}_j(x, y) + h(x, y)$, where $\hat{p}_j(x, y)$ is the inverse Fourier transform of $p_j(x, \xi)$ with respect to ξ. If P is an operator of finite order, it can be shown that $K(x, x - y)$ is a distribution and its analytic wave front is contained in the normal fibration of the diagonal. Hence pseudodifferential operator of finite order acts on distributions without enlarging their wave fronts.

The action of a pseudodifferential operator $P(x, D_x)$ can also be defined in another manner. Let ω be a domain in $\mathbb{R}^n$ and let U contain the set $\omega \times \{\xi, |\xi'| \leqslant k\xi_1\}$. Let $f \in A(\omega + iG)$, where G is the circular cone $\{x; x_1 > k|x'|\}$.

The values of $b(f)$ are canonically identified on $\omega \times S^{n-1}$ with a microfunction having support in $\omega \times G^0$. The microfunction $Pb(f)$ is defined in U, has a support in $\omega \times G^0$, and can be extended by zero beyond U to a microfunction defined in $\omega \times S^{n-1}$. Then there exists a function $g \in A(\omega + iG)$ such that $Pb(f) = b(g)$. The function g is defined in the following manner. Let Σ be a hyperplane in $\mathbb{C}^n$ on which $\operatorname{Im} z_1 = \varepsilon$. We define an operator P_Σ by the formula

$$P_\Sigma f(z) = \int_\Sigma K(z, z - \tilde{z}) f(\tilde{z})\, d\tilde{z}.$$

$P_\Sigma f \in A(\omega + iG)$ if $\varepsilon > 0$ is sufficiently small. We set $g = P_\Sigma f$; the microfunction $b(g)$ does not depend on the choice of ε.

The pseudodifferential operators of finite order constitute a ring sheaf $\mathscr{P}$ on $\mathbb{R}^n \times S^{n-1}$. It can be shown that such operators map $\mathscr{D}'(U)$ or $C^\infty(U)$ into itself.

If $P(x, D) = \sum p_j(x, D)$ and $Q(x, D) = \sum q_j(x, D)$ are pseudodifferential operators in a neighbourhood of a point (x_0, ξ_0), then their composition $R = P \cdot Q$ is a pseudodifferential operator and $R(x, D) = \sum r_j(x, D_x)$. Moreover,

$$r_j(x, \eta) = \sum_{k+l=|\alpha|+j} \frac{1}{\alpha!} \partial_\eta^\alpha p_k(x, \eta) D_x^\alpha q_l(x, \eta).$$

The operator $Q(x, D_x)$, formally adjoint to $P(x, D_x)$, is a pseudodifferential operator with symbol $Q(x, \eta) = \sum q_j(x, \eta)$, where

$$q_j(x, \eta) = \sum_{j+|\alpha|=k} \frac{(-1)^k}{\alpha!} \partial_\eta^\alpha D_x^\alpha p_j(x, \eta).$$

If $P(x, D_x)$ is a pseudodifferential operator of order m and if $p_m(x, \eta) \neq 0$ on a set $\Omega \subset T^*\mathbb{R}^n$, then there exists a unique psueodifferential operator $E(x, D)$ of finite order such that $PE = EP = I$.

Theorem 4.1. *Suppose that $P(x, D)$ and $Q(x, D)$ are pseudodifferential operators of order m and that their principal symbols coincide. Suppose further that the forms $dp_m(x, \eta)$ and $\eta\, dx$ are linearly independent at those points where $p_m(x, \eta) = 0$. Then we can find, locally, two invertible pseudodifferential operators $R(x, D_x)$ and $S(x, D_x)$ of finite order such that*

$$P(x, D)R(x, D) = S(x, D)Q(x, D).$$

This theorem, unlike the next one, is valid in the usual theory of pseudodifferential operators too.

Theorem 4.2. *Let $P(x, D)$ be a pseudodifferential operator of order m in a neighbourhood of the point (x_0, η_0), where $x_0 = 0$ and $\eta_0 = (0, \ldots, 0, 1)$. Let the principal symbol $p_m(x, \eta)$ of this operator be equal to η_1^m. Then the pseudodifferential equation $P(x, D)u = 0$ is microlocally equivalent to the equation $D_1^m u = 0$. In this way, one equation is transformed into the other by means of pseudodifferential operators of infinite order.*

For example, in the theory of microfunctions the equations

$$\frac{\partial^2 u}{\partial x_1^2} - \frac{\partial u}{\partial x_2} = 0, \qquad \frac{\partial^2 v}{\partial x_1^2} - i\frac{\partial v}{\partial x_2} = 0, \qquad \frac{\partial^2 w}{\partial x_1^2} = 0$$

are equivalent.

Corollary 4.1. *If $P(x, D)$ is an elliptic operator, then every solution u, a hyperfunction, of the equation $Pu = 0$ is an analytic function.*

Theorem 4.3. *Let $P(x, D)$ be a pseudodifferential operator such that the function $p_m(x, \xi)$ is real valued and is of principal type. Let $u \in B(\Omega)$ and $Pu \in A(\Omega)$. A closed set $F \subset \omega \times S^{n-1}$ can be the support of u if and only if F is a union of bicharacteristics.*

4.4. Fourier Integral Operators. Let U and V be two open sets in $\mathbb{R}^n \times S^{n-1}$. Let T be an analytic diffeomorphism from U to V that preserves the contact structure (or, in other words, a homogeneous canonical transformation of degree one of one open conical domain in $\mathbb{R}^n \times (\mathbb{R}^n \backslash 0)$ onto another). Such a T is known as a *contact* transformation. Then there exists an isomorphism (which is not unique) $\tilde{T}$ from $\mathscr{P}|_U$ onto $\mathscr{P}|_V$ (or T_1 from $C|_U$ onto $C|_V$), compatible with the structure of the ring sheaf on $\mathscr{P}$ (or with the structure of $\mathscr{P}$, the module on C), that preserves the subsheaves $\mathscr{D}'$ (or C^∞) and is also compatible with the action of T on the principal symbol.

In terms of formulae, to each contact transformation of a neighbourhood of the point (x_0, ξ_0) in $\mathbb{R}^n \times S^{n-1}$ onto the neighbourhood of (y_0, η_0) there corresponds a function $F(x, y)$ defined in the neighbourhood of (x_0, y_0) for which $dF(x_0, y_0) = (-\xi_0, \eta_0)$. Let Σ be the surface $\{(x, y); F(x, y) = 0\}$. Then $T = q \circ p^{-1}$, where

$$p\colon S^*_\Sigma\mathbb{R}^{2n} \to S^*\mathbb{R}^n; \qquad p(x, y, \xi, \eta) = (x, -\xi);$$
$$q\colon S^*_\Sigma\mathbb{R}^{2n} \to S^*\mathbb{R}^n; \qquad q(x, y, \xi, \eta) = (y, \eta).$$

If u is a microfunction in a neighbourhood of (x_0, ξ_0), then we can define the operator

$$\tilde{T}u(y) = \int K(x, y)u(x)\,dx,$$

where $K(x, y) = \delta(F(x, y))$. The kernel K is actually a distribution, and is known as the Fourier distribution associated with the phase function $F(x, y)\cdot\theta$ and $\tilde{T}$ as a microlocal Fourier integral operator. It can be shown that $\mathrm{WF}(u) \subset \mathrm{WF}_A(u) = S_\Sigma\mathbb{R}^n$. The operator $\tilde{T}$ can be examined in the space of distributions or microfunctions. $\tilde{T}$ preserves microdistributions and $\mathrm{WF}(\tilde{T}u) = T \circ \mathrm{WF}(u)$. The operator $\tilde{T}$ generates an isomorphism of the sheaf $\mathscr{P}$ under which $P \mapsto Q$, where $P\tilde{T} = \tilde{T}Q$.

References*

Arnol'd, V.I.

[1967] On the characteristic class present in the quantization condition. Funkts. Anal. Prilozh. *1*, No. 1, 1–14. English transl.: Funct. Anal. Appl. *1*, No. 1, 1–13 (1967), Zbl.175,203

[1974] Mathematical Methods in Classical Mechanics. Moscow: Nauka. English transl.: New York-Berlin-Heidelberg: Springer-Verlag 1978, Zbl.647.70001, Zbl.386.70001

Beals, R.

[1974] Spatially inhomogeneous pseudodifferential operators. II. Commun. Pure Appl. Math. *27*, 161–205, Zbl.283.35071

[1975] A general calculus of pseudodifferential operators. Duke Math. J. *42*, No. 1, 1–42, Zbl.343.35078

[1977] Characterisation of pseudodifferential operators and applications. Duke Math. J. *44*, No. 1, 45–57, Zbl.353.35088

Beals, R., Fefferman, C.

[1973] On local solvability of linear partial differential equations. Ann. Math., II. Ser. *97*, No. 3, 482–498, Zbl.256.35002

[1976] On hypoellipticity of second order operators. Commun. Partial. Differ. Equations *1*, No. 1, 73–85, Zbl.321.35025

Bolley, P.

[1977] Hypoellipticité partielle pour des opérateurs dégénérés nonfuchsiens. Commun. Partial. Differ. Equations *2*, No. 1, 1–30, Zbl.356.35016

Bolley, P., Camus, J., Helffer, B.

[1976] Sur une classe d'opérateurs partiellement hypoelliptiques. J. Math. Pures Appl., IX. Ser. *55*, 131–171, Zbl.294.35021

Bony, J.M.

[1976/77] Equivalence des diverses notions de spectre singulier analytique. Sémin. Goulaouic-Schwartz, Exp. No. 3, Zbl.367.46036

[1979/80] Propagation des singularités pour les équations aux dérivées partielles non linéaires. ibid., Exp. No. 22, Zbl.449.35006

[1981/82] Interaction des singularités pour les équations aux dérivées partielles non linéaires. ibid., Exp. No. 2, Zbl.498.35017

[1984] Propagation et interaction des singularités pour les solution des équations aux dérivées partielles non linéaires. Proc. Int. Congr. Math., Warszawa, 1983, No. *2*, 1133–1147, Zbl.573.35014

Bony, J.M., Schapira, P.

[1973] Propagation des singularités analytiques pour les solutions des équations aux dérivées partielles. Univ. Paris XI, *94*, 1–62. Appeared in Ann. Inst. Fourier *26*, No. 1, 81–140 (1976), Zbl.312.35064

Boulkhemair, A.

[1984] Opérateurs paradifférentiels et conjugaison par les opérateurs integraux de Fourier. Univ. Paris-Sud.

Boutet de Monvel, L.

[1966] Comportement d'un opérateur pseudodifferentiel sur une variété à bord. J. Anal. Math. *17*, 241–304, Zbl.161,79

[1971] Boundary problems for pseudodifferential operators. Acta Math. *126*, 11–51, Zbl.206,394

*For the convenience of the reader, references to reviews in Zentralblatt für Mathematik (Zbl.), compiled using the MATH database, have, as far as possible, been included in this bibliography.

[1974] Hypoelliptic operators with double characteristics and related pseudodifferential operators. Commun. Pure Appl. Math. *27*, 585–639, Zbl.294.35020

[1975] Propagation des singularités des solutions d'équations analogues à l'équation de Schrödinger. Lect. Notes Math. *459*, 1–14, Zbl.305.35088

[1979] Lacunas and transmissions. Ann. Math. Stud. *91*, 209–218, Zbl.445.35008

Boutet de Monvel, L., Grigis, A., Helffer, B.

[1976] Parametrixes d'opérateurs pseudodifférentiels à caracteristiques multiples. Astérisque, 34–35, 93–121, Zbl.344.32009

Boutet de Monvel, L., Trèves, F.

[1974] On a class of pseudodifferential operators with double characteristics. Invent. Math. *24*, 1–34, Zbl.281.35083

Bove, A., Lewis, J.E., Parenti, C.

[1983] Propagation of singularities for Fuchsian operators. Lect. Notes Math. *984*, Zbl.507.58002

[1985] Parametrix for a characteristic Cauchy problem. Ann. Sc. Norm. Super. Pisa, Cl. Sci., IV. Ser. *12*, No. 1, 1–42, Zbl.593.35050

Bronshtejn, M.D.

[1976] On analyticity of the solutions of equations that degenerate on a characteristic manifold. Usp. Mat. Nauk *31*, No. 3, 205–206, Zbl.334.35032

Calderon, A.P., Vaillancourt, R.

[1971] On the boundedness of pseudodifferential operators. J. Math. Soc. Japan *23*, 374–378, Zbl.214,390

Chazarain, J.

[1974a] Opérateurs hyperboliques à caracteristiques de multiplicité constante. Ann. Inst. Fourier *24*, No. 1, 173–202, Zbl.274.35045

[1974b] Propagation des singularités pour une classe d'opérateurs à caractéristiques multiples et resolubilité locale. ibid., 203–223, Zbl.274.35007

Dencker, N.

[1982a] On the propagation of polarization sets for systems of real principal type. J. Funct. Anal. *46*, No. 3, 351–372, Zbl.487.58028

[1982b] On the propagation of singularities for pseudodifferential operators of principal type. Ark. Mat. *20*, 23–60, Zbl.503.58031

Derridj, M., Zuily, C.

[1973] Régularité analytique et Gevrey pour des classes d'opérateurs elliptiques et paraboliques dégénérés du second order. Astérisque, 2–3, 371–381, Zbl.303.35027

Duistermaat, J.J.

[1973] Fourier integral operators. NY: Courant Inst. Math. Sci., Zbl.272.47028

Duistermaat, J.J., Guillemin, V.

[1975] The spectrum of positive elliptic operators and periodic bicharacteristics. Invent. Math. *29*, 39–79, Zbl.307.35071

Duistermaat, J.J., Hörmander, L.

[1972] Fourier integral operators. II. Acta Math. *128*, 183–269, Zbl.232.47055

Egorov, Yu.V.

[1969] On canonical transformations of pseudodifferential operators. Usp. Mat. Nauk *24*, No. 5, 235–236, Zbl.191,438

[1971] On solvability of differential equations with simple characteristics. ibid. *26*, No. 2, 183–198. English transl.: Russ. Math. Surv. *26*, No. 2, 113–130 (1972), Zbl.213,370

[1974] On sufficient conditions for local solvability of pseudodifferential equations of the principal type. Tr. Mosk. Mat. O.-Va *31*, 59–84. English transl.: Trans. Mosc. Math. Soc. *31*, 55–79 (1976), Zbl.309.35066

[1975] Subelliptic operators. Usp. Mat. Nauk *30*, No. 3, 57–104. English transl.: Russ. Math. Surv. *30*, No. 3, 55–105 (1975), Zbl.309.35068

[1984] Linear Differential Equation of Principal Type. Moscow: Nauka. English transl.: New York: Contemp. Sov. Math. 1986, Zbl.574.35001

[1985a] Lectures on Partial Differential Equations. Supplementary Chapters. Moscow: Izdat. Mosk. Univ., Zbl.615.35001

[1985b] On removable singularities in boundary conditions for differential equations. Vestn. Mosk. Univ., Ser. I, 1985, No. 6, 30–36. English transl.: Mosc. Univ. Math. Bull. *40*, No. 6, 35–42 (1985), Zbl.612.35003

Egorov, Yu.V., Popivanov, P.R.

[1974] On equations of principal type without solutions. Usp. Mat. Nauk *29*, No. 2, 172–189. English transl.: Russ. Math. Surv. *29*, No. 2, 176–194 (1974), Zbl.298.35056

Egorov, Yu.V., Rangelov, Ts.

[1978] On a class of pseudodifferential operators with multiple characteristics. Tr. Semin. Im. I. G. Petrovskogo *3*, 43–48. English transl.: Transl., II. Ser. Am. Math. Soc. *118*, 185–190 (1982), Zbl.458.35012

Eskin, G.

[1977] Parametrix and propagation of singularities for the interior mixed hyperbolic problem J. Anal. Math. *32*, 17–62, Zbl.375.35037

Farris, M.

[1981] Egorov's theorem on a manifold with diffractive boundary. Commun. Partial Differ. Equations *6*, 651–687, Zbl.474.58019

Fedosov, B.V.

[1974] Analytic formulae for the index of elliptic operators. Tr. Mosk. Mat. O.-Va *30*, 159–241. English transl.: Trans. Mosc. Math. Soc. *30*, 159–240 (1976), Zbl.349.58006

Fefferman, C.

[1983] The uncertainty principle. Bull. Am. Math. Soc., New Ser. *9*, No. 2, 129–206, Zbl.526.35080

Fefferman, C., Phong, D.H.

[1978] On positivity of pseudodifferential operators. Proc. Natl. Acad. Sci. USA *75*, 4673–4674, Zbl.391.35062

[1981] The uncertainty principle and sharp Gårding inequalities. Commun. Pure Appl. Math. *34*, 285–331, Zbl.458.35099

[1982] Symplectic geometry and positivity of pseudodifferential operators. Proc. Natl. Acad. Sci. USA *79*, 710–713, Zbl.553.35089

Flaschka, H., Strang, G.

[1971] The correctness of the Cauchy problem. Adv. Math. *6*, No. 3, 347–379, Zbl.213,373

Folland, G.B.

[1976] Applications of analysis on nilpotent groups to partial differential equations. Bull. Am. Math. Soc. *83*, No. 5, 912–930, Zbl.371.35008

Friedlander, F.G.

[1976] The wavefront set of a simple initial boundary value problem with glancing rays. Math. Proc. Camb. Philos. Soc. *79*, 145–159, Zbl.319.35053

Ganzha, E.I.

[1986] On hypoellipticity of a second-order differential operator with quadratic form of variable sign. Usp. Mat. Nauk *41*, No. 4, 215–216. English transl.: Russ. Math. Surv. *41*, No. 4, 175–176, Zbl.637.35024

Gårding, L.

[1971] Sharp fronts and lacunas. Actes Congr. Int. Math. 1970, *2*, 723–729, Zbl.222.35030

Gerard, C.

[1983] Réflexion du front d'onde polarisé des solutions de systèmes d'équations aux dérivées partielles. C. R. Acad. Sci., Paris, Ser. I *297*, 409–412, Zbl.555.35012

Grigis, A.

[1976] Hypoellipticité et paramétrix pour les opérateurs pseudodifférentiels à caractéristiques doubles. Astérisque 34–35, 183–205, Zbl.344.35019

[1979] Propagation des singularités pour des opérateurs pseudodifférentiels à caracteristiques doubles. Commun. Partial Differ. Equations *4*, No. 11, 1233–1262, Zbl.432.35082

Grigis, A., Sjöstrand, J.

[1985] Front d'onde analytique et sommes de carrés de champ de vecteurs. Duke Math. J. *52*, No. 1, 35–51, Zbl.581.35009

Grigis, A., Rothschild, L.P.

[1984] A criterion for analytic hypoellipticity of a class of differential operators with polynomial coefficients. Ann. Math., II. Ser. *118*, No. 3, 443–460, Zbl.541.35017

Grubb, G.

[1986] Functional calculus of pseudodifferential boundary problems. Prog. Math. *65*, Basel: Birkhäuser, Zbl. 622.35001

Grushin, V.V.

[1963] The extension of smoothness of solutions of differential equations of principal type. Dokl. Akad. Nauk SSSR *148*, No. 6, 1241–1244. English transl.: Sov. Math. Dokl. *4*, 248–252 (1963), Zbl.166,364

[1970] Pseudodifferential operators in $\mathbb{R}^n$ with bounded symbols. Funkts. Anal. Prilozh. *4*, No. 3, 37–50. English transl.: Funct. Anal. Appl. *4*, 202–212 (1971), Zbl. 223.35084

[1971a] On a class of elliptic pseudodifferential operators that degenerate on a submanifold. Math. Sb. *84*, 163–195. English transl.: Math. USSR, Sb. *13*, 155–185 (1971), Zbl.215,492

[1971b] A differential equation without solution. Mat. Zametki *10*, No. 2, 125–128. English transl.: Math. Notes *10*, 499–501 (1972), Zbl.223.35011

[1972] Hypoelliptic differential equations and pseudodifferential operators with operator-valued symbols. Mat. Sb., Nov. Ser. *88*, 504–521. English transl.: Math. USSR, Sb. *17*, 497–514 (1973), Zbl.243.35020

Grushin, V.V., Shananin, N.A.

[1977] Some theorems on singularities of solutions of differential equations with weighted principal symbols. Mat. Sb., Nov. Ser. *103*, 37–51. English transl.: Math. USSR, Sb. *32*, 32–44 (1977), Zbl.355.35015

Guillemin, V., Sternberg, S.

[1977] Geometric asymptotics. Providence, RI: Am. Math. Soc., Zbl.364.53011

Guillemin, V., Uhlmann, G.

[1981] Oscillatory integrals with singular symbols. Duke Math. J. *48*, No. 1, 251–287, Zbl.462.58030

Hanges, N.

[1979] Propagation of singularities for a class of operators with double characteristics. Inst. Adv. Stud., Princeton, 1977/78, Ann. Math. Stud. *91*, 113–126, Zbl.446.35087

Harvey, R., Polking, J.

[1970] Removable singularities of linear partial differential equations. Acta Math. *125*, 39–56, Zbl.214,100

Helffer, B.

[1976] Sur une classe d'opérateurs hypoelliptiques à caracteristiques multiples. J. Math. Pures Appl., IX. Ser. *55*, 207–215, Zbl.296.35019

[1977] Sur l'hypoellipticité des opérateurs pseudodifferentiels à caracteristiques multiples (perte de 3/2 dérivées) Bull. Soc. Math. Fr., Suppl., Mem. 51–52, 13–61, Zbl.374.35012

Helffer, B., Nourrigat, J.

[1979] Caractérisation des opérateurs hypoelliptiques homogénes in variants à gauche sur une groupe de Lie nilpotent gradué. Commun. Partial. Differ. Equations. *4*, No. 8, 899–958, Zbl.423.35040

Hörmander, L.

[1961] Hypoelliptic differential operators. Ann. Inst. Fourier *11*, 477–492, Zbl.99,301

[1963] Linear Partial Differential Operators. Berlin: Springer, Zbl.108,93

[1971] Fourier integral operator. I. Acta Math. *127*, 79–123, Zbl.212,466

[1975] A class of hypoelliptic operators with double characteristics. Math. Ann. *217*, 165–188, Zbl.306.35032

[1979a] Spectral analysis of singularities. Inst. Adv. Stud., Princeton, 1977/78, Ann. Math. Stud. *91*, 3–49. Zbl.446.47045

[1979b] Subelliptic operators. ibid., 127–208, Zbl. 446.35086

[1979c] The Weyl calculus of pseudodifferential operators. Commun. Pure Appl. Math. *32*, 359–443, Zbl.388.47032

[1983] The Analysis of Linear Partial Differential Operators. Vols. 1 and 2. Berlin-Heidelberg-New York-Tokyo: Springer-Verlag, Zbl.521.35001, Zbl.521.35002

[1985] The Analysis of Linear Partial Differential Operators. Vols. 3 and 4. Berlin-Heidelberg-New York-Tokyo: Springer-Verlag, Zbl.601.35001, Zbl.612.35001

Ivrii, V.Ya. (= Ivrij, V.Ya)

[1971] Differential equations with multiple characteristics and with no solutions. Dokl. Akad. Nauk SSR *198*, 279–282. English transl.: Sov. Math., Dokl. *12*, 769–772 (1971), Zbl.227.35032

[1979] Wave fronts of solutions of some pseudodifferential equations. Tr. Mosk. Mat. O.-Va *39*, 49–82. English transl.: Trans. Msoc. Math. Soc. 1981, No. 1, 49–86 (1981), Zbl.433.35077

[1980a] Wave fronts of solutions of boundary-value problems for a class of symmetric hyperbolic systems. Sib. Mat. Zh. *21*, No. 4, 62–71. English transl.: Sib. Math. J. 21, 527–534 (1981), Zbl.447.35055

[1980b] On the second term of the spectral asymptotic expansion of the Laplace-Beltrami operator on a manifold with boundary. Funkts. Anal. Prilozh. *14*, No. 2, 25–34. English transl.: Funct. Anal. Appl. *14*, 98–106 (1980), Zbl.435.35059

[1981] The propagation of singularities of solutions of non-classical boundary-value problems for second-order hyperbolic equations. Tr. Mosk. Mat. O.-Va *43*, 81–91. English transl.: Trans. Mosc. Math. Soc. 1983, No. 1, 87–99 (1983), Zbl. 507.35059

[1982] On precise spectral asymptotics for elliptic operators acting on vector bundles. Funkts. Anal. Prilozh. *16*, No. 2, 30–38. English transl.: Funct. Anal. Appl. *16*, 101–108 (1982), Zbl.505.35066

Kannai, Ya.

[1971] An unsolvable hypoelliptic differential equation. Isr. J. Math. *9*, 306–315, Zbl.211,406

[1973] Hypoelliptic ordinary differential operators. ibid. *13* (1972), No. 1–2, 106–134, Zbl.256.35021

Kashiwara, M., Kawai, T.

[1980] Second microlocalization and asymptotic expansions. Proc. Colloq., Les Houches 1979, Lect. Notes Phys. *126*, 21–76, Zbl.458.46027

Kashiwara, M., Schapira, P.

[1985] Microlocal study of sheaves. Astérisque *128*. Paris: Soc. Math. Fr. (1985), Zbl.589.32019

Kohn, J.J., Nirenberg, L.

[1965] An algebra of pseudodifferential operators. Commun. Pure Appl. Math. *18*, 269–305, Zbl.171,351

Kucherenko, V.V.

[1977] Asymptotics of the solution of the Cauchy problem for equations with complex characteristics. Itogi Nauki Tekh, Ser. Sovrem. Probl. Mat. *8*, 41–136. English transl.: J. Sov. Math. *13*, 24–81 (1980), Zbl.446.35091

Kumano-go, H.

[1976] A calculus of Fourier integral operators on $\mathbb{R}^n$ and the fundamental solution for an operator of hyperbolic type. Commun. Partial. Differ. Equations *1*, No. 1, 1–44, Zbl.331.42012

Landis, E.M., Olejnik, O.A.

[1974] Generalized analyticity and some related properties of solutions of elliptic anad parabolic equations. Usp. Mat. Nauk *29*, No. 2, 190–206. English transl.: Russ. Math. Surv. *29*, No. 2, 195–212 (1974), Zbl.293.35011

Lascar, B.

[1981] Propagation des singularités des solutions des équations aux derivées partielles non linéaires. Adv. Math., Suppl. Stud. *7B*, 455–482, Zbl.468.35002

Lascar, B., Lascar, R.

[1982] Propagation des singularités pour des équations hyperboliques à caracteristiques de multiplicité au plus double et singularités Masloviennes. II. J. Anal. Math. *41*, 1–38, Zbl.592.58053

Lascar, B., Sjöstrand, J.

[1982] Equation de Schrödinger et propagation des singularités pour des opérateurs pseudo-differentiels à caracteristiques réeles de multiplicité variable. Astérisque *95*, 167–207, Zbl.529.58033

[1985] Equation de Schrödinger et propagation des singularités pour des opérateurs pseudo-differentiels à caracteristiques réeles de multiplicité variable. II. Commun. Partial. Differ. Equations *10*, No. 5, 467–523, Zbl.589.58031

Lascar, R.

[1981] Propagation des singularités des solutions d'équations pseudodifferentieles à caracteristiques de multiplicités variables. Lect. Notes Math. *856*, Zbl.482.58031

Laurent, Y.

[1985] Théorie de la deuxième microlocalisation dans le domaine complex. Prog. Math. *53*, Boston: Birkhäuser, Zbl.561.32013

Lions, J.L., Magenes, E.

[1968] Problèmes aux limits non homogénes et applications, Vol.s 1, 2. Paris: Dunod, Zbl.165,108

Lychagin, V.V.

[1975] Local classification of non-linear first order partial differential equations. Usp. Mat. Nauk *30*, No. 1, 101–171. English transl.: Russ. Math. Surv. *30*, No. 1, 105–175 (1975), Zbl.308.35018

Martineau, A.

[1961] Les hyperfunctions de M. Sato. Sémin. Bourbaki 13 (1960/61), Exp. No. 214, 13p., Zbl.122,349

Maslov, V.P.

[1965a] The Theory of Perturbation and Asymptotic Methods. Moscow: Izdat. Moskov. Gos. Univ. French transl.: Paris: Dunod 1972, Zbl.247.47010

[1965b] The WKB method in the multi-dimensional case. Supplement to the book: Hoding, J.: Introduction to the Method of Phase Integrals. Moscow: Mir, 177–237

[1973] Operator Methods. Moscow: Nauka. English transl.: Moscow: Mir 1982, Zbl.288.47042

Maslov, V.P., Fedoryuk, M.V.

[1976] Quasi-Classical Approximation for Equations of Quantum Mechanics. Moscow: Nauka. English transl.: Dordrecht-Boston-London: D. Reidel Publ. Co. 1981, Zbl.449.58002

Melin, A.

[1971] Lower bounds for pseudodifferential operators. Ark. Math. *9*, 117–140, Zbl.211,171

Melin, A., Sjöstrand, J.

[1976] Fourier integral operators with complex phase functions and parametrix for an interior boundary value problem. Commun. Partial Differ. Equations *1*, No. 3, 313–400, Zbl.364.35049

Melrose, R.

[1976] Equivalence of glancing hypersurfaces. Invent. Math. *37*, 165–191, Zbl.354.53033

[1979] Differential boundary value problems of principal type. Inst. Adv. Stud., Princeton, 1977/78, Ann. Math. Stud. *91*, 81–112, Zbl.447.35053

[1981] Transformation of boundary problems. Acta Math. *147*, 149–236, Zbl.492.58023

Melrose, R., Sjöstrand, J.

[1978] Singularities of boundary value problems. I. Commun. Pure Appl. Math. *31*, 593–617, Zbl.368.35020

[1982] Singularities of boundary value problems. II. ibid. *35*, 129–168, Zbl.546.35083

Menikoff, A.

[1977a] On hypoelliptic operators with double characteristics. Ann. Sc. Norm. Super. Pisa, Cl. Sci. IV. Ser. *4*, 689–724, Zbl.376.35013

[1977b] Parametrices for subelliptic operators. Commun. Partial Differ. Equations *2*, 69–108, Zbl.354.35080

[1979] Hypoelliptic operators with double characteristics. Inst. Adv. Stud., Princeton 1977/78 Ann. Math. Stud. *91*, 65–79, Zbl.448.35029

Metivier, G.

[1976] Fonction spectrale et valeurs propres d'une class d'opérateurs non-elliptiques. Commun. Partial Differ Equations *1*, No. 4, 467–519, Zbl.376.35012

[1980] Hypoellipticité analytique sur des groupes nilpotents de rang 2. Duke Math. J. *47*, 195–221, Zbl.433.35015

Mishchenko, A.S., Sternin, B.Yu., Shatalov, V.E.

[1978] Lagrange Manifolds and the Method of Canonical Operator. Moscow: Nauka. English transl: Berlin-Heidelberg-New York: Springer-Verlag (1990), Zbl.727.58001

Nazajkinskij, V.E., Oshmyan, V.G., Sternin, B.Yu., Shatalov, N.E.

[1981] Fourier integral operators and the canonical operator. Usp. Mat. Nauk *36*, No. 2, 81–140. English transl.: Russ Math. Surv. *36*, No. 2, 93–161 (1981), Zbl.507.58043

Nirenberg, L.

[1973] Lectures on linear partial differential equations. Am. Math. Soc., Reg. Conf. Ser. Math. *17*, 1–58, Zbl.267.35001

Nirenberg, L., Trèves, F.

[1970] On local solvability of linear partial differential equations. I. Necessary conditions. II. Sufficient conditions. Commun. Pure Appl. Math. *23*, I. 1–38. II, 459–509, Zbl.191,391, Zbl.208,359

Olejnik,O.A., Radkevich, E.V.

[1971] Second-order equations with non-negative characteristic form. Itogi Nauki, Ser. Mat., Mat. Anal. 1969, Zbl.217,415

[1972] Analyticity of solutions of linear differential equations and systems. Dokl. Akad. Nauk SSSR *207*, No. 4, 785–788. English transl.: Sov. Math., Dokl. *13* (1972), 1614–1618 (1973), Zbl.266.35001

[1975] On analyticity of solutions of second-order linear partial differential equations. Tr. Semin. I. G. Petrovskogo *1*, 163–173. English transl.: Transl., II. Ser., Am. Math. Soc. *118*, 13–23 (1982), Zbl.354.35015

Parenti, C., Rodino, L.

[1980] Examples of hypoelliptic operators which are not microhypoelliptic. Boll. Unione Mat. Ital. IV. Ser B *17*, 390–409, Zbl.407.35074

Parenti, C., Segala, F.

[1981] Propagation and reflection of singularities for a class of evolution equations. Commun. Partial Differ. Equations *6*, 741–782, Zbl.496.35082

Popivanov, P.R.

[1975] On local solvability of a class of pseudodifferential equations with double characteristics. Tr. Semin. I. G. Petrovskogo *1*, 237–278. English transl.: Transl., II. Ser., Am. Math. Soc. *118*, 51–90 (1982), Zbl.323.35077

[1976] Local solvability of pseudodifferential equations with characteristics of multiplicity 2. Mat. Sb., Nov. Ser. *100*, 217–241. English transl.: Math. USSR, Sb. *29* (1976), 193–216 (1978), Zbl.376.35054

Popivanov, P.R., Iordanov, I.V.

[1983] Paradifferential Operators and Propagation of Singularities for Non-Linear Partial Differential Equations. Berlin: Akad. Wiss. der DDR, Zbl.518.35114

Rempel, S., Schulze, B.-W.

[1982a] Index Theory of Elliptic Boundary Problems. Berlin: Akademie Verlag, Zbl.504.35002

[1982b] Parametrices and boundary symbolic calculus for elliptic boundary problems without the transmission property. Math. Nachr. *105*, 45–149, Zbl.544.35095

Rothschild, L.P., Stein, E.M.

[1976] Hypoelliptic differential operators and nilpotent groups. Acta Math., *137*, 247–320, Zbl.346.35030

[1979] A criterion for hypoellipticity of operators constructed from vector fields. Commun. Partial Differ. Equations, *4*, 645–699, Zbl.459.35025

Sato, M.
[1970a] Hyperfunctions and partial differential equations. Proc. Int. Conf. Funct. Anal. Related Topics, Tokyo 1969, 91–94, Zbl.208,358
[1970b] Regularity of hyperfunction solutions of partial differential equations. Actes Congr. Int. Math. 1970, *2*, 785–794, Zbl.232.35004
Sato, M., Kawai, T., Kashiwara, M.
[1973] Microfunctions and pseudodifferential equations. Lect. Notes Math. *287*, 263–529, Zbl.277.46039
Schapira, P.
[1971] Hyperfonctions et problèmes aux limites elliptiques. Bull. Soc. Math. Fr. *99*, 113–141, Zbl.209,129
Schwartz, L.
[1950] Théorie des distributions. Paris: Hermann, Zbl.37,73
Seeley, R.T.
[1967] Complex powers of an elliptic operator. Proc. Symp. Pure Math. *10*, 288–307, Zbl.159,155
[1969] The resolvent of an elliptic boundary problem. Am. J. Math. *91*, 889–920, Zbl.191,118
Shananin, N.A.
[1982] On local non-solvability and non-hypoellipticity of (pseudo-)differential equations with weighted symbols. Mat. Sb., Nov. Ser. *119*, No. 4, 548–563. English transl.: Math. USSR, Sb. *47*, 541–556 (1984), Zbl.514.35089
Shubin, M.A.
[1978] Pseudodifferential Operators and Spectral Theory. Moscow: Nauka. English transl.: Berlin-Heidelberg-New York: Springer-Verlag (1987), Zbl.451.47064
Sjöstrand, J.
[1972] Operators of principal type with interior boundary conditions. Acta Math. *130*, 1–51 (1973), Zbl.253.35076
[1976] Propagation of singularities for operator with multiple involutive characteristics. Ann. Inst. Fourier *26*, No. 1, 141–155, Zbl.313.58021
[1979] Fourier integral operators with complex phase functions. Inst. Adv. Stud., Princeton 1977/78, Ann. Math. Stud. *91*, 51–64, Zbl.446.47046
[1982] Singularités analytiques microlocales. Astérisque *95*, 1–166, Zbl.524.35007
Sternin, B.Yu.
[1984] Differential equations of subprincipal type. Mat. Sb., Nov. Ser. *125*, No. 1, 38–69. English transl.: Math. USSR, Sb. *53*, 37–67 (1986), Zbl.587.58046
Tartakoff, D.S.
[1973] Gevrey hypoellipticity for subelliptic boundary-value problems. Commun. Pure Appl. Math. *26*, 251–312, Zbl.265.35023
[1978] Local analytic hypoellipticity for $\Box_b$ on non-degenerate Cauchy-Riemann manifolds. Proc. Natl. Acad. Sci USA, *75*, 3027–3028, Zbl.384.35020
Taylor, M.
[1979] Fourier integral operators and harmonic analysis on compact manifolds. Proc. Symp. Pure Math. *35*, 115–136, Zbl.429.35055
[1981] Pseudodifferential Operators. Princeton, NJ: Univ. Press, Zbl.453.47026
[1984] Noncommutative microlocal analysis. I. Mem. Am. Math. Soc. *313*, Zbl.554.35025
Trèves, F.
[1961] An invariant criterion of hypoellipticity. Am. J. Math. *83*, 645–668, Zbl.114,52
[1982] Introduction to Pseudodifferential and Fourier Integral Operators. New York-London: Plenum Press, Zbl.453.47027
Tulovskij, V.N.
[1979] The propagation of singularities for operators with characteristics of constant multiplicity. Tr. Mosk. Mat. O.-Va *39*, 113–134. English transl.: Trans. Mosc. Math. Soc. 1981, No. 1, 121–144 (1981), Zbl.474.47025

Uhlmann, G.
[1977] Pseudodifferential operators with involutive double characteristics. Commun. Partial Differ. Equations *2*, 713–779, Zbl.389.35047

Unterberger, A., Bokobza, J.
[1964] Les opérateurs de Calderon-Zygmund précisés. C. R. Acad. Sci., Paris *259*, 1612–1614, Zbl.143,369

Vasil'ev, D.G.
[1984] Two-term asymptotics of the spectrum of a boundary-value problem under interior reflection of general form. Funkts. Anal. Prilozh. *18*, No. 4, 1–13. English transl.: Funct. Anal. Appl. *18*, 267–277 (1984), Zbl.574.35032
[1986] The asymptotics of the spectrum of a boundary-value problem. Tr. Mosk. Mat. O.-Va *49*, 167–237. English transl.: Trans. Mosc. Math. Soc. 1987, 173–245 (1987), Zbl.623.58024

Vishik, M.I., Grushin, V.V.
[1969] On a class of degenerate elliptic equations of higher orders. Mat. Sb., Nov. Ser. *79*, No. 1, 3–36. English transl.: Math.USSR, Sb. *8*, 1–32 (1969), Zbl.177,143

Wasow, W.
[1965] Asymptotic Expansions for Ordinary Differential Equations. Pure Appl. Sc. *14*, New York-London-Sydney: Interscience Publishers, Zbl.133,353

Wenston, P.
[1975] On local solvability of linear partial differential equations not of principal type. Bull. Am. Math. Soc. *81*, No. 1, 215–218, Zbl.315.35028
[1977] A necessary condition for the local solvability of $P_m^2(x, D) + P_{2m-1}(x, D)$. J. Differ. Equations *25*, No. 1, 90–95, Zbl.356.35011

II. Linear Hyperbolic Equations

V.Ya. Ivrii

Translated from the Russian
by P.C. Sinha

Contents

Introduction

The present survey is devoted to the linear hyperbolic equations and systems. The concept of a hyperbolic equation first appeared in the case of a second-order equation

$$Pu = \sum_{i,j=0}^{n} a_{ij}\partial_i\partial_j u = 0, \tag{0.1}$$

with constant coefficients. It implied that the quadratic form $a(\xi) = \sum_{i,j=1}^{n} a_{ij}\xi_i\xi_j$ is hyperbolic, that is, its positive subspaces are of dimension 1 and the negative subspaces of dimension $n - 1$ (or vice-versa). But now such equations are referred to as strictly hyperbolic; a hyperbolic equation may have its quadratic form $a(\cdot)$ degenerate (parabolically degenerate, in old terminology). In this case, the equation (0.1) has solutions of the type of moving (plane) waves $u = v(\langle x, \xi \rangle)$, where $\langle x, \xi \rangle = x_0\xi_0 + \cdots + x_n\xi_n$, $a(\xi) = 0$ and v is an arbitrary function of a single variable. If $n = 1$, the general solution of (0.1) is a linear combination of two moving waves, and on the basis of this fact we can easily solve the Cauchy problem with data prescribed on any non-characteristic curve (that is, on a curve γ such that $a(N) \neq 0$, where N is the normal to γ), and even suitable mixed problems. Moreover, the Cauchy problem is uniquely solvable and there exists a triangle of dependence; the same is true for non-homogeneous equations. When lower terms are present, the Cauchy problem and the mixed problem can be solved, in principle at least, by the method of successive approximation, just as it is done in the case of the Cauchy problem for ordinary differential equations. Similar construction is available for the case of variable coefficients also, and again the non-characteristic Cauchy problem is uniquely solvable and there is a characteristic triangle of dependence (now curvilinear). These statements do not hold for equations of other types. Therefore, apart from the algebraic definition of hyperbolicity, it is possible to give a meaningful and close analytical definition: a given equation (system) is hyperbolic if some non-characteristic Cauchy problem for it is uniquely solvable for any smooth right-hand sides and initial data and if there exists a cone of dependence. This fact was already recognised in the last century, and from that time the algebraic definition of hyperbolicity was extended to systems of first order, to equations and systems of higher orders and with a large number of independent variables in such a way that it remained meaningfully close to the analytical definition.

By the above method of characteristics, a comparitively simple analysis can be given for the first-order systems

$$\sum_{i=0}^{n} A_i\partial_i u + Bu = f \tag{0.2}$$

in two independent variables, where $n = 1$, A_i and B are $d \times d$ matrices, and u and f are column vectors of size d. The method of characteristics works (possibly in conjuction with the method of successive approximations) if A_0 is a non-degenerate matrix, all eigenvalues of the matrix $C = A_0^{-1}A_1$ are real and if there exists a complete set of eigenvectors; when C is a variable matrix, these eigenvectors are required to depend smoothly on x. All these conditions are fulfilled when (0.2) is strictly hyperbolic, that is, when A_0 is a non-degenerate matrix and all the eigenvalues of C are real and distinct. We note that if C is a constant matrix and all it eigenvalues are real but there is no complete set of eigenvectors, the modified method of characteristics still works provided some additional conditions are imposed on the lower terms, that is, on B.

The next important class is that of equations of second order where the number of variables is greater than 2, and, in particular, the wave equation

$$\Box u \equiv (\partial_0^2 - \Delta)u = f, \qquad \Delta = \sum_{i=1}^{n} \partial_i^2. \tag{0.3}$$

The method of characteristics is no longer applicable to these equations, although under certain assumptions the solution can be found in the form of the integral of moving waves; by the Kirchoff's formula in the particular case of (0.3). But the method of energy integral works: (0.3) implies the energy identity

$$E(u, t) = E(u, t_0) + \int_{t_0}^{t} \int_{\mathbb{R}^n} 2 \operatorname{Re} \partial_0 \bar{u} \cdot f \, dx' \, dx_0, \tag{0.4}$$

where

$$E(u, t) = \int_{\mathbb{R}^n} \sum_{j=0}^{n} |\partial_j^2 u(t, x')|^2 \, dx',$$

$x' = (x_1, \dots, x_n)$, $dx' = dx_1 \dots dx_n$. From (0.4) there follows the energy inequality

$$E(u, t) \leqslant C\left[E(u, t_0) + \int_{t_0}^{t} \int_{\mathbb{R}^n} |f|^2 \, dx' \, dx_0\right] (t > t_0). \tag{0.5}$$

The unique solvability of the Cauchy problem can be established by means of (0.5) and analytic-functional arguments. For the general equation (0.1) the identity and inequality analogous to (0.4) and (0.5), respectively, hold under the assumption that (0.1) is hyperbolic and the hyperplanes $S_t = \{x_0 = t\}$ are spatially similar, that is, $a(x, N) > 0$, where $N = (1, 0, \dots, 0)$ is the normal to S_t. We have assumed here that the positive subspaces of the quadratic form $a(x, \cdot)$ are of dimension 1 and that for $|x'| > R$ the coefficients of the equation are constant or else some other conditions are fulfilled at infinity. Since any surface S can be reduced to the form $\{x_0 = 0\}$ by a change of coordinates, the Cauchy problem with data prescribed on a spatially similar surface is uniquely solvable. At the same time, it can be demonstrated that the Cauchy problem with data prescribed on a non-characteristic and non-spatially similar surface is not uniquely solvable. Thus for $n \geqslant 2$ not every non-characteristic Cauchy problem for a hyperbolic equation is uniquely solvable. Therefore the term "hyperbolicity" must be replaced by a more precise term "x_0-hyperbolicity" or "hyperbolicity in the direction N".

Petrovskij was the first to introduce the concept of a hyperbolic system of arbitrary order m. Namely, the system

$$Pu \equiv \sum_{|\alpha| \leqslant m} a_\alpha(x) D^\alpha u = f,$$

where a_α are $d \times d$ matrices, is called x_0-hyperbolic if the matrix $a_{(m,0,\ldots,0)}(x)$ is invertible and the characteristic equation

$$g(x, \xi) \equiv \det p(x, \xi) = 0 \quad \left(p(x, \xi) = \sum_{|\alpha| = m} a_\alpha(x) \xi^\alpha \right)$$

has only real roots ξ_0 for all $x, \xi' \in \mathbb{R}^n$. If, moreover, these roots are simple when $\xi' \neq 0$, the system is called strictly x_0-hyperbolic or, what has now become customary, Petrovskij hyperbolic. In a similar fashion, the Leray-Volevich hyperbolic and strictly hyperbolic systems can be defined (Leray [1954]).

Petrovskij established that the Cauchy problem for a strictly hyperbolic system is always well posed, that is, has always a unique solution. On the other hand, Lax and Mizohata proved that a non-characteristic and non-hyperbolic Cauchy problem is always ill posed. Thus the question whether the Cauchy problem for non-strictly hyperbolic equations and systems is well posed remains open. Although there is a vast number of works devoted to this question, it has not been conclusively resolved. In many cases, for the Cauchy problem to be well posed, conditions are required to be imposed on the lower terms. However, there are non-strictly hyperbolic equations and systems for which the Cauchy problem is well posed, no matter what the lower terms are; such equations and systems are referred to as regularly hyperbolic. At present, necessary and sufficient conditions are known for regular hyperbolicity of equations (but not systems) in an open domain.

By the well-posedness we meant hitherto the C^∞-well-posedness, that is, we meant the unique solvability for any right-hand sides and for any initial data belonging to C^∞. If C^∞ is replaced by a narrower space, the Gevrey class $G^{\{\varkappa\}}$, $1 < \varkappa < \infty$, it results in a weaker $G^{\{\varkappa\}}$-well-posedness. However, even for the $G^{\{\varkappa\}}$-well-posedness of the non-characteristic Cauchy problem, the system has to be necessarily hyperbolic. We remark that the condition imposed on the lower terms can be relaxed as $\varkappa$ becomes smaller; for $\varkappa < r/(r-1)$, where r is the maximal multiplicity of the roots of the characteristic equation, the hyperbolic Cauchy problem is always $G^{\{\varkappa\}}$-well-posed.

Since hyperbolicity is closely linked with well-posedness of the non-characteristic Cauchy problem, we shall devote a large part of our survey to the well-posedness question of the Cauchy problem.

Another topic of our discussion is the well-posedness of the mixed problem. This question is much more complicated because there are various types of lateral boundary (non-characteristic, (non) uniformly characteristic, characteristic only at separate points) and various types of boundary conditions, and that is why this question is not as thoroughly investigated. Thus no conclusive answer is available so far to the question as to under what conditions on boundary operators is the non-characteristic mixed problem for strictly hyperbolic equation of second order C^∞-well posed.

The third question which we deal with is closely connected with the preceding two. This is the question of the propagation of singularities (of wave fronts) of solutions of hyperbolic equations and systems as well as of the Cauchy problem and mixed problems for them. Although this problem is of comparatively recent origin, numerous results have already been achieved.

Because of the limitation of space as well as of the resources of the present author, many questions, closely related with the above three, remain untouched. Namely, characteristic Cauchy problem for hyperbolic equations and uniformization, hyperbolic equations where the coefficients of D_0^m degenerate at certain points (at these points the equation may not have hyperbolicity directions), the construction of oscillating solutions of hyperbolic equations and systems as well as of the Cauchy problem and mixed problems for them, the construction of parametrices of the Cauchy problem and mixed problems for hyperbolic equations and systems, the Maslov canonical operator and Fourier integral operators, the theory of scattering and related questions, the application of hyperbolic problems to obtaining precise spectral asymptotics, equations and systems obtained from hyperbolic equations and systems on replacing D_0 by a real or a complex parameter, non-linear equations and systems and the propagation of strong and weak breaks of their solutions, etc. Many of these questions deserve a separate survey of the same length as the present one, and the author hopes that they will be dealt with in other volumes published under the present series.

To conclude, we remark that the standard definition of hyperbolicity adopted in this survey is not the only possible one. First, non-hyperbolic equations such as $(\partial_0^2 + \Delta)u = 0$ may turn out to be abstractly-hyperbolic if they are examined in suitable function spaces of variables $x' \in \omega$ that are non-dense in $L_2(\omega)$. Second, non-standard hyperbolicity and non-standard characteristics arise when h – (pseudo) differential operators are studied. Finally, some equations, such as the Schrödinger equation $(i\partial_0 + x_1^2 - \partial_1^2)u = 0$, that are global with respect to the space variables can be taken to be hyperbolic if we examine only functions belonging to $\mathscr{S}'(\mathbb{R}^n)$ with respect to x' and give equal status to the variables x' and D' in the phase space, and if by the wave front we mean a cone in $\mathbb{R}^{2n}\backslash 0$ that shows in which direction the function g fails to be smooth and in which direction it does not decay fast enough at infinity.

The author sincerely thanks M.S. Agranovich for his well-wishing and constructive criticism.

Chapter 1
The Cauchy Problem

§ 1. C^∞-well-posedness of the Cauchy Problem

1.1. Basic Definitions and Notation. We shall use the following notation:

$$x = (x_0, x') = (x_0, x_1, \ldots, x_n),$$

$$\xi = (\xi_0, \xi') = (\xi_0, \xi_1, \ldots, \xi_n),$$

$$D = (D_0, D') = (D_0, D_1, \ldots, D_n), \qquad D_j = -i\frac{\partial}{\partial x_j}.$$

For $M \subset \mathbb{R}^{n+1}$, we denote by $\overline{M}$ its closure and by $\mathring{M}$ its interior, and set

$$M_t^\pm = \{x \in M, \pm(x_0 - t) \geqslant 0\},$$

$$M_t = \{x \in M, x_0 = t\}.$$

Let Ω be an open domain in $\mathbb{R}^{n+1}$, and let $X = \Omega \cap \{T_- \leqslant x_0 \leqslant T_+\}$, where $T_- < T_+$. We put $S_\pm = \Omega_{T_\pm}$. The necessity of considering such domains arises due to operators that are non-hyperbolic above the final hyperplane and below the initial hyperplane (the Tricomi operator is one such operator).

For a $d \times d$ matrix operator

$$P = \sum_{|\alpha| \leqslant m} a_\alpha(x) D^\alpha, \qquad a_\alpha \in C^\infty(\Omega),$$

we set

$$P_l(x, \xi) = \sum_{|\alpha| = l} a_\alpha(x) \xi^\alpha.$$

We call $p(x, \xi) = P_m(x, \xi)$ the *principal symbol* of P and $g = \det p$ the *characteristic symbol* of P.

We examine the Cauchy problem

$$Pu = f \qquad \text{in } X_T^+ \ (T \in [T_-, T_+)), \tag{1.1}$$

$$D_0^j u|_{X_T} = g_j \qquad (j = 0, \ldots, m-1). \tag{1.2}$$

Definition 1. The Cauchy problem (1.1)–(1.2) is said to be C^∞-*well posed* if the following conditions are fulfilled:

a) For any $f \in C^\infty(X_T^+)$ and any $g_j \in C^\infty(X_T)$ $(j = 0, \ldots, m-1)$ there exists a function $u \in C^\infty(X_T^+)$ satisfying (1.1)–(1.2).

b) From the fact that $u \in C^m(X_T^+)$ satisfies (1.1)–(1.2) and that $g_0 = \cdots = g_{m-1} = 0$, $f = 0$ in X_t^- for all $t > T$, it follows that $u = 0$ in X_t^-.

Remark. In a number of works, for instance in Ivrii and Petkov [1974], a formally weaker definition of C^∞-well-posedness is given. This makes sense in establishing necessary conditions for well-posedness. For example, all our results remain valid if we restrict ourselves in condition a) above to the zero initial data and to right-hand sides $f \in C_0^\infty(\Omega_T^+)$ and require that $u \in C^m(X_T^+)$.

Definition 2. Let $\Gamma = \Gamma(\hat{x})$ be an open convex cone in $\mathbb{R}^{n+1}$, with vertex $\hat{x} \in X_T^+ \setminus X_T$, such that $\Gamma \cap X_T^+ \Subset X_T^+$. We shall say that the uniqueness condition (U_Γ) holds if the fact that $u \in C^m(X_T^+)$ is a solution of (1.1)–(1.2), with $g_j|X_{T \cap \Gamma} = 0$ $(j = 0, \ldots, m-1)$ and $f = 0$ in Γ, implies that $u = 0$ in Γ.

The condition (U_Γ) expresses the fact that the perturbation is propagated with a finite speed, and this condition enables us not to stipulate beforehand that the Cauchy problem is non-characteristic.

Definition 3. The operator P is said to be *hyperbolic at a point x in the direction N* if

a) $g(x, N) \neq 0$,

and all the roots τ of the *characteritic equation*

b) $g(x, \xi + \tau N) = 0$

are real for all $\xi \in \mathbb{R}^{n+1}$. Moreover, if for any ξ, non-collinear with N, these roots are simple, then P is said to be *strictly hyperbolic at x in the direction N*.

In particular, P is known as *x_0-hyperbolic* (*strictly x_0-hyperbolic*) if it is hyperbolic (strictly hyperbolic) in the direction $N = dx_0 = (1, 0, \ldots, 0)$, that is, if the conditions in Definition 3 are satisfied and all the roots $\xi_0 = \lambda(x, \xi')$ of the characteristic equation

$$g(x, \xi_0, \xi') = 0 \tag{1.3}$$

are real (real and simple) for any $\xi' \in \mathbb{R}^n \setminus 0$.

1.2. Necessity of Hyperbolicity. L_2-well-posedness of the Cauchy Problem

Theorem 1 (Lax [1957], Mizohata [1961]; see also Ivrii and Petkov [1974]). *If the hyperplanes $\{x_0 = t\}$ are non-characteristic and the Cauchy problem* (1.1)–(1.2) *is C^∞-well posed, then the operator P is x_0-hyperbolic in X_T^+.*

Theorem 2 (Mizohata [1961]; see also Ivrii and Petkov [1974]). *Let $\hat{x} \in X_T^+ \setminus X_T$ and $g(\hat{x}, \cdot) \not\equiv 0$. If the Cauchy problem* (1.1)–(1.2) *satisfies the conditions* (1.4) *and* (U_Γ), *where $\Gamma = \Gamma(\hat{x})$, then P is hyperbolic at $\hat{x}$ in any direction N that is such that $\bar{\Gamma} \cap \{\langle x - \hat{x}, N \rangle = 0\} = \{\hat{x}\}$.*

The proofs of Theorems 1 and 2, like those of other necessary conditions for well-posedness, rest on constructing an asymptotic solution of the problem in question. This solution, constructed by assuming the invalidity of the desired condition, violates the *a priori* estimates that follow from the fact that the problem is well posed.

On the other hand, if P is strictly x_0-hyperbolic, the Cauchy problem (1.1)–(1.2) is C^∞-well posed in suitable domains.

Definition 4. Let $\mathbb{R}^{n+1} \Supset X$. Let the coefficients of P be infinitely smooth in a neighbourhood of $\bar{X}$, $\partial\Omega \in C^\infty$ and, finally, let P *be* x_0-hyperbolic in $\bar{X}$. If $g(x, v(x) + \tau N) \neq 0$ for any $x \in \sigma\Omega \cap \{T_- \leqslant x_0 \leqslant T_+\}$ and $\tau \geqslant 0$, where $N = dx_0$ and $v(x)$ is the outward normal to $\partial\Omega$ at x, then X is called a *lens of spatial type.*

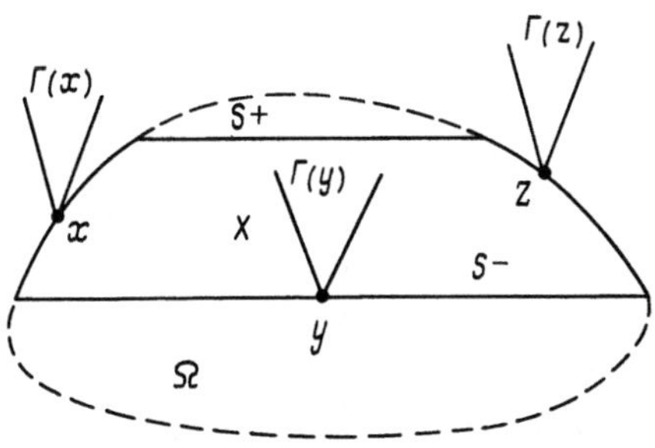

Fig. 1

For the wave equation, typical spatial type lens is shown in Fig. 1. Here the x_0-axis is directed upwards, $\Gamma(\bar{x}) = \{x, x_0 - \bar{x}_0 \geqslant |x' - \bar{x}'|\}$ is the light cone, directed upwards, with vertex $\bar{x}$, and the boundary of the domain Ω where it does not intersect the boundary of X is depicted by dotted lines. The spatial type lenses for the general hyperbolic operators look similar.

Theorem 3 (Petrovskij [1938], Leray [1954], Gårding [1957]). *Let P be a strictly x_0-hyperbolic operator in $X \Subset \mathbb{R}^{n+1}$, where X is a spatial type lens. Then the Cauchy problem* (1.1)–(1.2) *is C^∞-well posed for any $T \in [T_-, T_+)$. Moreover, for all $u \in C^\infty(\bar{X})$, $t > T$ and $k \geqslant l \geqslant 0$, the following Petrovskij* ***energy estimates*** *hold*:

$$\|u; t\|_{k+m-1,l+m-1} \leqslant C_{kl}\left(\|u; T\|_{k+m-1,l+m-1} + \int_T^t \|Pu; t'\|_{k,l}\, dt'\right), \tag{1.4}$$

where

$$\|u; t\|_{k,l} = \sum_{|\alpha| \leqslant k, \alpha_0 \leqslant l} \|D^\alpha u; L_2(X_t)\|.$$

The estimates (1.4) are established, for example, by applying the method of the alternating operator (Gårding [1957], Leray [1954]). After this Theorem 3 is proved by means of analytic-functional arguments. An operator Q is said to be a divisor operator if its order is 1 less than that of P and the characteristic roots of Q alternate with the characteristic roots of P[1]. The proofs, based on the construction of a symmetrizer, were suggested by Calderon, Mizohata and Yamaguti, although the main idea in an implicit form can be traced back to the classical work of Petrovskij.

Definition 1.5. The Cauchy problem (1.1)–(1.2) is said to be L_2-*well posed* if it is C^∞-well posed and the estimates (1.4) hold.

We mention the necessary conditions for the problem to be L_2-well posed.

[1] Here we are discussing scalar operators.

Theorem 4 (Ivrii and Petkov [1974]). (i) *If the estimate* (1.4) *holds for some k and l, then P is x_0-hyperbolic and the inequality*

$$\|p^{-1}(x,\xi)\| \leqslant C(-\operatorname{Im}\xi_0)^{-1}|\xi|^{1-m}$$
$$\forall x \in X, \qquad \zeta \in (\mathbb{C}_-\backslash\mathbb{R}) \times \mathbb{R}^n, \tag{1.5}$$

where $\mathbb{C}_\pm = \{z \in \mathbb{C}, \pm\operatorname{Im} z \geqslant 0\}$, holds.

(ii) *In particular, if P is a scalar operator, it is strictly x_0-hyperbolic.*

(iii) *If $m = 1, P = ID_0 - A(x, D')$ and $p(x,\xi) = I\xi_0 - a(x,\xi')$, then the matrix $a(x,\xi')$ is* ***uniformly diagonalizable****. That is, there exists a matrix $q(x,\xi')$, uniformly bounded on $X \times (\mathbb{R}^n\backslash 0)$ together with its inverse, such that $q^{-1}aq$ is a diagonal matrix.*

Thus in the case of scalar operators, strict hyperbolicity is a necessary and sufficient condition for the L_2-well-posedness. For matrix operators, the situation is more complicated, and this case has not been thoroughly investigated. If $p = p(\xi)$ (that is, if the principal part of P has constant coefficients), the inequality (1.5) is equivalent to L_2-well-posedness. If, however, $m = 1$ and the matrix $a(x,\xi')$ is *smoothly diagonalizable*, that is, if both the diagonalizing matrix $q(x,\xi')$ and its inverse are infinitely smooth in a neighbourhood of $\overline{X} \times (\mathbb{R}^n\backslash 0)$, then the Cauchy problem will be L_2-well posed. What is more, when $m = 1$, for the L_2-well posedness it is sufficient that the matrix a is *symmetrizable*, that is, there exists a Hermitian matrix $S(x,\xi')$ that is infinitely smooth and positive definite in the neighbourhood of $\overline{X} \times (\mathbb{R}^n\backslash 0)$ and is such that Sa is Hermitian (Friedrichs [1958]). However, as an example due to Strang [1967] shows, uniform diagonalization of a does not, in general, imply L_2-well-posedness.

There arise two problems that have attracted a large number of works.

Problem A. Obtain conditions on the *principal symbol of the operator* under which the Cauchy problem is C^∞-well posed for all lower terms (such operators are known as *regularly* or *strongly hyperbolic*).

Problem B. If P is not a *regularly hyperbolic* operator, find conditions on its lower terms under which the Cauchy problem is C^∞-well posed.

1.3. Operators with Constant Coefficients. Algebraic Properties of Hyperbolic Polynomials. Thanks to the Fourier transformation, a complete study can be made of the Cauchy problem for operators with constant coefficients.

Definition 6. An operator $P(D)$ with constant coefficients is said to be *Gårding hyperbolic in a direction N* if $g(N) \neq 0$ and there exists a C such that, for any $\xi \in \mathbb{R}^{n+1}$, all the roots τ of the polynomial $G(\xi + \tau N)$, where $G(\zeta) = \det P(\zeta)$, lie in the strip $|\operatorname{Im}\tau| \leqslant C$. In particular, $P(D)$ is said to be *Gårding x_0-hyperbolic* if it is hyperbolic in this sense in the direction $(1, 0, \dots, 0)$.

Theorem 5 (Gårding [1951]; see also Atiyah, Bott and Gårding [1970, 1973], and Hörmander [1963]). *In order that the non-characteristic Cauchy problem* (1.1)–(1.2) *for the operator $P(D)$, with constant coeffcients, be C^∞-well posed it is necessary* (*and sufficient if X is a spatial type lens*) *that $P(D)$ is Gårding x_0-hyperbolic.*

Proposition 1. *If $P(D)$ is Gårding hyperbolic in the direction N, it is hyperbolic in the same direction.*

The converse of this statement is not true. If the homogeneous operator $p(D)$ is hyperbolic in the direction N and $Q(D)$ is an operator of lower order, then the operator $p(D) + Q(D)$ may not be Gårding hyperbolic in the direction N. For instance, if $P = D_0^2$ and $Q = D_1$, then the operator $P + aQ$ will be Gårding x_0-hyperbolic only when $a = 0$. For admissible operators $Q(D)$ in the case of scalar operators, see Hörmander [1963]. Moreover, we have the following

Theorem 6 (Svensson [1970], Hörmander [1963]). *In order that the operator $p(D) + Q(D)$ be Gårding x_0-hyperbolic for any operator $Q(D)$ of a lower order, it is necessary and sufficient that p satisfies* (1.5).

In this case the Cauchy problem for any operator with principal symbol p will be L_2-well posed in any spatial type lens.

Until the end of this section, G will denote a scalar polynomial of degree M and g its principal part. We introduce the following sets: $h(M)$ is the set of homogeneous polynomials of degree M; $\operatorname{hyp}(N, M)$ and $\operatorname{hyp}^0(N, M)$ denote the set of homogeneous polynomials of degree M that are, respectively, hyperbolic and strictly hyperbolic in the direction N; $\operatorname{Hyp}(N, M)$ is the set of polynomials of degree M that are Gårding hyperbolic in the direction N. In view of Proposition 1 and Theorem 6, we have

$$\begin{aligned} &G \in \operatorname{Hyp}(N, M) \Rightarrow g \in \operatorname{hyp}(N, M), \\ &g \in \operatorname{hyp}^0(N, M) = G \in \operatorname{Hyp}(N, M). \end{aligned} \tag{1.6}$$

Evidently,

$$\begin{aligned} \operatorname{hyp}(-N, M) &= \operatorname{hyp}(N, M), \\ \operatorname{hyp}^0(-N, M) &= \operatorname{hyp}^0(N, M), \\ \operatorname{Hyp}(-N, M) &= \operatorname{Hyp}(-N, M), \end{aligned} \tag{1.7}$$

$$g \in \operatorname{hyp}(N, M) \Rightarrow g^{-1}(N)g(\cdot) \text{ is a polynomial with real coefficients,}$$

$$g \in \operatorname{hyp}(N, M) \Rightarrow \frac{\partial g}{\partial N} \in \operatorname{hyp}(N, M - 1), \tag{1.8}$$

$$g \in \operatorname{hyp}^0(N, M) \Rightarrow \frac{\partial g}{\partial N} \in \operatorname{hyp}^0(N, M - 1), \tag{1.9}$$

where $\frac{\partial g}{\partial N}(\xi) = \sum_j N_j \frac{\partial g}{\partial \xi_j}$. That is, the differentiation of a (strictly) hyperbolic homogeneous polynomial in the direction of hyperbolicity results in a (strictly) hyperbolic polynomial of order one less.

Let $\Gamma(g, N)$ be a connected component of the set $\{\xi\colon g(\xi) \neq 0\}$ containing N. Then (Atiyah, Bott and Gårding [1970, 1973], Hörmander [1963])

$$g \in \operatorname{hyp}(N, M) \Rightarrow \Gamma(g, N) \text{ is an open convex cone,} \tag{1.10}$$

$$g \in \text{hyp}(N, M), N' \in \Gamma(g, N) \Rightarrow g \in \text{hyp}(N', M),$$
$$g \in \text{hyp}^0(N, M), N' \in \Gamma(g, N) \Rightarrow g \in \text{hyp}^0(N', M), \tag{1.11}$$
$$G \in \text{Hyp}(N, M), N' \in \Gamma(g, N) \Rightarrow G \in \text{Hyp}(N', M).$$

That is, the property of (strict) hyperbolicity or Gårding hyperbolicity remains valid for all the elements of $\Gamma(g, N)$ simultaneously.

Let $\Gamma'(g, N) = \{x: \langle x, \xi \rangle \geqslant 0 \ \forall \xi \in \Gamma(g, N)\}$ be the *dual cone* to $\Gamma(g, N)$.

Theorem 7 (Atiyah, Bott and Gårding [1970, 1973], Hörmander [1963]). (i) *If* $G = \det P$ *and* $G \in \text{Hyp}(N, M)$, *then a fundamental solution* $E \in \mathscr{D}'(\mathbb{R}^{n+1})$ *of the operator* $P(D)$ *exists, is unique, and is such that* $\operatorname{supp} E \subset \{x: \langle x, N \rangle \geqslant 0\}$.

(ii) *The convex hull of the set* $\operatorname{supp} E$ coincides with $\Gamma'(g, N)$. *If, however,* $g \in \text{hyp}^0(N, M)$, *then* $\operatorname{sing\,supp} E \supset \partial \Gamma'(g, N)$.

For a more precise structure of supp E and sing supp E, and, in particular, of lacunae, we refer the reader to Atiyah, Bott and Gårding [1970, 1973].

We now discuss the structures of $\text{hyp}(N, M)$ and $\text{hyp}^0(N, M)$. Let $h_{\mathbb{R}}(M)$ denote the set of homogeneous polynomials of degree M with real coefficients.

Theorem 8 (Nuij [1968]). (i) *The sets* $\text{hyp}_{\mathbb{R}}(N, M) = \text{hyp}(N, M) \cap h_{\mathbb{R}}(M)$ *and* $\text{hyp}^0(N, M) = \text{hyp}^0(N, M) \cap h_{\mathbb{R}}(M)$ *consist of two connected components on which* $g(N) \gtrless 0$.

(ii) *The set* $\text{hyp}_{\mathbb{R}}(N, M)$ *is relatively closed in* $h_{\mathbb{R}}(M) \backslash \{g: g(N) = 0\}$; *the set* $\text{hyp}^0_{\mathbb{R}}(N, M)$ *is open in* $h_{\mathbb{R}}(M)$ *and dense in* $\text{hyp}_{\mathbb{R}}(N, M)$.

Thus every scalar hyperbolic polynomial can be approximated by strictly hyperbolic polynomials. For matrix operators, the situation is much more complicated (John [1978]). If $n \geqslant n(d)$, there does not exist any first-order strictly hyperbolic $d \times d$ matrix polynomial.

We shall now discuss the important concept of localization. To do this, we expand $t^M G(t^{-1}\xi + \xi)$ in ascending powers of t:

$$t^M G(t^{-1}\xi + \zeta) = t^r G_\xi(\zeta) + 0(t^{r+1})(t \to 0),$$

$G_\xi(\cdot) \not\equiv 0$. The number $r = r_\xi(G)$ is called the *multiplicity* of ξ *with respect to* G and the polynomial $G_\xi(\zeta)$ the *localization of* G *at* ξ.

In particular, $r_\xi(G) = 0$ if $g(\xi) \neq 0$, and $G_\xi(\zeta) = g_\xi(\zeta)$. If, however, $g(\xi) = 0$ and $dg(\xi) \neq 0$, then $r_\xi(g) = 1$, and $g_\xi(\zeta)$, $G_\xi(\zeta)$ are linear functions of ζ.

Theorem 9 (Atiyah, Bott and Gårding [1970, 1973]). *Let* $G \in \text{Hyp}(N, M)$. *Then*

a) $r_\xi(G) = r_\xi(g)$ *and* g_ξ *is the principal part of* G_ξ,

b) $g_\xi \in \text{hyp}(N, r_\xi)$, $G_\xi \in \text{Hyp}(N, r_\xi)$,

c) $\Gamma(g_\xi, N) \supset \Gamma(g, N)$.

Suppose that $g(x, \xi)$ is now a polynomial that depends smoothly on the parameter $x \in U$, where U is a domain in $\mathbb{R}^s$. Then the localization concept must take into account the dependence on x too.

Theorem 10. (i) *Let* $g(x, \cdot) \in \text{hyp}(N, M)$ *for any* $x \in U$, *where* U *is a neighbourhood of* x. *Then*

a) $\partial_x^\alpha \partial_\xi^\beta g(\hat{x}, \hat{\xi}) = 0$ *for* $|\alpha| + |\beta| < r = r_{\hat{\xi}}(g(\hat{x}, \cdot))$,

$$\left(\frac{\partial}{\partial N}\right)^r g(\hat{x}, \hat{\xi}) \neq 0;$$

b) *the polynomial*

$$g_{(\hat{x}, \hat{\xi})}(y, \eta) = \sum_{|\alpha|+|\beta|=r} \frac{1}{\alpha! \beta!} y^\alpha \eta^\beta (\partial_x^\alpha \partial_\xi^\beta g)(\hat{x}, \hat{\xi})$$

in the $n + s + 1$ *variables* (y, η) *is hyperbolic in the direction* $\tilde{N} = (0, N)$; $g_{(\hat{x}, \hat{\xi})}(\cdot)$ *is said to be the* ***localization of*** g ***at the point*** $(\hat{x}, \hat{\xi})$.

(ii) *Let* $g(x, \cdot) \in \text{hyp}(N, M)$ *for any* $x \in U \cap \{x_1 \geqslant 0\}$. *Then*

c) $\partial_x^\alpha \partial_\xi^\beta g(\hat{x}, \hat{\xi}) = 0$ for $2\alpha_1 + |\alpha'| + |\beta| < r = r_{\hat{\xi}}(g(\hat{x}, \cdot))$, $\alpha' = (\alpha_z, \ldots, \alpha_s)$,

$$\left(\frac{\partial}{\partial N}\right)^r g(\hat{x}, \hat{\xi}) \neq 0.$$

The polynomial $g = \xi_0^p (\xi_0^2 - x_1 \xi_1^2)^q$ demonstrates that under conditions (ii) the statement a) does not, in general, hold.

Finally, for polynomials depending on a parameter the following important results hold.

Theorem 11 (Bronshtejn [1979]). *Let* $g(x, \cdot) \in \text{hyp}(N, M)$ *for all* $x \in U$, $N = (1, 0, \ldots, 0)$. *If* $\xi_0 = \lambda_j(x, \xi')$ *are the characteristic roots of the polynomial* g, *then the estimates*

$$\left|\frac{\partial \lambda_j}{\partial x_k}\right| \leqslant C|\xi'|, \qquad \left|\frac{\partial \lambda_j}{\partial \xi_l}\right| \leqslant C, \qquad C = C_K,$$

hold on any compact set $K \Subset U$. (Note that there exists a nowhere dense closed set $\Sigma \subset U \times \mathbb{R}^n$ such that outside Σ the roots λ_j have constant multiplicity depending, possibly, on the connected components $U \times \mathbb{R}^n \setminus \Sigma$; in particular, λ_j are infinitely smooth in the exterior of Σ.)

In § 1.4–§ 1.9, we shall assume, without repeating it each time, that P is a scalar x_0-hyperbolic operator with a real principal symbol. We start by discussing Problem A which has been solved till the present time for the case of an open domain X.

1.4. Regularly and Completely Regularly Hyperbolic Operators. We note first that if the Cauchy problem is C^∞-well posed then a linear map

$$\mathscr{L}: C^\infty(X_T^+) \times (C^\infty(X_T))^m \ni (f, g_0, \ldots, g_{m-1}) \to u \in C^\infty(X_T^+)$$

is defined that associates with the right-hand side and the initial data a solution, and for any $l \in \mathbb{Z}^+$ and any compact set $K \Subset X$ there is a $q \in \mathbb{Z}^+$ such that $\mathscr{L}$ can be extended to a continuous map acting from $C_0^q(K_T^+) \times (C_0^q(K_T))^m$ into $C^l(K_T^+)$. In view of this, we have

Definition 7. A pair of numbers $(l, q) \in \mathbb{Z}^+ \times \mathbb{Z}^+$ is known as the *well-posedness index of the Cauchy problem at a point* $\hat{x} \in X_T^+$ if there exists a K-neighbourhood of this point such that $\mathscr{L}$ can be extended to a continuous map acting from

$$C_0^q(K_T^+) \times (C_0^q(K_T))^m \text{ to } C^l(K_T^+).$$

Definition 8. An operator P is called *regularly hyperbolic in* X if for any point $\hat{x} \in X$ we can find a neighbourhood K and a number $\hat{T}$, with the property that $\hat{T} < \hat{x}_0$ if $\hat{x} \in \mathring{X} \cup S_+$ and $\hat{T} = \hat{x}_0$ if $\hat{x} \in \hat{S}_-$, such that the Cauchy problem (1.1)–(1.2) for the operator $P + Q$, where Q is any differential operator of order not exceeding $m - 1$ with infinitely smooth coefficients, is C^∞-well posed in K_T^+ for $T \geqslant \hat{T}$.

If the well-posedness index at every point $\hat{x}$ can be chosen independently of Q, then P is said to be *completely regularly hyperbolic* in X.

It is a simple matter to give a description of completely regularly hyperbolic operators.

Theorem 12. *In order that P, an x_0-hyperbolic operator in X, be completely regularly hyperbolic in X, it is necessary and sufficient that*

$$d_{x,\xi}p(x, \xi) \neq 0 \tag{1.12}$$

for $(x, \xi) \in X \times (\mathbb{R}^{n+1}\backslash 0)$. *Moreover, the Cauchy problem for P is C^∞-well posed in any spatial type lens $\tilde{X} \Subset X$, and the following estimates hold.*

(i) *If* $\tilde{T}_- = T_-$, $\tilde{T}_+ < T_+$ *and* $\varkappa \in [1, 2]$, $l \geqslant s$, *then*

$$\|u; \tilde{X}_t\|_{l+m-1,s+m-1} \leqslant C\|u; X_T\|_{l+m,s+m-1} + C\left(\int_T^t \|Pu; \tilde{X}_{t'}\|_{l,s}^{\varkappa}(t' - T)^{(\varkappa/2)-1}\, dt'\right)^{1/\varkappa} \tag{1.13}$$

uniformly in t, $T \in [\tilde{T}_-, \tilde{T}_+]$, $T \leqslant t$.

(ii) *If* $\tilde{T}_- > T_-$, $\tilde{T}_+ = T_+$ *and* $\varkappa \in [2, \infty]$, $l \geqslant s$, *then*

$$\left(\int_T^t \|u; \tilde{X}_{t'}\|_{l+m-1,s+m-1}^{\varkappa}(T_+ - t')^{(\varkappa/2)-1}\, dt'\right)^{1/\varkappa} \leqslant C\left(\|u; \tilde{X}_T\|_{l+m-1,s+m-1} + \int_T^t \|Pu; X_{t'}\|_{l,s}\, dt'\right) \tag{1.14}$$

uniformly in t, $T \in [\tilde{T}_-, \tilde{T}_+]$, $T \leqslant t$; *if* $\varkappa = \infty$, *the left-hand side of* (1.14) *has to be replaced by*

$$\|u; \tilde{X}_t\|_{l+m-1,s+m-1}(T_+ - t)^{1/2}.$$

The necessity of condition (1.12) has been established by Ivrii and Petkov [1974] and its sufficiency by Ivrii [1976a] by the method of alternating operator.

Corollary. *If P is completely regularly hyperbolic in X, then it is strictly hyperbolic in X and the multiplicity of its characteristic roots does not exceed 2 if* $(x, \xi') \in (S_+ \cup S_-) \times (\mathbb{R}^n\backslash 0)$.

Example 1. The operator $P_\pm = -D_0^2 \pm x_0 D_1^2$ is completely regularly hyperbolic in $\overline{\mathbb{R}}^\pm \times \mathbb{R}$.

A point $\bar{\rho} = (\bar{x}, \bar{\xi}) \in X \times (\mathbb{R}^{n+1} \backslash 0)$ is said to be *critical* if $d_{x,\xi} p(\bar{\rho}) = 0$. At every critical point, we introduce a *fundamental matrix* (the Hessian) of the symbol p:

$$F_p = \begin{pmatrix} p_{\xi x} & p_{\xi\xi} \\ -p_{xx} & -p_{x\xi} \end{pmatrix},$$

where $p_{\xi\xi}$, $p_{x\xi} = {}^t p_{\xi x}$ and p_{xx} are $(n+1) \times (n+1)$ matrices of second-order derivatives, and ${}^t(\cdot)$ denotes the transpose matrix. The fundamental matrix has the following properties (Ivrii and Petkov [1974], Hörmander [1977]):

a) $H\{p, q\}(\bar{\rho}) = -F_p(\bar{\rho}) H_q(\bar{\rho})$,
where $H_q = (\partial q/\partial \xi)(\partial/\partial x) - (\partial q/\partial x)(\partial/\partial \xi)$ is the *Hamiltonian field* generated by q and $\{p, q\} = H_p q$ are the *Poisson brackets.*

b) The fundamental matrix is transformed by a canonical diffeomorphism into a similar matrix; thus its eigenvalues are symplectic invariants.

c) The matrices F_p and $-F_p$ are similar.

d) If the symbol p is hyperbolic in X and $\bar{\rho} = (\bar{x}, \bar{\xi})$ is its critical point, then all the eigenvalues of the matrix $F_p(\bar{\rho})$ are either pure imaginary or there are two non-zero simple real eigenvalues $\mu(\bar{\rho})$ and $-\mu(\bar{\rho})$ and the remaining eigenvalues are pure imginary.

e) Moreover, if $\bar{\xi}_0$ is a root, with multiplicity r, of the polynomial $p(\bar{x}, \xi_0, \bar{\xi}')$, then $r \geqslant 2$. If $\bar{x} \in \mathring{X}$ and $r \geqslant 3$ or if $\bar{x} \in S_+ \cup S_-$ and $r \geqslant 4$, then $F_p(\bar{\rho}) = 0$. If $\bar{x} \in S_+ \cup S_-$ and $r = 3$, then either $F_p(\bar{\rho}) = 0$ or rank $F_p(\bar{\rho}) = 2$ and $F_p(\bar{\rho})$ has two non-zero real eigenvalues.

f) Let $Q(z) = \frac{1}{2}\sigma(F_p(\bar{\rho})z, z)$, the localization of the symbol p at the point $\bar{\rho}$, be a quadratic form in $z \in V = \mathbb{R}^{n+1} \times \mathbb{R}^{n+1}$, where σ is a bilinear *symplectic form*, $\sigma((x, \xi), (y, \eta)) = \langle x, \eta \rangle - \langle y, \xi \rangle$. If $r = 2$, $\bar{x} \in X$, p is x_0-hyperbolic in X and $p_{\xi_0 \xi_0}(\bar{\rho}) < 0$, then V can be decomposed into a direct sum of symplectic orthogonal subspaces V_j that are invariant under $F_p(\bar{\rho})$ and on each of which Q has one of the following canonical forms:

(a) $Q = \frac{1}{2}\omega(x_1^2 + \xi_1^2)$, $\omega > 0$;

(b) $Q = \mu x_1 \xi_1$, $\mu > 0$;

(c) $Q = \xi_1^2$;

(d) $Q = -\xi_1^2$;

(e) $Q = 0$;

(f) $Q = -\xi_1^2 + 2\xi_1\xi_2 + x_2^2$.

Moreover, there is exactly one subspace of the type (b) $\cup$ (d) $\cup$ (f); the subspaces of the types (a) – (e) are two dimensional, while those of the type (f) four dimensional. To the subspace of the type (a) there correspond eigenvalues $\pm i\omega$ of the fundamental matrix and to the type (b) the eigenvalues $\pm\mu$; the eigenvalue zero corresponds to the remaining types of subspaces.

f′) If, however, $r = 3$, $\bar{x} \in S_+ \cup S_-$ and p is x_0-hyperbolic in X, then in the decomposition of V there may be subspaces of types (b) and (e) only; moreover, the number of subspace of type (b) cannot exceed 1.

Definition 9. An x_0-hyperbolic operator P with real principal symbol p is called *effectively hyperbolic* if at any critical point ρ its fundamental matrix has non-zero real eigenvalues $\pm\mu$, $\mu > 0$.

At critical points ρ, we also introduce the *subprincipal symbol* of the operator by the formula

$$p^s = P_{m-1} + \frac{i}{2}\sum_{j=0}^{n} p_{x_j\xi_j};$$

this is also invariant.

Ivrii and Petkov [1974] conjectured that an operator is regularly hyperbolic if and only if it is effectively hyperbolic, and established its validity in one direction.

Theorem 13. *If the operator P is regularly hyperbolic, it is effectively hyperbolic.*

Corollary. *Let the operator P be regularly hyperbolic in X. The multiplicity of its characteristic roots does not exceed 2 if $(x, \xi') \in \mathring{X} \times (\mathbb{R}^n\backslash 0)$ and if $(x, \xi') \in (S_+ \cup S_-) \times (\mathbb{R}^n\backslash 0)$, the multiplicity does not exceed 3.*

Theorem 14. *Let P be effectively hyperbolic in X. Suppose that one of the following three additional conditions is satisfied*:

(i) $X = \mathring{X}$, $S_+ = S_- = \varnothing$.

(ii) *p can be factorized in the neighbourhood of each of its critical points ρ, that is, $p = p_1 p_2$ in the neighbourhood of ρ, where $p_1, p_2 \in C^\infty$* and $p_1(\rho) = p_2(\rho) = 0$.

(iii) *p can be factorized in the neighbourhood of each of its critical points $\rho = (x, \xi)$ such that ξ_0 is a characteristic root of multiplicity* 3.

Then P is regularly hyperbolic in X and the Cauchy problem is C^∞-well posed in any spatial type lens $\tilde{X} \Subset X$.

Note that under assumption (ii) all the eigenvalues of $F_p(\rho)$ are zero except, possibly, the two $\pm\{p_1, p_2\}(\rho)$.

Under assumption (ii) (as also under certain other conditions when $n = 1$), Theorem 14 has been proved in Ivrii [1976a], while under assumption (i) in Iwasaki [1983a, 1984] and in Melrose [1983]. Under assumption (iii), it can be proved by combining the results of Ivrii [1976a] and Iwasaki [1983a, 1984]. The proofs carried out in Ivrii [1976a] and Iwasaki [1983a, 1984] are based on the transformation of the operator P to an operator with "nice" lower terms by means of integro-pseudodifferential operators and on the energy estimates for the resulting operator. The proof of Melrose [1983] is based on the study of operators that are strictly hyperbolic for $x_0 \neq 0$ but degenerate at $x_0 = 0$; the Cauchy problem for the general operators is reduced to the one for degenerate operators by means of arguments based on strictly hyperbolic approximation and the uniqueness theorem. As a by product of the proofs, we have

Theorem 15. *Suppose that the hypotheses of Theorem 14 are fulfilled. Let $\bar{x} \in X$ and K be a sufficiently small neighbourhood of $\bar{x}$ that depends only on p. If the inequality*

$$|Im\ p^s(\rho)| \leqslant (L - n - 2)|\mu(\rho)|,$$

where $\pm\mu$ are non-zero real eigenvalues of $F_p(\rho)$, holds at any critical point $\rho \in K \times (\mathbb{R}^{n+1}\backslash 0)$, then any pair $(q + m, q + L)$ will be a well-posedness index in K.

The following assertion well supplements Theorems 13 and 14.

Theorem 16 (Ivrii and Petkov [1974]). *Let $(\bar{x}, \bar{\xi}) \in X \times (\mathbb{R}^{n+1}\backslash 0)$ be a critical point of p. If $F_p(\bar{x}, \bar{\xi})$ has non-zero real eigenvalues $\pm\mu$, then for the pair (l, q) to be a well-posedness index, it is necessary that*

$$|\mathrm{Im}\ p^s(\bar{x}, \bar{\xi})| \leqslant Cn(q + m - l)|\mu|,$$

where C is an absolute constant.

Example 2. The operator

$$P = -D_0^2 + x_0^2 D_n^2 + \sum_{j=1}^{n} \omega_j(D_j^2 + x_j^2 D_n^2) + \sum_{j=k+1}^{n-1} D_j^2 \quad (0 \leqslant k \leqslant n - 1, \omega_j > 0)$$

is regularly hyperbolic in $\mathbb{R}^{n+1}$.

Example 3. The operator

$$P_\pm = (-D_0^2 \pm x_0 D_1^2)D_0$$

is regularly hyperbolic in $\bar{\mathbb{R}}^\pm \times \mathbb{R}$.

Example 4. The operators

$$P_1 = -D_0^2 + \sum_{j=1}^{k} \omega_j(D_j^2 + x_j^2 D_n^2) + \sum_{j=k+1}^{n-1} D_j^2 \quad (0 \leqslant k \leqslant n - 1, \omega_j > 0),$$

$$P_2 = -D_0^2 + D_0 D_1 + x_1^2 D_n^2 + \sum_{j=2}^{k} \omega_j(D_j^2 + x_j^2 D_n^2) + \sum_{j=k+1}^{n-1} D_j^2$$

$$(1 \leqslant k \leqslant n - 1, \omega_j > 0),$$

$$P_3 = -D_0^2 + x_1 D_0 D_n + D_1^2 + \sum_{j=2}^{k} \omega_j(D_j^2 + x_j^2 D_n^2) + \sum_{j=k+1}^{n-1} D_j^2$$

$$(1 \leqslant k \leqslant n - 1, \omega_j > 0)$$

are x_0-hyperbolic in $\mathbb{R}^{n+1}$ but are not regularly hyperbolic in X if $X \cap \{x_s = \cdots = x_k = 0\} \neq \varnothing$, where $s = 1$ for P_1, P_3 and $s = 2$ for P_2.

We now turn to Problem B.

1.5. Operators with Characteristics of Constant Multiplicity

Definition 10. An operator P is said to be an *operator with characteristics of constant multiplicity* (locally) in X if in the neighbourhood of every point $\bar{\rho} \in \{p = 0\} \subset X \times (\mathbb{R}^{n+1}\backslash 0)$ its principal symbol can be expressed in the form $p = eq^r$. Here $e(\bar{\rho}) \neq 0$, q is a real symbol, $q(\bar{\rho}) = 0$, and $H_q(\bar{\rho}) \neq 0$ is non-collinear with $\xi\dfrac{\partial}{\partial\xi}$, and $r \in \mathbb{Z}^+\backslash 0$ and depends, in general, on the connected component of the characteristic set $\{p = 0\}$.

Definition 11. An operator P, x_0-hyperbolic in X, is said to be an *operator with characteristic roots of constant multiplicity* if all its characteristic roots $\xi_0 = \lambda_j(x, \xi')$ are of constant multiplicity $r_j \in \mathbb{Z}^+\backslash 0$, $r_1 + \cdots + r_s = m$ (and then they depend smoothly on $(x, \xi') \in X \times (\mathbb{R}^n\backslash 0)$).

Proposition 2. (i) *An x_0-hyperbolic operator is an operator with characteristic roots of constant multiplicity in X if and only if its principal symbol can be represented in the form*

$$p = a_0(x)p_1^{r_1} \dots p_s^{r_s}, \tag{1.15}$$

where $a_0(x)$ does not vanish, $r_j \in \mathbb{Z}^+\backslash 0$, and where p_j are strictly hyperbolic polynomials in X no two of which vanish simultaneously.

(ii) *The operator P, x_0-hyperbolic in X, is an operator with characteristics of constant multiplicity in X if and only if its principal symbol can be represented in the form* (1.15), *where p_j are completely regularly hyperbolic polynomials in X no two of which vanish simultaneously.*

Thus for hyperbolic operators, Definitions 10 and 11 coincide if $S_+ = S_- = \varnothing$ but Definition 10 is somewhat more general when $S_+ \cup S_- \neq \varnothing$. Let us formulate the Levi-Strang-Flaschka condition that turns out to be a necessary and sufficient condition for well-posedness. Without loss of generality, we may assume that in Definitions 10 and 11 q is a homogeneous positive symbol in ξ of degree 1.

Definition 12. An operator P with characteristics of constant multiplicity r in a neighbourhood of the point $\bar{\rho} = (\bar{x}, \bar{\xi}) \in X \times (\mathbb{R}^{n+1}\backslash 0)$ is said to satisfy the *Levi-Strang-Flaschka condition* (L) in the neighbourhood of this point if

$$P = \sum_{k=0}^{r} A_k Q^{r-k}, \tag{1.16}$$

where Q is a pseudodifferential operator of order 1 with principal symbol q and A_k are pseudodifferential operators of order $m - r$, to within a pseudodifferential operator with symbol equal to zero in the neighbourhood of $\bar{\rho}$ (Taylor [1981], Trèves [1980]).

Remarks. a) If condition (L) is satisfied in $\mathring{X} \times (\mathbb{R}^{n+1}\backslash 0)$, it is also satisfied in $X \times (\mathbb{R}^{n+1}\backslash 0)$.

b) If $q_{\xi_0}(\bar{\rho}) \neq 0$, we may assume, without loss of generality, that $q = \xi_0 - \lambda(x, \xi')$ and $Q = D_0 - \lambda(x, D')$, where $\lambda(x, D')$ is a pseudodifferential operator with respect to x' with real principal symbol (which is a characteristic root of P). Then we may assume, without loss of generality, that in condition (L) A_k are differential operators with respect to x_0 and pseudodifferential operators with respect to x' of order $m - r$.

c) If P is x_0-hyperbolic in X and $q_{\xi_0}(\bar{\rho}) = 0$, then $\bar{x} \in S_+ \cup S_-$, $q_{\xi_0\xi_0}(\bar{\rho}) \neq 0$ and $H_q(\bar{\rho}) \neq 0$ is collinear with H_{x_0}. In view of the Weierstrass basic theroem, we may then assume, without loss of generality, that q is a polynomial in ξ_0 of degree 2, where the coefficient of ξ_0^2 is 1, q is positive homogeneous in ξ of degree 2, Q and A_k's are, respectively, differential operators of order 2 with respect to x_0 and pseudodifferential operators of order $m - r + k$ with respect to x'.

d) In view of Proposition 2, the x_0-hyperbolic operators that have characteristics of constant multiplicity and satisfy condition (L) can be characterized in the following inductive manner. Suppose that such operators have already been characterized in (1.15) for $\max_j r_j < r$, and let $p' = \prod_{j:z_j=r} p_j$, $p = p'p''$. Then P satisfies condition (L) if and only if it can be expressed in the form $P = P'P'' + A$, where P' and P'' are differential operators with principal symbols p' and p'', respectively, P'' satisfies condition (L) and A is a differential operator of order $m - r$.

e) Let $d_\xi q(\bar{\rho}) \neq 0$. Then condition (L) is satisfied in a neighbourhood of the point $\bar{\rho}$ if and only if for all (or for some, it is immaterial) real functions $\varphi \in C^\infty$, that satisfy the *eikonal equation*

$$q(x, d\varphi(x)) = 0 \tag{1.17}$$

in a neighbourhood of $\bar{x}$ and are such that $d\varphi(\bar{x}) = \bar{\xi}$, and for any $f \in C^\infty$

$$P(e^{i\varphi\rho} f) = 0(\rho^{m-r}) \quad \text{as} \quad \rho \to +\infty \tag{1.18}$$

in a certain neighbourhood of $\bar{x}$.

The real functions φ satisfying (1.17) are known as *phase functions.*

Theorem 17. *Let P be an x_0-hyperbolic operator in X having characteristics of constant multiplicity. For the Cauchy problem for P to be C^∞-well posed, it is necessary* (*and sufficient if X is a spatial type lens*) *that condition* (L) *holds.*

The necessity of condition (L) has been established by Flaschka and Strang [1971] and its sufficiency by Chazarain [1974] by constructing a parametrix for the Fourier integral operator. It was done under the assumption that P has characteristic roots of constant multiplicity. However, the necessity of condiion (L) in the general case of the operator having characteristics of constant multiplicity follows from this result in view of Remark a), and the sufficiency can be easily established, in view of Remark d), by means of the estimates (1.4), (1.11) and (1.12). For operators with characteristic roots of constant multiplicity, we need only the estimate (1.4). We note that when $r \geqslant 4$ the proof of Flaschka and Strang [1971] does not go through unless condition (U_r) is assumed to hold (see the note to the Russian translation of this work). But we can do away with this condition if we follow the method of Ivrii [1976b].

1.6. Irregularly Hyperbolic Operators of General Form. Let us formulate, first of all, the most general necessary conditions on the lower terms. The following assertion well supplements Theorem 13.

Theorem 18 (Ivrii and Petkov [1974], Hörmander [1977]). *Let P be an x_0-hyperbolic operator in X with a real principal symbol p. Let $\rho \in X \times (\mathbb{R}^{n+1}\setminus 0)$ be a critical point of p. Suppose that all the eigenvalues of $F_p(\rho)$ are pure imaginary. In order that the Cauchy problem for P be C^∞-well posed, it is necessary that the following two conditions are satisfied:*

a) $\operatorname{Im} p^s(\rho) = 0$,

b) $|p^s(\rho)| \leqslant \frac{1}{2} Tr^+ F_p(\rho)$,

where $Tr^+ F_p(\rho) = \sum |\mu_j|$ and $\pm i\mu_j$ are all the non-zero eigenvalues of $F_p(\rho)$ (with regard to their multiplicity).

The *Ivrii–Petkov–Hörmander conditions* a) and b) turn out to be very precise.

Theorem 19 (Ivrii [1977], Hörmander [1977]). *Let P be an x_0-hyperbolic operator in X, with real principal symbol p. Suppose that the following conditions are fulfilled:*

c) $\Sigma = \{p = dp = 0\}$ *is a C^∞-manifold;*

d) $\rho \in \Sigma \Rightarrow T_\rho \Sigma = \operatorname{Ker} F_p(\rho)$, $\operatorname{Ker} F_p^2(\rho) \cap \operatorname{Ran} F_p^2(\rho) = 0$, $\operatorname{Spec} F_p(\rho) \subset i\mathbb{R}$ *(then there exist $\varepsilon = \pm 1$ and a $z^0(\rho) \in \operatorname{Ker} F_p(\rho)$ such that*

$$\sigma(z^0, z) = 0, \qquad z \notin \operatorname{Ker} F_p(\rho) \Rightarrow \varepsilon\sigma(F_p(\rho)z, z) > 0).$$

e) *It is possible to choose a $z^0 \in C^\infty(\Sigma)$. If the condition* a) *and the condition* b′) $|p^s(\rho)| < \frac{1}{2} Tr + F_p(\rho)$ *(in contrast to* b), *the inequality here is strict) are fulfilled, then the Cauchy problem for P is C^∞-well posed in any spatial type lens $\tilde{X} \Subset X$.*

Theorem 20 (Hörmander [1977]). *Let P be x_0-hyperbolic in X. Assume that the following conditions, apart from condition* c), *are also satisfied:*

f) $\rho \in \Sigma \Rightarrow T_\rho \Sigma = \operatorname{Ker} F_p(\rho) \subset \operatorname{Ran}(\rho)$ *(that is, Σ is an involutive manifold).*

g) $\rho \in \Sigma \Rightarrow p^s(\rho) = 0$.

Then the Cauchy problem for P is C^∞-well posed in any spatial type lens $\tilde{X} \Subset X$.

Remarks. Theorem 19 remains valid under more general conditions (Ivrii [1977], Iwasaki [1984]). Under conditions of Theorems 19 and 20 very strong *a priori* estimates hold (Ivrii [1977], Hörmander [1977]) that are established by the method of the divisor operator; in comparison to strictly hyperbolic operators, the loss in smoothness equals $\frac{1}{2}$ and 1, respectively. For model operators, we can prove that the Cauchy problem is C^∞-well posed also under the conditions a) and b) (but not b′)), but the loss in smoothness as compared to strictly hyperbolic operators will be 1. Under the hypotheses of Theorem 20, the condition g) turns out, in view of Theorem 18, to be necessary and sufficient for the Cauchy problem to be C^∞-well posed.

Example 4′. Let P_l be one of the operators of Example 4, and let $X_T^+ \cap \{x_s = \cdots = x_k = 0\} \neq \varnothing$. In order that the Cauchy problem for the operator $P_l + cD_n (c \in \mathbb{C})$ be C^∞-well posed, it is necessary (and sufficient if X is a spatial type lens) that $c \in \mathbb{C}$ and $|c| \leqslant \overline{\omega} \equiv \sum_j \omega_j$.

1.7. Necessary Conditions for the Cauchy Problem to be Well Posed. We do not have results, apart from Theorem 18, that are both general and precise enough. All the same, the following result is fairly general and in a number of cases proves to be precise enough.

Theorem 21. *Let $\bar{x} \in X_T^+$, $\bar{\xi} = (0, \ldots, 0, 1)$ and $q = (q_0, \ldots, q_n) \in [0, 1)^n$. Let*

$$\partial_\xi^\alpha \partial_x^\beta p(\bar{x}, \bar{\xi}) = 0$$

for (1.19)

$$|\alpha| + \langle \beta - \alpha, q \rangle < r(1 - q_0)$$

and

$$\partial_{\xi_0}^r p(\bar{x}, \bar{\xi}) \neq 0. \tag{1.20}$$

If the Cauchy problem for P is C^∞-well posed in X_T^+, then

$$\partial_\xi^\alpha \partial_x^\beta P_s(\bar{x}, \bar{\xi}) = 0 \tag{1.21}$$

for

$$|\alpha| + \langle \beta - \alpha, q \rangle < r(1 - q_0) - (m - s).$$

By results a) and b) of Theorem 10, we have

Corollary. *If $\bar{\xi}_0$ is a characteristic root of multiplicity r, then in order that the Cauchy problem be C^∞-well posed, it is necessary to impose conditions on all the members of P which are of order greater than $m - \gamma r$, where $\gamma = \frac{1}{2}$ if $\bar{x} \in \mathring{X}$ and $\gamma = \frac{1}{3}$ if $\bar{x} \in S_+ \cup S_-$.*

In a particular case, Theorem 21 was proved by Ivrii and Petkov [1974]; for $r \geqslant 4$, it was assumed that condition (U_Γ) holds. However, the method of Ivrii [1976c] enables us to prove it in its present form. By the same method, Mandai recently established a more general theorem. In his formulation, Mandai used Newton's polyhedra instead of half-spaces in $\mathbb{Z}^{+(2n+2)}$. We note that both Theorem 21 and Mandai's theorem are non-invariant under the change of coordinates.

Example 5. In order that the Cauchy problem for the operators

$$P_1 = D_0^2 - x_0^{2k} x_1^{2l} D_1^2 + a_0(x) D_0 + a_1(x) D_1 + a_2(x) \quad (0 \in X_T^+ \subset \mathbb{R}^2),$$

$$P_2 = D_0^2 \mp x_0^{2k+1} x_1^{2l} D_1^2 + b_0(x) D_0 + b_1(x) D_1 + b_2(x)$$

$$(0 \in X_T^+ \subset \mathbb{R}^\pm \times \mathbb{R}) \quad (k, l \in \mathbb{Z}^+)$$

be C^∞-well posed, it is necessary (and sufficient if X_T^+ is a spatial type lens) that $a_1 = x_0^{k-1} x_1^l \tilde{a}_1$ and $b_1 = x_0^k x_1^l \tilde{b}_1$ with $\tilde{a}_1, \tilde{b}_1 \in C^\infty$.

1.8. Degenerate Hyperbolic Equations. Among non-strictly hyperbolic equations we identify degenerate equations that are either strictly hyperbolic or have characteristic roots of constant multiplicity in the exterior of initial or final or intermediate hyperplanes. For a number of years only these equations were investigated, and there is a vast literature devoted to them. We mention one of the most specific results.

Theorem 22 (Olejnik [1970]). *Let*

$$P = D_0^2 - \sum_{i,j=1}^{n} D_i a_{ij} D_j + \sum_{j=0}^{n} b_j D_j + c,$$

where $a_{ij} = a_{ji}$. *Let the quadratic form*

$$a(x, \xi') = \sum_{i,j=1}^{n} a_{ij}\xi_i\xi_j$$

be positive definite for $x_0 \neq T_0 \in [T_-, T_+]$. *If the lower symbol* $b(x, \xi') = \sum_{j=1}^{n} b_j\xi_j$ *satisfies the inequality*

$$|b(x, \xi')|^2 \leqslant c_0(x_0 - T_0)^{-1}\frac{\partial}{\partial x_0}a(x, \xi') + Ca(x, \xi'),$$

then the Cauchy problem for P is C^∞-well posed in any spatial type lens.

More recent works on this topic are devoted to equations of order greater than 2 and to equations with infinite order of gluing of characteristic roots. A typical example of the latter is the operator

$$P = D_0^2 - \lambda^2(x_0)D_1^2 + b(x)D_1 + c(x),$$

where the function λ has a zero of infinite order at $x_0 = 0$. By Theorem 21, in order that the Cauchy problem for P be C^∞-well posed, it is necessary that b has a zero of infinite order at $x_0 = 0$. But neither this necessary condition nor the sufficient condition following from Theorem 22 are precise.

Example 6 (Yagdzhyan [1980]). Let $X = \mathbb{R}^+ \times \mathbb{R}$,

$$P = D_0^2 - \exp(-2x_0^{-s})D_1^2 + Bx_0^{-s+l+1}(-\ln x_0)^k \exp(-x_0^{-s})D_1,$$

and $0 < s$, $l, k \in \mathbb{R}$, $B \in \mathbb{C}$. A necessary and sufficient condition for the Cauchy problem to be C^∞-well posed is $\{l > 0 \text{ or } l = 0, k \leqslant 0\}$ if $B \notin \mathbb{R}$ and $\{l > -s \text{ or } l = -s, k \leqslant 0\}$ if $B \in \mathbb{R}$.

1.9. Second-Order Equations in Two Variables. These equations are the simplest. Assuming the coefficients to be analytic and $S_+ \cup S_- = \varnothing$, Nishitani [1984a] carried out a complete analysis of C^∞-well-posedness of the Cauchy problem for these equations. Namely, let the operator

$$P = D_0^2 - A(x)D_1^2 + B(x)D_1 + c(x)D_0 + R(x)$$

be x_0-hyperbolic in a neighbourhood of 0 (that is, $A(x) \geqslant 0$), and let $A(x)$, $B(x)$ be analytic functions, with $A(0) = 0$ and $A \not\equiv 0$. Then, in view of the Weierstrass basic theorem, $A(x) = x_1^{2l}\tilde{A}(x)E(x)$ in the neighbourhood of 0, where

$$\tilde{A}(x) = \prod_{\mu=1}^{2m} (x_0 - t_\mu(x_1))$$

is the Weierstrass polynomial in x_0 and $E(0) > 0$. Let us expand the function $t_\mu(x_1)$ in Puiseux series for $\pm x_1 > 0$. Then

$$\operatorname{Re} t_\mu(x_1) = \sum_{j \geqslant 0} c^{\pm}_{\mu,j}(\pm x_1)^{j/p(\mu)},$$

where $c^{\pm}_{\mu,j} \in \mathbb{R}$ and $p(\mu) \in \mathbb{Z}^+ \setminus 0$.

For an arbitrary function $f(x)$, which is analytic in a neighbourhood of 0, we introduce the functions $f_\mu(x) \equiv f(x_0 + \operatorname{Re} t_\mu(x_1), x_1)$ and for them define the Newton polyhedra at the points $(0, \pm 0)$. To do this, we expand f_μ in series

$$f_\mu(x) = \sum_{i,j \geqslant 0} c^{\pm}_{\mu,i,j} x_0^i (\pm x_1)^{j/p(\mu)}$$

and denote by $N_\pm(f_\mu)$ the convex hull of the set

$$\bigcup_{\{c^{\pm}_{\mu,i,j} \neq 0\}} \{(i, j/p(\mu)) + \overline{\mathbb{R}}^{+2}\}.$$

If $f \equiv 0$, we set $N_\pm(f_\mu) = \varnothing$.

Theorem 23. *In order that the Cauchy problem for P be C^∞-well posed in some neighbourhood of* 0, *it is necessary and sufficient that $B(x) = x_1^l \tilde{B}(x)$, where $\tilde{B}$ is analytic and*

$$N_\pm(x_0 \tilde{B}_\mu) \subset \tfrac{1}{2} N_\pm(\tilde{A}_\mu) \textit{ for all } \mu = 1, \ldots, 2m.$$

We now examine first-order systems. In § 1.10 and § 1.11, $P = ID_0 - A(x, D')$ denotes a $d \times d$ matrix operator,

$$A(x, D') = \sum_{j=1}^{n} a_j(x) D_j + a'(x),$$

and $p(x, \xi) = I\xi_0 - a(x, \xi')$ and $a(x, \xi')$ are principal symbols of P and A, respectively, and $g = \det p$.

1.10. Systems with Characteristic Roots of Constant Multiplicity. Let $\xi_0 = \lambda(x, \xi')$ be a characteristic root with the following properties:

a) In a neighbourhood of the point $(\bar{x}, \bar{\xi}') \in X \times (\mathbb{R}^n \setminus 0)$ the root $\lambda(x, \xi')$ is of constant multiplicity r.

b) The matrix $\tilde{a}(x, \xi') = a(x, \xi') - I\lambda(x, \xi')$ is of a constant rank $d - r + \varkappa$ in the neighbourhood of $(\bar{x}, \bar{\xi}')$.

Since for $\varkappa = 0$, the corresponding part of the Jordan's normal form of the matrix a is diagonal and it does not spoil the well-posedness, we should consider only the case $\varkappa \geqslant 1$. Let $\varphi(x)$ be a *phase function*, that is, a real function satisfying the *eikonal equation*

$$\varphi_{x_0}(x) = \lambda(x, d_{x'}\varphi(x)), \qquad d_{x'}\varphi(\bar{x}) = \bar{\xi} \tag{1.22}$$

in a neighbourhood of $\bar{x}$.

Let us first examine the necessary conditions for C^∞-well-posedness that were established by Petkov for $r = 2, 3$. Let $R_k(x, \xi')$ and $L_k(x, \xi')$ $(k = 1, \ldots, r - \varkappa)$ be smooth vector functions that constitute a basis in Ker $\tilde{a}$ and Ker $\tilde{a}^*$, respectively; when $\varkappa = r - 1$, we shall omit the index $k = 1$. Since the identity $\langle L, R\rangle = 0$ is valid for $\varkappa = r - 1$, $r \geqslant 2$, we can find a smooth vector function $h(x, \xi')$ such that $\tilde{a}h = R$. If, moreover,

$$\langle L_1(x, d_{x'}\varphi), P(R_1(x, d_{x'}\varphi))\rangle = 0, \tag{1.23}$$

a smooth vector function $H(x)$ can be found such that

$$\tilde{a}(x, d_{x'}\varphi)H(x) = P(R(x, d_{x'}\varphi)).$$

If $r = 3$ and $\varkappa = 1$, we may assume, without less of generality, that $\langle L_j, R_k\rangle = \delta_{j2}\delta_{k2}$.

Theorem 24 (Petkov [1975]). *If conditions* a), b) *and* (1.22) *are satisfied, $r = 2, 3$ and $\varkappa \geqslant 1$, then in order that the Cauchy problem for P be C^∞-well posed, it is necessary that* (1.23) *is satisfied in a neighbourhood of $\bar{x}$. Furthermore, for $r = 3$, $\varkappa = 2$, it is necessary that*

$$\langle L(x, d_{x'}\varphi), P(h(x, d_{x'}\varphi)) + H(x)\rangle = 0, \tag{1.24}$$

$$\langle L(x, d_{x'}\varphi), P(H(x))\rangle = 0 \tag{1.25}$$

in a neighbourhood of $\bar{x}$, while for $r = 3$, $\varkappa = 1$ it is necessary that

$$\mathscr{L}_{1,2} \cdot \mathscr{L}_{2,1} \equiv 0 \tag{1.26}$$

in a neighbourhood of $\bar{x}$, where $\mathscr{L}_{j,k} = \langle L_j(x, d_{x'}\varphi), P(R_k(x, d_{x'}\varphi))\rangle$.

We can put these necessary conditions for well-posedness in an almost equivalent form as we did with the Levi-Strang-Flaschka conditions for scalar operators. Let $\varkappa = r - 1$ or $\varkappa = 1$; in the latter case suppose that $R_1 \in \operatorname{Ran} \tilde{a}$. We introduce

Condition (L). For any real function φ satisfying (1.22) and any $f \in C_0^\infty$ there are scalar functions $g_k(x)$ and vector functions $v_k(x)$ such that $g_k(\bar{x}) = 0$ and

$$P\left[\left(f(x)R_1(x, d_{x'}\varphi) + \sum_{k=2}^{r-1} g_k(x)R_k(x, d_{x'}\varphi) + \sum_{k=1}^{\varkappa} v_k(x)\rho^{-k}\right)e^{i\varphi\rho}\right]$$
$$= 0(\rho^{-\varkappa}) \qquad \text{as } \rho \to +\infty$$

in a neighbourhood of $\bar{x}$.

It can be demonstrated that condition (L) is equivalent to (1.23) for $r = 2$, $\varkappa = 1$, to (1.23), (1.22), (1.25) for $r = 3$, $\varkappa = 2$, and to (1.23) and (1.26) for $r = 3$, $\varkappa = 1$; $\mathscr{L}_{1,2} \equiv 0$ or $\mathscr{L}_{2,1} \equiv 0$ in the neighbourhood of x.

Theorem 25 (Petkov [1978]). *Let all the characteristic roots $\lambda_j(x, \xi')$ of P be of constant multiplicity r_j in $X \times (\mathbb{R}^n\backslash 0)$. Let condition* b) *hold for any j, with $\varkappa_j \in \{0, 1, r_j - 1\}$, and at each point $(x, \xi') \in X \times (\mathbb{R}^n\backslash 0)$ let condition* (L) *hold for any j, with $\varkappa_j \geqslant 1$. Then the Cauchy problem for P is C^∞-well posed in any spatial type lens.*

The case where only $r = 2$ is constant while $\varkappa = 0, 1$ may be variables has been discussed in Demay [1977]. Let ${}^{co}p$ denote the matrix of cofactors of p, $l = p^s \cdot {}^{co}p - \frac{i}{2}\{p, {}^{co}p\}$, $l' = \{{}^{co}p \cdot p^s - \frac{i}{2}\{{}^{co}p, p\}$, where p^s is the subprincipal symbol of P, and let

$$\{f, g\} = \sum_{j=0}^{n} \left(\frac{\partial f}{\partial \xi_j}\frac{\partial g}{\partial x_j} - \frac{\partial f}{\partial x_j}\frac{\partial g}{\partial \xi_j}\right)$$

denote the matrix Poisson brackets.

Theorem 26. *Let all the characteristic roots $\lambda_j(x, \xi')$ of P be of constant multiplicity $r_j = 1, 2$ in $X \times (\mathbb{R}^n \backslash 0)$. Assume that there are smooth matrices $s_j(x, \xi')$ and $s_j'(x, \xi')$ such that*

$$\begin{aligned} \xi_0 = \lambda_j(x, \xi'), \qquad r_j = 2 \Rightarrow l(x, \xi) &= p(x, \xi)s_j(x, \xi'), \\ l'(x, \xi) &= s_j'(x, \xi')p(x, \xi). \end{aligned} \tag{1.27}$$

Then the Cauchy problem for P is C^∞-well posed in any spatial type lens.

1.11. Necessary Conditions for Regular Hyperbolicity of First-Order Systems in Two Variables. We conclude the present section with the only necessary condition, known to us, for the Cauchy problem to be C^∞-well posed for systems with variable coefficients and with characteristics of variable multiplicity.

Theorem 27 (Petkov and Kutev [1976]). *Let $n = 1$. Let the operator*

$$P = ID_0 + a(x)D_1 + b(x)$$

be x_0-hyperbolic in a neighbourhood of $\bar{x} \in \mathbb{R}^2$. Let $\bar{\lambda}$ be an eigenvalue of $a(\bar{x})$ having multiplicity $r \geqslant 2$ with $\operatorname{rank}(a(\bar{x}) - \bar{\lambda}I) = d - 1$. Set $\bar{p} = (\bar{x}, -\bar{\lambda}, 1)$. For regular hyperbolicity of P in the domain $X \ni \bar{x}$, it is necessary that the matrix $F_g(\bar{p})$ has a pair of non-zero real eigenvalues. Otherwise, all its eigenvalues are zero, and for the Cauchy problem for P to be C^∞-well posed in $X_T^+ \ni \bar{x}$, it is necessary that

$$\langle L_1, a'R_0\rangle - \langle L_0, a'R_1\rangle - 2i\langle L_0, b(\bar{x})R_0\rangle = 0,$$

where R_0, R_1, L_0, and L_1 are vectors such that

$$\bar{a}R_0 = 0, \qquad \bar{a}^*L_0 = 0, \qquad \bar{a}R_1 = R_0, \qquad \bar{a}^*L_1 = L_0,$$

$$\bar{a} = a(\bar{x}) - \bar{\lambda}I, \qquad a' = \left(\frac{\partial}{\partial x_0} + \bar{\lambda}\frac{\partial}{\partial x_1}\right)a(\bar{x}).$$

§ 2. Well-posedness of the Cauchy Problem in Gevrey Classes

2.1. The Main Definitions. As we have seen in the preceding section, the Cauchy problem is C^∞-well posed for non-strictly hyperbolic equations and systems only when, in general, some conditions on the lower terms are fulfilled.

If we desire the Cauchy problem to be well posed for any hyperbolic operator with analytic coefficients, we must give up C^∞-well-posedness and replace C^∞ by some narrower class of functions but strive for this new class to be as wide as possible.

Definition 13. Let $\varkappa \in [1, \infty)$ and $l \in \mathbb{Z}^+ U_\infty$. By an *inductive Gevrey class* $G^{l(\varkappa)}(X)$ we mean the set of functions $u \in C^{l,\infty}_{x_0,x'}(X)$ such that for any compact set $K \Subset X$ there are quantities $R = R_K$, $C = C_K$ for which the following inequalities hold:

$$|D^\alpha u(x)| \leqslant CR^{|\alpha|}(|\alpha|!)^\varkappa, \qquad \forall x \in K, \qquad \alpha \in \mathbb{Z}^{+(n+1)}: \alpha_0 \leqslant l. \tag{1.28}$$

By a *projective Gevrey class* $G^{l\langle\varkappa\rangle}(X)$ we mean the set of functions $u \in C^{l,\infty}_{x_0,x'}(X)$ such that for any compact set $K \Subset X$ and any $R > 0$ there is a $C = C_{K,R}$ for which the inequalities (1.28) hold.

Evidently, $G^{l\langle\varkappa\rangle} \subset G^{l(\varkappa)} \subset G^{l\langle\varkappa_1\rangle}$ for $\varkappa < \varkappa_1$, and $G^{l\{\varkappa\}} \subset G^{l_1\{\varkappa\}}$ for $l > l_1$, and all the embeddings are proper (throughout, $\{\cdot\}$ will mean either $(\cdot)$ or $\langle\cdot\rangle$). We denote by $G^{\infty(1)}$ and $G^{\infty\langle1\rangle}$, respectively, the space of analytic functions and that of entire analytic functions.

Definition 14. Let the coefficients of the operator P belong to $G^{l\{\varkappa\}}(X)$. The Cauchy problem (1.1)–(1.2) is said to be $G^{l\{\varkappa\}}$*-well posed* if the following conditions are satisfied:

Existence condition a). For any $f \in G^{l\{\varkappa\}}(X_T^+)$, $g_j \in G^{\{\varkappa\}}(X_T)$ $(j = 0, \ldots, m-1)$ there exists a $u \in G^{l+m\{\varkappa\}}(X_T^+)$ satisfying (1.1)–(1.2).

Uniqueness condition b) of Definition 1.

The Cauchy problem (1.1)–(1.2) is said to be *locally* $G^{l\{\varkappa\}}$*-well posed* if the following conditions are satisfied:

Existence condition a′). For any $f \in G^{l\{\varkappa\}}(X_T^+)$, $g_j \in G^{\{\varkappa\}}(X_T)$, $(j = 0, \ldots, m-1)$ there exists a neighbourhood X' of X_T and a $u \in G^{l+m\{\varkappa\}}(X_T'^+)$ satisfying (1.1)–(1.2) in $X_T'^+$.

Uniqueness condition b) of Definition 1 in each of the domains X' mentioned in the preceding condition.

We shall examine the $G^{(\varkappa)}$-well-posedness, $G^{\langle\varkappa\rangle}$-well-posedness and local $G^{(\varkappa)}$-well-posedness. Since $G^{l\langle\varkappa\rangle}$, in contrast to $G^{l(\varkappa)}$, are countably-normed spaces and since in conditions a) and a′) above the requirement $u \in G^{l+m\{\varkappa\}}(X_T^+)$ is, apparently, meaningfully equivalent to the requirement that $u \in C^m(X_T^+)$, the local $G^{\langle\varkappa\rangle}$-well-posedness has no new effects in comparison to the $G^{\langle\varkappa\rangle}$-well-posedness. Because till now we have not come across even a single example where the (local) $G^{l\{\varkappa\}}$-well-posedness depends substantially on l, we take some liberty and omit the letter l in the well-posedness notation.

2.2. Necessity and Sufficiency of Hyperbolicity. As we already mentioned, for the non-characteristic Cauchy problem to be well posed in Gevrey classes the hyperbolicity is a necessary condition.

Theorem 28 (Ivrii [1976b], Komatsu [1977], Nishitani [1978]). *Let the coefficients of the operator P belong to $G^{\infty\{\varkappa\}}$. Assume that the hyperplanes $\{x_0 = t\}$*

are non-characteristic. If $\varkappa > 1$ *and the Cauchy problem* (1.1)–(1.2) *is* $G^{\{\varkappa\}}$*-well posed, then* P *is* x_0*-hyperbolic in* X_T^+. *If, however, the Cauchy problem* (1.1)–(1.2) *is locally* $G^{\{\varkappa\}}$*-well posed, then* P *is* x_0*-hyperbolic in* $X_T'^+$, *where* X' *is some neighbourhood of* X_T.

What is more, if $\{\varkappa\} = (1)$ and the Cauchy problem (1.1)–(1.2) is $G^{(1)}$-well posed, then P is x_0-hyperbolic in X_T^+.

At the same time, by the Cauchy-Kovalevskaya theorem, for $\{\varkappa\} = (1)$ any non-characteristic Cauchy problem is locally $G^{(1)}$-well posed.

On the other hand, the hyperbolicity is a sufficient condition for the Cauchy problem to be well posed in a suitable Gevrey class.

Theorem 29 (Bronshtejn [1980]). *Let* P *be* x_0*-hyperbolic in* X, $S_+ = S_- = \varnothing$, *and let* X_T^+ *be a spatial type lens. Let* $\varkappa^* = r/(r-1)$, *where* r *is the maximum multiplicity of characteristic roots of* P. *If the coefficients of* P *are in* $G^{\infty(\varkappa)}(X)(G^{\infty\langle\varkappa\rangle}(X))$ *and* $\varkappa \in [1, \varkappa^*)(\varkappa \in (1, \varkappa^*])$, *then the Cauchy problem* (1.1)–(1.2) *is* $G^{(\varkappa)}$*-well posed* ($G^{\langle\varkappa\rangle}$*-well posed*). *If, however,* $\varkappa = \varkappa^*$, *then this problem is locally* $G^{(\varkappa)}$*-well posed.*

As confirmed by examples (see § 2.3 and § 2.4), in none of these assertions can the conditions imposed on $\varkappa$ be, in general, relaxed.

When $S_+ \cup S_- \neq \varnothing$, we have a less precise result.

Theorem 30 (Ivrii [1975]). *Let the operator* P *be* x_0*-hyperbolic in* X *and* X_T^+ *be a spatial type lens. Let the characteristic symbol* g *of the operator be analytic in* $X \times (\mathbb{R}^{n+1}\setminus 0)$, *and, finally, let the coefficients of* P *belong to* $G^{\infty\{\varkappa\}}(X)$. *Then the assertions of Theorem* 29 *remain valid for* $\varkappa^* = (2r-2)/(2r-3)$.

Remark. For $r = 2$, both the formulae for $\varkappa^*$ coincide.

We now examine the following problem. Obtain conditions for the Cauchy problem to be well posed in a given Gevrey class. We shall assume below, without mentioning it each time, that P is x_0-hyperbolic in X and that X_T^+ is a spatial type lens.

2.3. Operators with Constant Coefficients. For such operators the well-posedness conditions can be easily obtained by means of the Fourier transformation.

Theorem 31 (Hörmander [1963]). *Let* $P(D)$ *be a (matrix) operator with constant coefficients and* $g(N) \neq 0$, $N = (1, 0, \ldots, 0)$. *Then for the Cauchy problem for* P *to be* $G^{(\varkappa)}$*-well posed, it is necessary and sufficient that for any* $\alpha > 0$ *there exists a* $C = C_\alpha$ *such that*

$$\forall \xi' \in \mathbb{R}^n \; G(\zeta_0, \xi') = 0 \Rightarrow |\operatorname{Im} \zeta_0| \leqslant \alpha |\xi'|^{1/\varkappa} + C, \tag{1.29}$$

where $G(\zeta) = \det P(\zeta)$. *For the problem to be* $G^{\langle\varkappa\rangle}$*-well posed or locally* $G^{(\varkappa)}$*-well posed, it is necessary and sufficient that* (1.29) *holds for suitable* α *and* C.

This result and the Seidenberg-Tarski theorem together give the following

Corollary. *Let $g \in \text{hyp}(N, M)$ but $G \notin Hyp(N, M)$. Then the Cauchy problem for P is $G^{(\varkappa)}$-well posed for $\varkappa \in [1, \varkappa^*]$ (and for these values only) and is $G^{\langle\varkappa\rangle}$-well posed as well as locally $G^{(\varkappa)}$-well posed for $\varkappa \in [1, \varkappa^*]$ (and for these values only), where $\varkappa^* = \varkappa^*(G)$ is a rational number, $\varkappa^* \geqslant r/(r-1)$. When $G \in \text{Hyp}(N, M)$, we can take $\varkappa^* = \infty$.*

We note that the well-posedness conditions (1.29) may be replaced by the following conditions:

$$\begin{gathered} \forall \xi \in \mathbb{R}^{n+1}, \qquad \tau \in \mathbb{C} \\ G(\xi + \tau N) = 0 \Rightarrow |\text{Im}\, \tau| \leqslant \alpha |\xi|^{1/\varkappa} + C. \end{gathered} \tag{1.29'}$$

Therefore for any $N \neq 0$, we can introduce the classes of polynomials

$$\text{Hyp}(N, M, \varkappa)$$

and

$$\text{Hyp}(N, M, \langle\varkappa\rangle) = \bigcap_{\varkappa' < \varkappa} \text{Hyp}(N, M, \varkappa').$$

These classes decrease as $\varkappa$ increases, and we can establish for them Properties (1.7) and (1.11). Theorem 7 can also be proved, but now the fundamental solution lies in $\mathscr{D}'_{\{\varkappa\}}$, the space of generalized functions over $G^{\infty\{\varkappa\}} \cap C_0$. But Theorem 31 will become valid only after the localization concept is modified.

In what follows, we shall examine only scalar operators.

2.4. Operators with Characteristics of Constant Multiplicity. We can prove for such operators also necessary and sufficient conditions for the problem to be well posed in Gevrey classes. Suppose that in a neighbourhood of the point $\bar\rho = (\bar x, \bar\xi) \in X \times (\mathbb{R}^{n+1} \backslash 0)$ the principal symbol of P is of the form $p = aq^r$, where $a(\bar\rho) \neq 0$, $q(\bar\rho) = 0$, and where $H_q(\bar\rho) \neq 0$ and is non-collinear with $\xi \dfrac{\partial}{\partial \xi}$, $r \in \mathbb{Z}^+ \backslash 0$. We shall say that the *generalized Levi condition* ($L'_\varkappa$) is satisfied in this neighbourhood if the representation (1.16) holds to within a pseudodifferential operator with symbol zero in the neighbourhood of $\bar\rho$; here Q is a first-order pseudodifferential operator with principal symbol q and A_k are pseudodifferential operators of order not exceeding $m - r + k/\varkappa$. If the orders of A_k are strictly less than $m - r + k/\varkappa$ for $k \geqslant 1$, we shall say that the *generalized Levi condition* ($L_\varkappa$) is satisfied. What is more, just as in § 1.5, all the operators can be taken to be differential operators with respect to x_0 and pseudodifferential with respect to x'; Remark a) of § 1.5 remains valid. If $d_\xi q(\bar\rho) \neq 0$ and φ is a real function that satisfies the eikonal equation in a neighbourhood of $\bar x$ and is such that $d\varphi(\bar x) = \bar\xi$, then the condition ($L_\varkappa$) means precisely that

$$\begin{aligned} P(\exp i(\varphi\rho + \psi\rho^{1/\varkappa})\cdot h) &= \rho^{m - r(1 - 1/\varkappa)} \exp i(\varphi\rho + \psi\rho^{1/\varkappa}) \\ &\times \left(\sum_{|\alpha| = r} \frac{1}{\alpha!}\, \partial_\xi^\alpha P(x, d\varphi)(d\psi)^\alpha h + o(1) \right) \text{ as } \rho \to +\infty \end{aligned} \tag{1.30}$$

in the neighbourhood of $\bar{x}$ for any real function $\psi \in C^\infty$ and any $h \in C^\infty$. Similarly, condition $(L'_\varkappa)$ signifies precisely that

$$P(\exp i(\varphi\rho + \psi\rho^{1/\varkappa})\cdot h) = 0(\rho^{m-r(1-1/\varkappa)} \text{ as } \rho \to +\infty \tag{1.31}$$

in the neighbourhood of $\bar{x}$ for any real function $\psi \in C^\infty$ and any $h \in C^\infty$.

Finally, the implications $(L) \Rightarrow (L_\varkappa) \Rightarrow (L'_\varkappa) \Rightarrow (L_{\varkappa'})$ turn out to be valid for $\varkappa > \varkappa'$, and it can easily be shown that of all these conditions substantially distinct are only the conditions (L), $(L_{i/j})$ with $1 \leqslant j < i \leqslant r$, $i, j \in \mathbb{Z}^+$.

Theorem 32 (Ivrii [1976b], Komatsu [1977]). *Let the operator P be x_0-hyperbolic in X and have characteristics of constant multiplicity. Let the coefficients of P belong to $G^{\infty(\varkappa)}(X)(G^{\infty\langle\varkappa\rangle}(X))$, and let X_T^+ be a spatial type lens. Then for the Cauchy problem (1.1)–(1.2) to be $G^{(\varkappa)}$-well posed ($G^{\langle\varkappa\rangle}$-well posed), it is necessary and sufficient that the condition $(L_\varkappa)$ (condition $(L'_\varkappa)$) is satisfied in X_T^+. For the problem to be locally $G^{(\varkappa)}$-well posed, it is necessary and sufficient that the condition $(L'_\varkappa)$ is satisfied in $X_T'^+$, where X' is some neighbourhood of X_T.*

2.5. Necessary Conditions for Well-posedness in the Gevrey Classes. The following result is analogous to Theorem 21.

Theorem 33 (Ivrii [1976c]). *Let the operator P have analytic coefficients. Let $\bar{x} \in X_T^+$, $\bar{\xi} = (0, \ldots, 0, 1)$ and $q = (q_0, \ldots, q_n) \in [0, 1)^{n+1}$. Assume that conditions (1.19) and (1.20) are satisfied.*

(i) *If the Cauchy proboem (1.1)–(1.2) is $G^{(\varkappa)}$-well posed, then*

$$\partial_\xi^\alpha \partial_x^\beta P_s(\bar{x}, \bar{\xi}) = 0 \tag{1.32}$$

for $|\alpha| + \langle\beta - \alpha, q\rangle \leqslant r(1 - q_0) - (m - s)\varkappa/(\varkappa - 1)$, $s < m$.

(ii) *If $q_0 = 0$, $\bar{x} \in X_T$ and the Cauchy problem (1.1)–(1.2) is locally $G^{(\varkappa)}$-well posed, then*

$$\partial_\xi^\alpha \partial_x^\beta P_s(\bar{x}, \bar{\xi}) = 0 \tag{1.33}$$

for $|\alpha| + \langle\beta - \alpha, q\rangle \langle r(1 - q_0) - (m - s)\varkappa/(\varkappa - 1)$, $s < m$.

(iii) *If $q_0 > 0$, $\bar{x} \in X_T$ and the Cauchy problem (1.1)–(1.2) is locally $G^{(\varkappa)}$-well piosed, then (1.32) holds.*

(iv) *If the Cauchy problem (1.1)–(1.2) is $G^{\langle\varkappa\rangle}$-well posed, then (1.33) holds.*

It follows from this theorem and from a), b) of Theorem 10 that for the Cauchy problem to be $G^{(\varkappa)}$-well posed or to be locally $G^{\langle\varkappa\rangle}$-well posed for $\bar{x} \in X_T$ ($G^{\langle\varkappa\rangle}$-well posed), it is necessary to impose suitable conditions on all the members of P of order not lower (greater) than $m - \gamma r(\varkappa - 1)/\varkappa$, where $\gamma = \frac{1}{2}$ if $\bar{x} \in \mathring{X}$ and $\gamma = \frac{1}{3}$ if $\bar{x} \in S_+ \cup S_-$.

Definition 15. Let the operator P have analytic coefficients. The Cauchy problem is said to be *$G^{\{\varkappa\}}$-regular* (*locally $G^{\{\varkappa\}}$-regular*) if it is $G^{\{\varkappa\}}$-well posed (locally $G^{\{\varkappa\}}$-well posed) for the operaor $P + Q$, where Q is any operator, with coefficients in $G^{\infty\{\varkappa\}}$, of order not greater than $m - 1$.

Theorem 34 (Ivrii [1976c]). *Let the operator P have analytic coefficients, and let $\bar{x} \in X_T^+$, $\bar{\xi} = (0, \ldots, 0, 1)$ and $q = (q_0, \ldots, q_n) \in [0, 1)^{n+1}$. Assume that $\partial_\xi^\alpha \partial_x^\beta p(\bar{x}, \bar{\xi}) = 0$ for $|\alpha| + \langle \beta - \alpha, q \rangle < \sigma$.*

(i) *If the Cauchy problem* (1.1)–(1.2) *is $G^{(\varkappa)}$-regular, then $\varkappa < \varkappa^* = \sigma/(\sigma - 1)$.*

(ii) *If $q_0 = 0, \bar{x} \in X_T$ and the Cauchy problem* (1.1)–(1.2) *is locally $G^{(\varkappa)}$-regular, then $\varkappa \leqslant \varkappa^*$.*

(iii) *If $q_0 > 0, \bar{x} \in X_T$ and the Cauchy problem* (1.1)–(1.2) *is locally $G^{(\varkappa)}$-regular, then $\varkappa < \varkappa^*$.*

(iv) *If the Cauchy problem* (1.1)–(1.2) *is $G^{\langle \varkappa \rangle}$-regular, then $\varkappa \leqslant \varkappa^*$.*

The following result has no analogue in the C^∞-well-posedness theory.

Theorem 35 (Irvii [1976c]). *Let the operator P have analytic coefficients, and let $\bar{x} \in X_T^+$, $\bar{\xi} = (0, \ldots, 0, 1)$ and $q = (q_0, \ldots, q_n) \in [0, 1)^{n+1}$.* Assume that

$$\partial_\xi^\alpha \partial_x^\beta p(\bar{x}, \bar{\xi}) = 0$$

for

$$|\alpha| + \langle \beta - \alpha, q \rangle \leqslant \sigma, \qquad (\alpha, \beta) \neq (\bar{\alpha}, \bar{\beta}),$$

where $\bar{\alpha} = \bar{\beta} = (\sigma, 0, \ldots, 0)$, and that

$$\partial_{x_0}^\sigma \partial_{\xi_0}^\sigma p(\bar{x}, \bar{\xi}) \neq 0.$$

If the Cauchy problem (1.1)–(1.2) *is $G^{\langle \varkappa \rangle}$-regular, then $\varkappa < \varkappa^* = r/(r - 1)$.*

By Theorem 34 and the results a) and b) of Theorem 10, for the Cauchy problem to be $G^{(\varkappa)}$-regular or locally $G^{(\varkappa)}$-regular for $\bar{x} \in X_T (G^{\langle \varkappa \rangle}$-regular), it is necessary that $\varkappa < \varkappa^* (\varkappa \leqslant \varkappa^*)$, where $\varkappa^* = r/(r - 2)$ if $\bar{x} \in X$ and $\varkappa^* = r/(r - 3)$ if $\bar{x} \in S_+ \cup S_-$.

Remark. If the corresponding inequality $\varkappa \leqslant \varkappa^*$ or $\varkappa < \varkappa^*$ does not hold in any one of the assertions (i)–(iv) of Theorem 34 or of Theorem 35, then for the Cauchy problem to be suitably well posed it is necessary that $P_{m-1}(\bar{x}, \bar{\xi}) = 0$.

2.6. Examples. Since fairly general results, apart from Theorems 29 and 30, concerning well-posedness in Gevrey classes of the Cauchy problems for operators with characteristics of variable multiplicity are not available, we confine ourselves to examining a few examples that are, in our view, quite instructive.

Example 7 (Ivrii [1978]). Let $0 \in X_T \subset \mathbb{R}^2$, and let

$$P = D_0^2 - x_1^{2\mu} D_1^2 + a(x) x_1^\nu D_1,$$

where $\mu, \nu \in \mathbb{Z}^+$, $\nu < \mu$, and a is analytic in X, with $a(0) \neq 0$.

For the Cauchy problem to be locally $G^{(\varkappa)}$-well posed or to be $G^{\langle \varkappa \rangle}$-well posed, a necessary and sufficient condition is $\varkappa \leqslant \varkappa^* = (2\mu - \nu)/(\mu - \nu)$. If $\mu > 1$, a necessary and sufficient condition for the $G^{(\varkappa)}$-well-posedness is $\varkappa < \varkappa^*$. If, however, $\mu = 1$, then a sufficient condition for the $G^{(\varkappa)}$-well-posedness is $\varkappa < \varkappa^* = 2$ and a necessary condition is $\varkappa \leqslant 2$. Thus the question whether the Cauchy problem for P is $G^{(2)}$-well posed for $\mu = 1$ and $\nu = 0$ remains open.

Example 8. Let $P = P_1 + cD_n$ be the operator as in Examples 4 and 4′, and $c \in \mathbb{C}\backslash[-\overline{\omega}, \overline{\omega}]$. Let $X_T \cap \{x_1 = \cdots = x_k = 0\} \neq \emptyset$. We recall that the condition that $c \in [-\overline{\omega}, \overline{\omega}]$ is necessary and sufficient for the Cauchy problem for P to be C^∞-well posed.

It can easily be shown that a necessary and sufficient condition for the $G^{(\varkappa)}$-well-posedness is $\varkappa < 2$, while for the $G^{\langle\varkappa\rangle}$-well-posedness or for the local $G^{(\varkappa)}$-well-posedness it is $\varkappa \leqslant 2$.

Example 9 (Ivrii [1978]). Let (i) $0 \in X_T$ and

$$P = D_0^2 - x_0^{2\mu} D_1^2 + a(x_0) x_0^\nu D_1,$$

or let (ii) $0 \in S_+$ and

$$P = D_0^2 - (-x_0)^{2\mu} D_1^2 + a(x_0) x_0^\nu D_1,$$

where ν, $2\mu \in \mathbb{Z}^+$, $\nu < \mu - 1$, and a is analytic on $[T_-, T_+]$, with $a(0) \neq 0$.

Then a necessary and sufficient condition for the $G^{(\varkappa)}$-well-posedness, as well as for the local $G^{(\varkappa)}$-well-posedness in case (i), is the inequality $\varkappa < \varkappa^* = (2\mu - \nu)/(\mu - \nu - 1)$, while for the $G^{\langle\varkappa\rangle}$-well-posedness it is $\varkappa \leqslant \varkappa^*$.

Example 10 (Ivrii [1978]). Let $n = 1$, and let

$$p(x, \xi) = \prod_{j=1}^{s} (\xi_0 - \lambda_j x_0^\mu \xi_1)^{r_j},$$

where μ, $r_j \in \mathbb{Z}^+\backslash 0$, $\lambda_j \neq \lambda_k$ for $j \neq k$ and $0 \in X_T^+$. We set $r = \max r_j$, $\varkappa_1 = r/(r-1)$, $\varkappa_2 = m/(m - 1 - 1/\mu)$, where $m = r_1 + \cdots + r_s$, and, finally, $\varkappa^* = \min(\varkappa_1, \varkappa_2)$.

If $\varkappa_1 < \varkappa_2$, a necessary and sufficient condition for the $G^{(\varkappa)}$-regularity of the Cauchy problem is $\varkappa < \varkappa^*$, while for the $G^{\langle\varkappa\rangle}$-regularity or local $G^{(\varkappa)}$-regularity for $0 \in X_T$ it is $\varkappa \leqslant \varkappa^*$. If, however, $\varkappa_1 > \varkappa_2$, then a necessary and sufficient condition for the $G^{(\varkappa)}$-regularity or for the local $G^{(\varkappa)}$-regularity for $0 \in X_T$ is $\varkappa < \varkappa^*$; a necessary condition for the $G^{\langle\varkappa\rangle}$-regularity is $\varkappa \leqslant \varkappa^*$ and a sufficient condition is $\varkappa < \varkappa^*$. If $\varkappa = \varkappa^*$, the Cauchy problem is $G^{\langle\varkappa\rangle}$-well posed for any operator whose principal symbol is p and coefficients are analytic and depend only on x_0. Finally, if $\varkappa_1 = \varkappa_2$, a necessary and sufficient condition for the $G^{(\varkappa)}$-regularity or for the local $G^{(\varkappa)}$-regularity for $0 \in X_T$ or for the $G^{\langle\varkappa\rangle}$-regularity is $\varkappa < \varkappa^*$.

Example 11 (Ivrii [1978]). Let (i) $0 \in X_T$ and

$$p(x, \xi) = \xi_0^{r_1} \prod_{j=2}^{s} (\xi_0^2 - \lambda_j^2 x_0^{2\mu} |\xi'|^2)^{r_d},$$

or let (ii) $0 \in S_+$ and

$$p(x, \xi) = \xi_0^{r_1} \prod_{j=2}^{s} (\xi_0^2 - \lambda_j^2 (-x_0)^{2\mu} |\xi'|^2)^{r_d},$$

where 2μ, $r_j \in \mathbb{Z}^+\backslash 0$, $\lambda_j \neq \lambda_k$ for $j \neq k$ and $\lambda_j > 0 \,\forall j$.

Then all the conclusions of the preceding example remain valid.

Remark. If $\nu \geqslant \mu$ in Example 7 or $\nu \geqslant \mu - 1$ in Example 9, then the Cauchy problem is C^∞-well posed. If in Examples 10 and 11 $r = 1$ and $m = 2$, $\mu = \frac{1}{2}$, 1 or $m = 3$, $\mu = \frac{1}{2}$, then the Cauchy problem is C^∞-regular. In all the Examples 7 to 11, the fact that the Cauchy problem is C^∞-well posed (regular) implies that it is $G^{\{\varkappa\}}$-well posed (regular) provided that the coefficients of the operator belong to the corresponding Gevrey class.

Example 12 (Yagdzhyan [1978]). Let $0 \in X_T = S_-$ and let

$$P = D_0^2 - \exp(-2x_0^{-1})D_1^2 + Bx_0^l \exp(-\beta x_0^{-1})D_1,$$

where $B \in \mathbb{C}\backslash 0$, $l \in \mathbb{R}$, $\beta \leqslant 1$. Then for $\beta < 1$ and $\varkappa < (2 - \beta)/(1 - \beta)$, and for $\beta = 1$ and any $\varkappa$, the Cauchy problem is $G^{(\varkappa)}$- and $G^{\langle\varkappa\rangle}$-well posed; at the same time, if $\beta = 1$, it is C^∞-well posed not for all l (see Example 6).

Example 13. Assume that $\beta = 0$, $l \in \mathbb{Z}^+$ in the preceding example. Then a necessary and sufficient condition for the $G^{(\varkappa)}$-well-posedness, the local $G^{(\varkappa)}$-well-posedness or the $G^{\langle\varkappa\rangle}$-well-posedness is that $\varkappa \leqslant 2$.

§ 3. Propagation of Singularities

We mean by singularities various types of wave fronts (C^∞, $G^{\{\varkappa\}}$ or analytic). It should be kept in mind that the hyperbolicity of an operator is a concept that is local in x but global ξ, while at the same time the wave front is a microlocal concept, that is, is local in both x and ξ. Therefore for describing the propagation of singularities the hyperbolicity of the operator is not as important a property as are its other microlocal properties, such as microhyperbolicity. However, hyperbolicity is a strong constraint too (for instance, the localization of a hyperbolic operator is itself hyperbolic), and hyperbolic operators cannot belong to several microlocal types. We confine our attention to only those microlocal types that are admissible for hyperbolic operators.

3.1. Propagation of C^∞-Singularities. We first recall the basic concepts of the microlocal analysis (Hörmander [1971], Duistermaat and Hörmander [1972], Taylor [1981], Trèves [1980]).

Definition 16. Let $u \in \mathscr{D}'(\Omega)$. The *wave front* of u is a closed conical set $\mathrm{WF}(u) \subset T^*\Omega\backslash 0$: $(\bar{x}, \bar{\xi}) \notin \mathrm{WF}(u)$ is there is a function $\varphi \in C_0^\infty(\Omega)$, with $\varphi(\bar{x}) = 1$, and a conical neighbourhood Γ of $\bar{\xi}$ such that

$$(1 + |\xi|)^s F(\varphi u) \in L_2(\Gamma) \tag{1.34}$$

for all $s \in \mathbb{R}$; here $F = F_{x\to\xi}$ denotes the Fourier transform. For each $s \in \mathbb{R}$, the same condition defines $\mathrm{WF}^s(u)$, the *wave front of order s*. We recall that

$$\mathrm{WF}(Pu) \subset \mathrm{WF}(u) \subset \mathrm{WF}(Pu) \cup \mathrm{Char}\, P, \tag{1.35}$$

$$\mathrm{WF}^s(Pu) \subset \mathrm{WF}^{s+m}(u) \subset \mathrm{WF}^s(Pu) \cup \mathrm{Char}\, P, \tag{1.35'}$$

where Char $P = \{g \equiv \det p = 0\} \subset T^*\Omega\backslash 0$ is the *characteristic set of the operator* P and m is the order of P.

We start by examining the propagation of singularities of symmetric hyperbolic systems for which the Cauchy problem is undoubtedly well posed. Let

$$P = ID_0 - A(x, D'), \tag{1.36}$$

where A is a differential or a classical pseudodifferential operator (Taylor [1976, 1979, 1981]) with principal symbol $a(x, \xi')$; $a(x, \xi')$ is a Hermitian matrix.

Theorem 36 (Ivrii [1979c]). *Let* $\Omega = (T_-, T_+) \times \omega$ *and* $T \in (T_-, T_+)$. *Let* $u \in C(T_-, T_+), \mathscr{D}'(\omega))$, $Pu \in C^\infty(\Omega)$ *and* $u|_{\Omega_T} = v$. *Then*

$$\mathrm{WF}(u)|_{\Omega_T} \subset \iota^{-1}\mathrm{WF}(v) \cup (N^*\Omega_T\backslash 0), \tag{1.37}$$

where ι: Char $P|_{\Omega_T} \to T^*\Omega_T$ *is the natural map, and* $N^*\Omega_T$ *is the normal bundle of* Ω_T. *Moreover,*

$$\mathrm{WF}(u)|_{\Omega_T} \subset \iota^{-1}\mathrm{WF}(v), \tag{1.37'}$$

if P *is a differential operator.*

The same assertions hold for WF^s *too.*

Theorem 36 enables us to reduce the study of the propagation of singularities of solutions to the Cauchy problem to that of the propagation in the whole space.

Suppose now that $P = P(x, D)$ is a classical pseudodifferential operator of first order. Assume that the following is satisfied:

Condition a). The matrix $p(x, \xi)$, the principal symbol of P, is Hermitian.

Theorem 37 (Ivrii [1979c]). *Suppose that condition* a) *is satisfied. Let* $T^*\Omega\backslash 0 \ni V$ *be an open conical set. Let* $\varphi \in C^\infty(V)$ *be a real positive definite homogeneous function (in* ξ*) of degree* 0 *such that*

$$\langle\{p, \varphi\}(\rho)v, v\rangle > 0 \qquad \forall v \in \operatorname{Ker} P(\rho)\backslash 0, \quad \rho \in \operatorname{Char} P \cap V. \tag{1.38}$$

If

$$\mathrm{WF}(Pu) \cap V \cap \{\varphi \leqslant 0\} \cap \operatorname{Char} P = \varnothing, \tag{1.39}$$

$$\mathrm{WF}(u) \cap \partial V \cap \{\varphi \leqslant 0\} \cap \operatorname{Char} P = \varnothing, \tag{1.40}$$

then

$$\mathrm{WF}(u) \cap V \cap \{\varphi \leqslant 0\} \cap \operatorname{Char} P = \varnothing. \tag{1.41}$$

The same conclusion holds for WF^s *too.*

This theorem bears a strong resemblance to the uniqueness theorem for the Cauchy problem in a spatial type lens. But instead of the spatial type lens, a subset of $\mathbb{R}^{n+1}_x$, we now have $V \cap \{\varphi \leqslant 0\}$, which is a subset of the phase space, and instead of the base of the lens we have the set $\partial V \cap \{\varphi \leqslant 0\}$, while the statements of the type $v = 0$ in X are now replaced by the statements of the type $v \sim 0$ in W (that is, $\mathrm{WF}(v) \cap W = \varnothing$).

Assume that the following is satisfied:

Condition b). The condition (1.38) holds for $\varphi = x_0$.

Then for any real function the condition (1.38) can be rewritten in the equivalent form

$$(H_\varphi^k H_{x_0}^{r-k} g)(\rho) \cdot (H_{x_0}^r g)^{-1}(\rho) > 0$$
$$\forall k = 1, \ldots, r(\rho) = \dim \operatorname{Ker} p(\rho), \qquad \rho \in \operatorname{Char} P \cap V, \tag{1.42}$$

where

$$(H_{x_0}^r g)(\rho) \neq 0, \qquad (\partial_\xi^\alpha \partial_x^\beta g)(\rho) = 0$$

for $|\alpha| + |\beta| < r = r(\rho)$ when conditions a) and b) are satisfied.

Theorem 37 can be stated in an equivalent form. Let $\rho \in \operatorname{Char} P$ and let $g\rho(\cdot)$, the localization of symbol g at ρ (see b) of Theorem 10), be a polynomial on $T_\rho T^*\Omega$. It can easily be shown that this polynomial is hyperbolic in the direction $H_{x_0}(\rho)$ if conditions a) and b) hold. Assume that $T_\rho T^*\Omega \supset \Gamma(\rho) = \Gamma(g_\rho(\cdot)$, and $H_{x_0}(\rho))$, containing $H_{x_0}(\rho)$, is a connected component of the set $\{\zeta, g_\rho(\zeta) \neq 0\}$. Under these assumptions (1.42) is equivalent to the inclusion $H_\varphi(\rho) \subset \Gamma(\rho)$. Finally, let

$$\Gamma^\sigma(\rho) = \{z \in T_\rho T^*\Omega \colon \forall \zeta \in \Gamma(\rho) \sigma(z, \zeta) \geqslant 0\}$$

be the *dual cone*, where σ is a symplectic form.

Definition 17. A *generalized bicharacteristic of the operator* P is a connected Lipschitz curve $\gamma = \{\rho = \rho(t) \in \operatorname{Char} P\}$ such that $\dfrac{d\rho}{dt} \in \Gamma^\sigma(\rho)$ for almost all t.

We shall analyse below the notion of the generalized bicharacteristic in various particular cases.

We can clearly take $t = x_0$ for a parameter on the generalized bicharacteristic. We now assume that $T^*\Omega \backslash 0 \Supset V$ is an open conical set and that $\bar\rho \in \operatorname{Char} P \cap V$. Let $K^\pm(\bar\rho, V)$ be the union of all generalized bicharacteristics issuing from $\bar\rho$ in the direction of increasing (decreasing) x_0 and lying in V. It can easily be shown that

$$K^\pm(\bar\rho, V) \text{ are closed sets,} \tag{1.43}$$

$$\varlimsup_{\rho \to \bar\rho} K^\pm(\rho, V) \subset K^\pm(\bar\rho, V), \tag{1.44}$$

that is, for any W-neighbourhood of $K^\pm(\bar\rho, V)$ we can find a W_0-neighbourhood of $\bar\rho$ such that

$$\rho \in W_0 \cap \operatorname{Char} P \Rightarrow K^\pm(\rho, V) \subset W, \tag{1.45}$$

$$\forall \rho, \rho', \rho'' \in V \quad \rho' \in K^\pm(\rho, V),$$
$$\rho'' \in K^\pm(\rho', V) \Rightarrow \rho'' \in K^\pm(\rho, V),$$
$$\forall \rho, \rho' \in V \quad \rho' \in K^\pm(\rho, V) \Leftrightarrow \rho \in K^\mp(\rho', V). \tag{1.46}$$

In other words, the relations $\rho' \in K^{\pm}(\rho, V)$ between the points $\rho, \rho' \in V$ are transitive and symmetric.

We can now reformulate Theorem 37 in the following equivalent form.

Theorem 38 (Wakabayashi [1984]). *Suppose that conditions* a) *and* b) *above hold. Let* $T^*\Omega\backslash 0 \ni V$ *be an open conical set and* $\bar{\rho} \in \operatorname{Char} P \cap V$. *If*

$$\operatorname{WF}(Pu) \cap K^{\pm}(\bar{\rho}, V) = \varnothing, \tag{1.47}$$

$$\operatorname{WF}(u) \cap K^{\pm}(\bar{\rho}, V) \cap \partial V = \varnothing, \tag{1.48}$$

then $\operatorname{WF}(u) \cap K^{\pm}(\bar{\rho}, V) = \varnothing$.

The same conclusions hold for WF^s *too.*

Like Theorem 37, this theorem also resembles the uniqueness theorem for the Cauchy problem, only now in a characteristic conoid the role of which is played by $K^{\pm}(\bar{\rho}, V)$.

Theorems 36, 37 and 38 enable us to estimate from above the wave front of the fundamental solution of the Cauchy problem. Namely, let $E = E(x, y) \in C^{\infty}(X_y, \mathscr{D}'(\Omega_x))$ satisfy the conditions

$$P(x, D_x)E = I\delta(x - y), \tag{1.49}$$

$$\operatorname{supp} E \subset \{x_0 \geqslant y_0\}. \tag{1.50}$$

If P is a symmetric x_0-hyperbolic differential operator of first order and $\mathbb{R}^{n+1} \ni X$ is a spatial type lens, then such a function exists, is unique and satisfies the dual system in y.

Theorem 39. *Let* $\mathbb{R}^{n+1} \ni X$ *be a spatial type lens and* P *a first-order differential operator satisfyng* (1.47) *and* (1.48). *Then*

$$\begin{aligned} {}^T\operatorname{WF}(E) \subset \operatorname{diag}(T^*X\backslash 0) \cup \{(\rho, \rho') \in (\operatorname{Char} P \cap T^*X\backslash 0)^2,& \\ \rho \in K^{+}(\rho', T^*X\backslash 0)\},& \end{aligned} \tag{1.51}$$

where ${}^TM = \{(x, \xi, y, \eta)\colon (x, \xi, y, -\eta) \in M\}$, $M^2 = M \times M$ *and* $\operatorname{diag} M^2 = \{(\rho, \rho), \rho \in M\} \subset M^2$ *is the diagonal of* M^2.

It turns out that Theorems 36, 37, 38 and 39 remain valid for other types of operators, described in § 1, too. Namely, suppose that one of the following conditions holds. P is strictly hyperbolic or else has constant coefficients and is Gårding hyperbolic; the scalar operator P is completely regularly hyperbolic or else has characteristic roots or at least characteristics of constant multiplicity and satisfies the Levi-Strang-Flaschka condition or the hypotheses of one of the Theorems 4, 19 or 20; the first-order matrix operator P satisfies the hypotheses of one of the Theorems 25 or 26. Then the conclusions of Theorems 36, 37, 38 or 39 remain valid if we carry out the following modifications.

(i) In Theorem 36, we should now set

$$v = D_0^j u|_{\Omega_T}; \qquad j = 0, \dots, m - 1.$$

(ii) If P is hyperbolic only in X, then we should now regard u as an element of $\mathscr{D}'(\mathring{X})$ and replace $\mathrm{WF}(u)$ by $\overline{\mathrm{WF}(u)} \cap (T^*X\backslash 0)$; this concerns Theorems 36 and 39.

(iii) In Theorem 37, we replace condition (1.38) by (1.42).

(iv) In the statements concerning wave fronts of finite order, we should write $\mathrm{WF}^s(Pu)$, $\bigcup_{j=0}^{m-1} \mathrm{WF}^{s+m-j-1}(v_j)$ and $\mathrm{WF}^{s+m-L-1}(u)$. Here $L = 0$ for strictly and completely regularly hyperbolic operators, $L = \frac{1}{2}$ for operators satisfying the conditions of Theorem 19, $L = 1$ for operators satisfying the conditions of Theorem 20, $L = r - 1$ for scalar operators with characteristics of constant multiplicity, $L = \varkappa$ and $L = 1$ for operators satisfying, respectively, the conditions of Theorems 25 and 26, etc.

It should be kept in mind that if we choose $t = x_0$ for the parameter, then the generalized bicharacteristic will be a Lipschitz curve only if $x_0 \in (T_-, T_+)$. In order that the curve be Lipschitz near $x_0 = T_\pm$, we must take for the parameter on it $t = -(T_+ - x_0)^{1/2}$ and $t = (x_0 - T_-)^{1/2}$, respectively.

In conclusion, we remark that Theorems 37 and 38 remain valid also for more general operators than the ones mentioned above. In particular, the conditions imposed on the lower terms that guarantee the desirable propagation of singularities are far weak than the well-posedness conditions for the Cauchy problem (Ivrii [1979b]).

3.2. The Geometry of the Propagation of Singularities. In this section, we shall discuss the structure of the sets $K^\pm(\rho)$. We first note that if in a neighbourhood of ρ the characteristic symbol of the operator is of the form $g = eg_1^{s_1} \dots g_k^{s_k}$, where $e(\rho) \neq 0$, g_j are real and $g_j(\rho) = 0$, $s_j \in \mathbb{Z}^+\backslash 0$, then φ satisfies (1.42) for g if and only if it satisfies the same for each g_j. In particular, if C is a connected component of the set $\{g = 0\}$ and if $g = eg_1^s$ in a neighbourhood of C, $e \neq 0$ on C and $\rho \in C$, then the sets $K^\pm(\rho)$, constructed for g and g_1, coincide.

If $g(\rho) = 0$ and $g_{\xi_0}(\rho) \neq 0$, then (1.42) implies precisely that $\dfrac{d\varphi}{dx_0} > 0$, where $\dfrac{d}{dx_0}$ denotes the differentiation along the *bicharacteristic* of g, that is, the curve $\gamma = \{\rho = \rho(t)\}$ along which

$$\frac{d\rho}{dt} = H_g(\rho), \qquad g(\rho) = 0. \tag{1.52}$$

Therefore if the symbol g is strictly (or completely regularly) x_0-hyperbolic in X, then $K^\pm(\rho)$ coincides with $\gamma_g^\pm(\rho)$, the bicharacteristic of g issuing from ρ in the direction of increasing (decreasing) x_0. Thus we obtain a unique description of the propagation of singularities for x_0-hyperbolic operators with characteristics of constant multiplicity. What is more, we can do away with the hyperbolicity condition and establish the following

Theorem 40. *Let $T^*\Omega\backslash 0 \ni V$ and V_1 be open conical sets. Suppose that $g = eg_1^s$ in V, where $e \neq 0$, $s \in \mathbb{Z}^+\backslash 0$ and g_1 is a* ***real symbol of the principal type,*** *that is,*

$$g_1 = 0 \Rightarrow H_{g_1} \neq 0 \text{ and is non-collinear with } \xi \frac{\partial}{\partial \xi}. \tag{1.53}$$

Suppose further that P is a matrix operator such that $g_1(\rho) = 0 \Rightarrow \dim \operatorname{Ker} p(\rho) = s$ or P is a scalar operator satisfying the Levi-Strang-Flaschka condition or, finally, P is a first-order matrix operator satisfying the hypotheses of Theorem 25 or 26.

a) *If $V \supset \gamma$ is a segment of the bicharacteristic of the symbol g_1 and $\mathrm{WF}(Pu) \cap \gamma \neq \varnothing$, then either $\mathrm{WF}(u) \cap \gamma = \varnothing$ or $\mathrm{WF}(u) \supset \gamma$.*

b) *If $V \supset \gamma$ is a segment of the bicharacteristic of g_1 and the endpoints of γ do not lie in $\overline{V}_1 \subset V$, then we can find a function $u \in \mathscr{D}'(\Omega)$ such that $\mathrm{WF}(Pu) \cap V_1 = \varnothing$ and $\mathrm{WF}(u) \cap V_1 = \gamma_c \cap V_1$, where $M_c = \{(x, \lambda\xi), (x, \xi) \in M, \lambda > 0\}$ is the conical hull of M.*

Therefore we have the following precise description of the wave front of the fundamental solution of the Cauchy problem.

Theorem 41. *If the hypotheses of Theorem 40 are satisfied in $V = T^*X\backslash 0$ and if P is x_0-hyperbolic in X and X is a spatial type lens, then*

$$^T\mathrm{WF}(E) = \operatorname{diag}(T^*X\backslash 0)^2 \cup \{(\rho, \rho') \in (\operatorname{Char} P \cap T^*X\backslash 0)^2, \rho \in \gamma_g^+(\rho')\}. \tag{1.54}$$

A curve γ containing its limit points and in a neighbourhood of which the hypotheses of Theorem 40 are satisfied and which is a bicharacteristic of the symbol g_1 will be called a *regular bicharacteristic*. A closed set $M \subset V$ will be called a *limiting bicharacteristic* in $V_1 \Subset V$ if there is an open set V_2, $V_1 \Subset V_2 \Subset V$, and a sequence of regular bicharacteristics γ_k, with endpoints not in V_2, such that $\overline{\lim}_{k\to\infty} (\gamma_k \cap \overline{V}_1) = M \cap \overline{V}_1$, that is, if $M \cap \overline{V}_1$ is the smallest closed set M' such that $\overline{\lim}_{k\to\infty} (\gamma_k \cap \overline{V}_1) \subset M'$. From the arguments based on the Baire's category theorem it follows that assertion b) of Theorem 41 remains valid for limiting bicharacteristic also. In many cases, this result enables us to show that Theorem 39 is precise.

It can easily be shown that if g is an x_0-hyperbolic symbol, then there exists a nowhere dense closed subset Σ in $C = \{g = 0\}$ such that the multiplicity of the characteristics of g is constant in $C\backslash\Sigma$, although the same depends, in general, on the connected component of $C\backslash\Sigma$. In view of Theorems 40 and 41, the main interest lies in the propagation of singularities in a neighbourhood of Σ.

We first assume that g can be factorized, that is,

$$g = eg_1^{s_1} \dots g_k^{s_k}, \tag{1.55}$$

where $e \neq 0$ and g_j are real symbols of the principal type. Suppose that one of the following conditions holds:

$$k = 2, \qquad g_1 = g_2 = 0 \Rightarrow \{g_1, g_2\} \neq 0; \tag{1.56}$$

$$k = 2, \qquad g_1 = g_2 = \{g_1, g_2\} = 0 \Rightarrow (H^2_{g_1} g_2)\cdot(H^2_{g_2} g_1) < 0;$$
$$k \geqslant 3, \qquad \forall j\; g_j = 0 \Rightarrow g_{j,\xi_0} > 0 \tag{1.57}$$

Suppose further that for any set $\mathscr{K} \subset \{1, \ldots, k\}$, the facts that $g_j(\rho) = 0\,\forall j \in \mathscr{K}$ and that $\lambda = (\lambda_j)_{j\in\mathscr{K}}$ is a non-zero vector with non-negative components imply $C_{\mathscr{K}}\lambda \neq 0$, where

$$C_{\mathscr{K}} = (\{g_i, g_j\})_{i,j\in\mathscr{K}} \tag{1.58}$$

(this condition will be fulfilled if, in particular,

$$\forall i, j \qquad g_i = g_j = 0, \qquad i < j \Rightarrow \{g_i, g_j\} > 0,$$
$$g_j = 0 \Rightarrow g_{j,\xi_0} > 0).$$

Then it can be shown that any generalized bicharacteristic of g is a *polygonal bicharacteristic*. That is, the curve γ, parametrized by x_0, consists of segments of bicharacteristics of g_j, the endpoints (and only the endpoints) of these segments lying in Σ; at points of Σ a "transition" to bicharacteristics of other factors may take place; $\Sigma \cap \gamma$ is a locally finite set. Thus in the above-mentioned cases, and in some other cases as well (Ivrii [1979c]), $K^\pm$ turn out to be "bicharacteristic trees" with branches at points of Σ. We note that in (1.57) and (1.58) we require definite signs and not merely non-vanishing of the quantities concerned. Otherwise there may take place infinitely many transitions on a finite interval for the polygonal bicharacteristics and there may appear generalized bicharacteristics lying entirely in Σ and being bicharacteristics of the linear combinations of the factors (with non-negative coefficients) but not of the factors themselves.

If, however, the factors are in involution, that is, if

$$g_i = g_j = 0 \Rightarrow \{g_i, g_j\} = 0 \quad \text{and}$$
$$\forall i_1 < \cdots < i_l \qquad g_{i_1} = \cdots = g_{i_l} = 0 \Rightarrow \tag{1.59}$$
$$H_{g_{i_1}}, \ldots, H_{g_{i_l}}, \qquad \xi\partial/\partial\xi \text{ are linearly independent}$$

and if $g_1(\rho) = \cdots = g_k(\rho) = 0$, then $K^\pm(\rho)$ are *bicharacteristic sheets*, that is, pieces of a k-dimensional manifold $Y \ni \rho$ spanned over the vector fields $H_{g_j} (j = 1, \ldots, k)$ and bound by $(k-1)$-dimensional manifolds $Y_i \ni \rho$ spanned over the vector fields $H_{g_j} (j = 1, \ldots, k, j \neq i)$; for $K^\pm(\rho)$ we take the piece on which $x_0 \gtrless x_0(\rho)$. In the general case, the dimension of the bicharacteristic sheet $K^\pm(\rho)$ equals the number of factors that vanish at the point ρ; the same factors also vanish on $K^\pm(\rho)$ but the other factors do not vanish on $K^\pm(\rho)$.

Assume that the factors are divided into different groups in such a manner that factors belonging to the same group are in involution but those belonging to distinct groups are not. If $\{g_i, g_j\} > 0$ when g_i belongs to the group with a smaller number than g_j, then $K^\pm(\rho)$ are trees consisting of bicharacteristic sheets of different dimensions corresponding to distinct groups of factors.

The geometry of the propagation of singularities becomes much more complicated when the characteristic symbol g cannot be factorized, and to analyse the behaviour of bicharacteristics becomes a far more difficult task. However, a complete analysis of the general situation (at a point) is still possible if the multiplicity of characteristics does not exceed 2. These are the situations that turn out to be the general case for narrower classes of systems of chrystaloptics and the theory of elasticity in anisotropic media.

Let Σ be a *stationary set* of g, that is, let

$$\Sigma = \{dg = 0\} \subset \{g = 0\}. \tag{1.60}$$

To start with, we assume that at each point $\rho \in \Sigma$ the fundamental matrix $F_g(\rho)$ has two non-zero real eigenvalues $\pm\mu(\rho)$ (that is, g is effectively hyperbolic). Then it can be shown that for any point $\bar{\rho} \in \Sigma$ in some neighbourhood V of Σ there exist two C^∞-curves $\gamma_j(\bar{\rho}) = \{\rho = \rho_j(t)\}$ $(j = 1, 2)$ such that $\rho_j(0) = \bar{\rho}$, $\rho_j(t) \notin \Sigma$ for $t \neq 0$ and such that γ_j is a bicharacteristic of g for $t \neq 0$ (but, of course, this parametrization does not coincide with the one in (1.52)), and $\frac{d\rho_j}{dt}(0)$ is a non-zero eigenvector of $F_g(\bar{\rho})$ corresponding to the eigenvalue $(-1)^j\mu(\bar{\rho})$.

What is more, for a parameter on γ_j outside $x_0 = T_\pm$ we may take $t = x_0$. We denote by $\gamma_j^\pm(\bar{\rho})$ that part of $\gamma_j(\bar{\rho})$ on which $x_0 \gtrless x_0(\bar{\rho})$. Suppose, finally, that $\Sigma_j = \gamma_j(\Sigma)$ and $\Sigma_j^\pm = \gamma_j^\pm(\Sigma)$. Then we have the following results. Σ_j are closed conical sets, $\Sigma_1 \cap \Sigma_2 = \Sigma$, $\Sigma_1 \cup \Sigma_2 \subset C$, and $\Sigma_1 \cup \Sigma_2 = C$ if and only if g can be factorized. If Σ is a C^∞-manifold of dimension d, then Σ_j are C^∞-manifolds of dimension $d + 1$. If $\rho \in V \cap C \backslash (\Sigma_1^\mp \cup \Sigma_2^\mp)$, then $\gamma_g^\pm(\rho)$ has no limit points in Σ and $K^\pm(\rho) = \gamma_g^\pm(\rho)$; if $\rho \in \Sigma$, then $K^\pm(\rho) = \gamma_1^\pm(\rho) \cup \gamma_2^\pm(\rho)$, and, finally, if $\rho \in \Sigma^\mp \backslash \Sigma$, then $\rho \in \gamma_j^\mp(\bar{\rho})$ for some point $\bar{\rho} \in \Sigma$ and $K^\pm(\rho) = \gamma_j(\rho, \bar{\rho}) \cup K^\pm(\bar{\rho})$, where $\gamma_j(\rho, \bar{\rho})$ is a segment of $\gamma_j(\bar{\rho})$ between the points ρ and $\bar{\rho}$.

Thus under the above assumptions, generalized bicharacteristics are regular bicharacteristics not passing through points of Σ and polygonal bicharacteristics having branches at points of Σ.

We now assume that the hypotheses c)–d) of Theorem 19 hold with $p = g$. We have the following results.

a) If $\rho \in V \cap C \backslash \Sigma$, V is a neighbourhood of Σ, then $\gamma_g(\rho)$ has no limit points in Σ, and, moreover, its distance from Σ remains all the time of the same order; $K^\pm(\rho) = \gamma_g^\pm(\rho)$.

b) If $\rho \in \Sigma$, then any generalized bicharacteristic passing through ρ lies in Σ and satisfies the inclusion

$$\frac{d\rho}{dt} \in \Gamma^\sigma(\rho) \cap \operatorname{Ker} F_g(\rho). \tag{1.61}$$

Let us replace condition e) of Theorem 19 by a stronger condition, namely,

e′) $\sigma|_\Sigma$ has a constant defect d (then $d \geqslant 1$).

Then the manifold Σ is fibered into d-dimensional bicharacteristic C^∞-submanifolds $\Lambda = \Lambda(\rho) \ni \rho$: $\rho \in \Lambda \Rightarrow T_\rho\Lambda = \operatorname{Ran} F_g(\rho) \cap \operatorname{Ker} F_g(\rho)$; and in this case $\sigma|_\Lambda = 0$. Let $\rho \in \Lambda$. Then the restriction of the localization $g_\rho(\cdot)$ to $\operatorname{Ker} F_g^2(\rho)$ is a well defined quadratic form on the factor space $\operatorname{Ker} F_g^2(\rho)/\operatorname{Ker} F_g(\rho)$. Since there exists a unique isomorphism between this factor space and $T_\rho^*\Lambda$, we obtain a symbol h defined on $T^*\Lambda$ which, by statement b) of Theorem 10, is strictly x_0-hyperbolic of second order. Then $K^{\pm}(\rho)$ lie in $\Lambda(\rho)$ and are conoids of dependence for h.

In particular, in the general case $d = 1$, the *bicharacteristic manifolds* (they are the limiting bicharacteristics as well as the generalized bicharacteristics) are the curves $\gamma = \{\rho = \rho(t)\}$ defined by the inclusions

$$\frac{d\rho}{dt} \in \operatorname{Ker} F_g(\rho) \cap \operatorname{Ran} F_g(\rho). \tag{1.62}$$

Since now through every point $\rho \in \Sigma$ there passes exactly one generalized bicharacteristic, assertions a) and b) of Theorem 40 hold for generalized bicharacteristics in this case and the inclusion (1.51) turns into an equality.

Now there arises the question as to how precise is the description of the propagation of singularities obtained above. We have in mind the inclusion (1.51) in the first place. The existing results (Ivrii [1979a, b], Uhlmann [1982]) are not exhaustive nor even somewhat general, but enable us to presume that in the general case (1.51) turns into an equality for almost all lower terms, although for some lower terms there may be abnormally poor propagation of singularities. What is more, if g can be factorized into two factors, not in involution, then for almost all lower terms all the singularities, and not only the typical ones, are propagated in a prescribed manner (Ivrii [1979a]).

3.3. The Construction of a Parametrix. The length of the present survey paper does not permit us to give an account of the construction of a parametrix for various cases, and we confine ourselves to a cursory look at the basic results.

For strictly hyperbolic operators and for diagonalizable hyperbolic systems of first order, with characteristic roots of constant multiplicity, the parametrix was constructed by many authors by means of the oscillatory integrals of a certain type. The investigation of these integrals and operators led Hörmander [1971] to the development of the theory of Fourier integral operators (and distributions). Similar reasonings connected, however, not with the construction of the parametrix but rather with the global construction of individual asymptotic solution of the Cauchy problem with oscillatory initial data had earlier led Maslov [1965, 1973, 1976, 1983] to the construction of a canonical operator. The further development of these two theories ran parallel to a great extent, although each one of them has its own deep individual features.

It should be noted that not the fundamental solution $E(x, y)$ is a Fourier integral distribution in the strict sense but a distribution $U(x, y)$ such that

$$P(x, D_x)U = 0, \tag{1.63}$$

$$D_0^j U|_{x_0=y_0} = i\delta_{j,m-1}\delta(x - y)I \quad (j = 0, \ldots, m - 1). \tag{1.64}$$

If the coefficient of D_0^m in P is 1, then

$$E(x, y) = \begin{cases} U(x, y) & x_0 \geqslant y_0 \\ 0 & x_0 < y_0; \end{cases} \tag{1.65}$$

$E(x, y)$ is a Fourier integral distribution in the sense of Melrose and Uhlmann [1979b], that is, is a Fourier integral distribution with a Lagrange manifold having self-intersections.

For scalar operators with characteristic roots of constant multiplicity that satisfy the Levi-Strang-Flaschka condition a parametrix was constructed in Chazarain [1974] by means of Fourier integral operators, while for first-order systems satisfying the hypotheses of Theorem 25 in Petkov [1978]. For various classes of effectively hyperbolic operators the parametrix has been constructed in Kucherenko and Osipov [1983], Alinhac [1978], Yoshikawa [1977, 1978a, b, 1980]. For operators with factorizable principal symbols for the case when the factors are in involution and also for operators satisfying the conditions of Theorem 20, parametrices have been constructed in Sjöstrand and Uhlmann [1979a, b]. For diagonalizable first-order systems the parametrices have been constructed in the form of series in Kucherenko [1974]. A more detailed study led Kumano-go and his students to the construction of the theory of Fourier integral operators with multiphase functions (Kumano-go and Taniguchi [1979], R. Lascar [1981]).

3.4. Propagation of Analytic Singularities and Gevrey Singularities. We first define the wave fronts $\mathrm{WF}^{\{\varkappa\}}(u)$, $\varkappa \in [1, \infty)$, corresponding to the Gevrey classes.

Definition 18. Let $u \in \mathscr{D}'(\Omega)$. Then $(\bar{x}, \bar{\xi}) \notin \mathrm{WF}^{(\varkappa)}(u)$ if there exists a function $\varphi \in G^{(\varkappa)} \cap C_0(\Omega)$, with $\varphi(\bar{x}) = 1$, and a concial neighbourhood Γ of $\bar{\xi}$ such that the inequality

$$|F(\varphi u)(\xi)| \leqslant C \exp(-A|\xi|^{1/\varkappa}) \qquad \text{for} \quad \xi \in \Gamma \tag{1.66}$$

holds for some C and $A > 0$.

Similarly,

Definition 19. Let $u \in \mathscr{D}'(\Omega)$. Then $(\bar{x}, \bar{\xi}) \notin \mathrm{WF}^{\langle\varkappa\rangle}(u)$ if there exists a function $\varphi \in G^{\langle\varkappa\rangle} \cap C_0(\Omega)$, with $\varphi(\bar{x}) = 1$, and a conical neighbourhood Γ of $\bar{\xi}$ such that (1.66) holds for any $A > 0$ and suitable $C = C_A$.

The extension of these definitions to the case $\varkappa = 1$ is not automatic. For distributions there are three equivalent definitions of the analytic wave front (or of the essential spectrum or of singular spectrum); these are by Sato, by Bros and Iagolnitzer, and by Hörmander. Here we present the one given by Bros and Iagolnitzer.

Definition 20. Let $u \in \mathscr{D}'(\Omega)$. Then $(\bar{x}, \bar{\xi}) \notin \mathrm{WF}_a(u) \equiv \mathrm{WF}^{(1)}(u)$ if there exists a function $\varphi \in C_0^\infty(\Omega)$, $\varphi = 1$ in a neighbourhood of $\bar{x}$, and a conical neighbourhood Γ of $\bar{\xi}$ such that the inequalities

$$|F(\varphi u)(x, \xi, \lambda)| \leqslant C_N(1 + |\xi|)^{-N} e^{-A\lambda} \quad \forall \xi \in \Gamma, \qquad \lambda \in [0, \gamma|\xi|], \tag{1.67}$$

hold for suitable constants $A > 0$, $\gamma > 0$ and C_N, and for all $N \in \mathbb{Z}^+$; here

$$(Fv)(x, \xi, \lambda) = \int \exp(-i < y, \xi > - \lambda|x - y|^2) v(y)\, dy$$

is the *Fourier–Bros–Iagolnitzer transformation.* We note that $\mathrm{WF}^{\{\varkappa\}}(u)$ and $\mathrm{WF}(u)$ can also be defined in terms of this transformation.

It can easily be shown that $\mathrm{WF}^{\{\varkappa\}}(u)$ are closed conical subsets of $T^*\Omega\backslash 0$, and that $\mathrm{WF}^{(\varkappa)}(u) \supset \mathrm{WF}^{\langle\varkappa_1\rangle}(u) \supset \mathrm{WF}^{(\varkappa_1)}(u) \supset \mathrm{WF}(u)$ for $1 \leqslant \varkappa < \varkappa_1$. What is more, $\mathrm{WF}^{\{\varkappa\}}$ have basic properties of the usual wave fronts: $\pi_x \mathrm{WF}^{\{\varkappa\}}(u) = \operatorname{sing\,supp}_{\{\varkappa\}}(u)$, etc.

For $\varkappa > 1$, the propagation of singularities for hyperbolic operators was investigated by Wakabayashi [1983a]. The propagation of analytic singularities was studied by many authors, but, in our view, Wakabayashi [1983b] succeeded in putting these results for hyperbolic operators in the most convenient form.

Theorem 42. *Let P be an x_0-hyperbolic operator in Ω whose coefficients lie in $G^{\infty\{\varkappa\}}(\Omega)$. Assume that at least one of the following three conditions is fulfilled:*

(i) *$\varkappa^* = r/(r-1)$, where r is the maximum multiplicity of the characteristic roots of P.*

(ii) *P is a scalar operator with characteristics of constant multiplicity and the condition $(L'_{\varkappa^*})$ holds.*

(iii) *The coefficients of P are constant and* (1.29) *holds with $\varkappa = \varkappa^*$.*

We have the following results.

(i) *If $\{\varkappa\} = (\varkappa)$, $\varkappa \in [1, \varkappa^*]$ or if $\{\varkappa\} = \langle\varkappa\rangle$, $\varkappa \in (1, \varkappa^*]$, $Pu \in G^{\infty\{\varkappa\}}(\Omega)$ and $v = (D_0^j u|_{\Omega_T}, j = 0, \dots, m-1)$ are the Cauchy data, then*

$$\mathrm{WF}^{\{\varkappa\}}(u)|_{\Omega_T} \subset \imath^{-1}\mathrm{WF}^{\{\varkappa\}}(v); \tag{1.37''}$$

(ii) *If $\{\varkappa\} = (\varkappa)$, $\varkappa \in [1, \varkappa^*)$ or if $\{\varkappa\} = \langle\varkappa\rangle$, $\varkappa \in (1, \varkappa^*]$, $\rho \in \operatorname{Char} P \cap V$, and $T^*\Omega\backslash 0 \Supset V$ is an open conical set and*

$$\mathrm{WF}^{\{\varkappa\}}(Pu) \cap K^{\pm}(\bar\rho, V) = \varnothing, \tag{1.47'}$$

$$\mathrm{WF}^{\{\varkappa\}}(u) \cap K^{\pm}(\bar\rho, V) \cap \partial V = \varnothing, \tag{1.48'}$$

then $\mathrm{WF}^{\{\varkappa\}}(u) \cap K^{\pm}(\bar\rho, V) = \varnothing$.

Corollary. *Let P be a scalar operator with characteristics of constant multiplicity in Ω and with coefficients in $G^{\infty\{\varkappa\}}(\Omega)$. Let P satisfy the condition $(L'_{\varkappa^*})$ with either $\{\varkappa\} = (\varkappa)$, $\varkappa \in [1, \varkappa^*)$ or $\{\varkappa\} = \langle\varkappa\rangle$, $\varkappa \in (1, \varkappa^*]$. Then the assertions* a) *and* b) *of Theorem* 40 *remain valid for $\mathrm{WF}^{\{\varkappa\}}$ too.*

Chapter 2
Mixed Problems for Hyperbolic Operators

§ 1. Well-posedness of Mixed Problems

1.1. Preliminary Remarks. In the present section we consider the question of the well-posedness of mixed problems. By well-posedness we mean the C^∞-well-posedness as well as the L_2-well-posedness in two of its substantially distinct variants, namely, in a strong sense and in a weak sense. As we already remarked, the well-posedness question for the mixed problem is far more complicated than the one for the Cauchy problem, because it depends on three objects. Namely, on the operator P itself, on the boundary values and on the boundary surface itself. What is more, the class of admissible operators P gets somewhat enlarged because on the other side of the boundary surface or below the initial or above the final hyperplanes the operator may fail to be hyperbolic.

Let ω be an open domain in $\mathbb{R}^n$ with a C^∞-boundary $\partial\omega$. Let $X = [T_-, T_+] \times \bar{\omega}$ be a cylindrical domain with respect to x_0 with lateral boundary $S = [T_-, T_+] \times \partial\omega$. Let $v(x)$ denote the inward normal to S and $S_\pm = \{T_\pm\} \times \bar{\omega}$ the base of X, $-\infty \leqslant T_- < T_+ < +\infty$. Put $N = dx_0$. We shall represent $\partial\omega$ in a local coordinate system by the equation $x_n = 0$ and ω by the inequality $x_n > 0$.

We examine the following mixed problem:

$$Pu = f \quad in\ X_T^+, \tag{2.1}$$

$$B_k u = \varphi_k \quad (k = 1, \dots, \mu) \quad \text{on } S_T^+, \tag{2.2}$$

$$D_0^j u|_{X_T} = g_j \quad (j = 0, \dots, m-1), \tag{2.3}$$

where P and B_k are scalar differential operators, with C^∞-coefficients, of order m and m_k, respectively, $0 \leqslant m_k \leqslant m - 1$.

We also consider the mixed problem

$$Pu = f \quad in\ X_T^+, \tag{2.1$'$}$$

$$Bu = \varphi \quad on\ S_T^+, \tag{2.2$'$}$$

$$u|_{X_T} = g, \tag{2.3$'$}$$

where P is a $d \times d$ matrix differential operator of first order with C^∞-coefficients and $B = B(x)$ is a $\mu \times d$ C^∞-matrix.

Thus right from the beginning we exclude from our study those cases where the lateral surface has corners or edges or where different types of boundary conditions are prescribed on its different parts.

Before defining a C^∞-well posed mixed problem, we remark that for no mixed problem with $\mu > 0$ (that is, other than the Cauchy problem) does infinite smoothness of the right-hand side, the boundary and initial data imply infinite smoothness of its solution unless compatibility conditions of infinite order are satisfied. Namely, we say that

$$\{f, \varphi_k(k = 1, \dots, \mu), g_j(j = 0, \dots, m-1)\} \in C^\infty$$

satisfy *compatibility conditions of infinite order* if there is a $u \in C^\infty$ such that (2.1)–(2.3) are satisfied to within functions having zero of infinite order for $x_0 = T$.

Definition 1. The mixed problem (2.1)–(2.3) is said to be C^∞-*well posed* if the following two conditions are satisfied:

a) The problem (2.1)–(2.3) has a solution $u \in C^\infty(X_T^+)$ for any

$$\{f, \varphi_k(k = 1, \dots, \mu), g_j(j = 0, \dots, m-1)\} \in C^\infty$$

that satisfy the compatibility conditions of infinite order.

b) The fact that $u \in C^m$ satisfies, for any $t > T$, (2.1)–(2.3) with $g_0 = \cdots = g_{m-1} = 0$ and with $f = 0$, $\varphi_k = 0$ $(k = 1, \ldots, \mu)$ for $x_0 < t$, implies that $u = 0$ for $x_0 < t$.

For the problem (2.1′)–(2.3′), the definitions of compatibility conditions of infinite order and of C^∞-well-posedness are analogous. In both cases, we shall assume that the hyperplanes $\{x_0 = t\}$ are non-characteristic, that is,

$$g(x, N) \neq 0 \qquad \forall x \in X, \tag{2.4}$$

where g is the characteristic symbol of P.

It can easily be shown that for the mixed problem (2.1)–(2.3) (or (2.1′)–(2.3′)) to be C^∞-well posed it is necessary that P is x_0-hyperbolic in X_T^+; the proof coincides word for word with the corresponding proof for the Cauchy problem.

Therefore we shall assume that P is x_0-hyperbolic in X. Then for $x \in S$, $\xi \in \mathbb{R}^{n+1}$, $\tau \in \mathbb{C}_- \setminus \mathbb{R}$, the polynomial $g(x, \xi + \tau N + \sigma \nu(x))$ (in σ) has no real roots σ. What is more, the degree of this polynomial, by b) of Theorem 10, is independent of $(\xi, \tau) \in \mathbb{R}^{n+1} \times (\mathbb{C}_- \setminus \mathbb{R})$; we denote it by $M(x)$. Thus the number of roots σ of this polynomial with $\operatorname{Im} \sigma \gtrless 0$ is also independent of $(\xi, \tau) \in \mathbb{R}^{n+1} \times (\mathbb{C}_- \setminus \mathbb{R})$; we denote it by $M^\pm(x)$, and set $M^+(x) + M^-(x) = \tilde{M}(x)$. Moreover, if M is independent of $x \in S$, then $M^\pm(x)$ depends solely on the connected component of S.

Using the method of Ivrii and Petkov [1974], we can easily prove that for the *a priori* estimate, which is a consequence of the C^∞-well-posedness of (2.1)–(2.3) or of (2.1′)–(2.3′), to hold it is necessary that

$$\mu \geqslant M^+(x) \qquad \text{at every point } x \in S_T^+. \tag{2.5}$$

If the number of boundary conditions is too small, then an asymptotic solution violating this estimate is constructed.

On the other hand, the fact that the problem (2.1)–(2.3) is solvable implies *a priori* estimate for the adjoint problem

$$P^*v = h \quad \text{in } X_T^+, \tag{2.6}$$

$$C_k v = \psi_k \; (k = 1, \ldots, \mu') \quad \text{on } S_T^+, \tag{2.7}$$

$$\operatorname{supp} v \Subset X \cap \{T \leqslant x_0 < T_+\}, \tag{2.8}$$

that is, for the problem which is such that

$$(Pu, v)_X - (u, P^*v)_X = \sum_{k=1}^{\mu} (B_k u, \beta_k v)_s - \sum_{k=1}^{\mu'} (\gamma_k u, C_k v)_s \tag{2.9}$$

for any $u, v \in C^\infty(X)$, with $D_0^j u|_{S_-} = 0$ and $D_0^j v|_{S_+} = 0$ $(j = 0, \ldots, m-1)$. Here $(\cdot, \cdot)_M$ denotes the scalar product in $L_2(M)$, and β_k and γ_k are certain differential operators of order $m - m_k - 1$ and $m - m_k' - 1$, respectively, in which m_k' is the order of C_k. From this *a priori* estimate one can deduce by the same methods the inequality

$$\mu' \geqslant M^-(x), \qquad x \in S_T, \tag{2.5'}$$

that is, for the solvability of the problem the number of boundary conditions must not be too big.

We remark that the adjoint problem does not always exist. However, if the original problem is, for instance, normal (see below), the adjoint problem does exist.

Similar arguments apply also to the problem (2.1')–(2.3') but now (2.7) and (2.9) are to be replaced, respectively, by

$$Cu = \psi \quad \text{on } S_T^+, \tag{2.7'}$$

$$(Pu, v)_X - (u, P^*v)_X = (Bu, \beta v)_s - (\gamma u, Cv)_s,$$

where C, β and γ are, respectively, $\mu' \times d$, $\mu \times d$ and $\mu' \times d$ matrices.

Evidently, a necessary condition for (2.1')–(2.3') to be C^∞-well posed is that

$$\operatorname{rank} B(x) = \mu, \quad \forall x \in S_T^+. \tag{2.10}$$

If in a neighbourhood of $\bar{x} \in S_T^+$ rank $p(x, v(x))$ is constant and equals s, then in this neighbourhood the adjoint problem is defined with $\mu' = d - s - \mu$, and (2.5), (2.5') imply

$$M^+(\bar{x}) \leqslant \mu \leqslant M^+(\bar{x}) + d - s - \tilde{M}(\bar{x}). \tag{2.5''}$$

Hence, if $\tilde{M}(x) \equiv d - s$ (which, for instance, holds when S is non-characteristic), we obtain

$$\mu = M^+, \tag{2.5'''}$$

a condition that is necessary for the problem (2.1')–(2.3') to be C^∞-well posed.

We know that the mixed boundary problem for an equation can be reduced to that for a first-order system, and, moreover, since the non-characteristic mixed boundary problem (that is a mixed problem with a non-characteristic lateral boundary) is reduced to a similar problem having the same number of boundary conditions. In this reduction the C^∞-well-posedness of the original problem is equivalent to the C^∞-well-posedness of the resulting problem. Therefore (2.5''') is a necessary condition, under the assumption that the boundary surface is non-characteristic, for both the problems (2.1)–(2.3) and (2.1')–(2.3') to be C^∞-well posed. And all the arguments remain valid provided that the solution is required to exist and to be unique in the same class of functions; for instance, in C^l $(m \leqslant l \leqslant \infty)$.

Unfortunately, the arguments of this section have not been carried out accurately in any of the works known to the present author.

In the case of characteristic boundary, the number of boundary conditions may depend on the lower terms of P and on the required smoothness of the solution, as, for instance, for the operator

$$P = D_0^2 - x_1 D_1^2 + aD_1, \quad a \in \mathbb{C}. \tag{2.11}$$

In order to overcome this difficulty, the concept of a uniformly characteristic boundary is introduced. A surface $S = \{\varphi = 0\}$, with $\alpha\ \varphi \neq 0$, is said to be *uniformly characteristic* if for small α the surfaces $\{\varphi = \alpha\}$ are characteristic, and of the same multiplicity. This concept is very important on account of the Maxwell system.

1.2. Operators with Constant Coefficients. In this and the next section, we follow Sakamoto [1970a, b, 1980].

In the present section $P = P(D)$ and $B_k = B_k(D)$ are operators with constant coefficients, $X = \mathbb{R}^n \times \overline{\mathbb{R}}^+ = \{x \in \mathbb{R}^{n+1}, x_n \geqslant 0\}$ is a half-space and $S = \mathbb{R}^n = \{x \in \mathbb{R}^{n+1}, x_n = 0\}$ a hyperplane.

Let the operator $P(D)$ be Gårding x_0-hyperbolic. This is a necessary and sufficient condition for the C^∞-well-posedness of the non-characteristic Cauchy problem, and, hence, a necessary condition for the C^∞-well-posedness of the mixed problem. We assume that the hyperplanes $\{x_0 = t\}$ are non-characteristic. Then the polynomial $G\,(\xi_0, \xi'', \xi_n)$ has no real roots ξ_n for $\xi'' = (\xi_1, \ldots, \xi_{n-1}) \in \mathbb{R}^{n-1}$ and for $\operatorname{Im} \xi_0 < -C$; recall that $G(\zeta) = \det P(\zeta)$.

Assume that the hyperplane S is non-characteristic, that is, $g(v) \neq 0$, where v is the inward normal to S. Then, for $\xi'' \in \mathbb{R}^{n-1}$ and $\operatorname{Im} \xi_0 < -C$, $G(\xi_0, \xi'', \xi_n)$ *has* $M^\pm$ roots ξ_n with $\operatorname{Im} \xi_n \gtrless 0$. Denoting these roots by $\lambda_j^\pm(\xi_0, \xi'')$ $(j = 1, \ldots, M^\pm)$ and setting

$$G_\pm(\xi) = \prod_{j=1}^{M^\pm} (\xi_n - \lambda_j^\pm(\xi_0, \xi'')),$$

we have $G(\xi) = cG_+(\xi)G_-(\xi)$, $c \neq 0$.

We perform (formal, for the time being) Fourier transformation with respect to $x'' \to \xi'' \in \mathbb{R}^{n-1}$ and Fourier–Laplace transformation with respect to $x_0 \to \xi_0$, $\operatorname{Im} \xi_0 < -C$. Then (2.1)–(2.3) goes over to a boundary-value problem for an ordinary differential equation with constant coefficients:

$$P(\xi_0, \xi'', D_n)\hat{u} = F, \tag{2.12}$$

$$B_k(\xi_0, \xi'', D_n)\hat{u}|_{x_n=0} = \Phi_k \quad (k = 1, \ldots, \mu), \tag{2.13}$$

where the right-hand side F of (2.12) is determined by the right-hand side of (2.1) and by the initial data, while Φ_k are determined by the boundary conditions and the initial data.

The general solution of (2.12), with $F \equiv 0$, that decays as $x_n \to +\infty$ is of the form

$$\hat{u} = \sum_{j=1}^{M^+} \frac{c_j}{2\pi i} \int_\gamma e^{ix_n\xi_n} \xi_n^{j-1} G_+^{-1}(\xi)\, d\xi_n, \tag{2.14}$$

where γ is a closed contour in $\mathbb{C}$ that goes round all the roots $\lambda_k^+(\xi_0, \xi'')$ $(k = 1, \ldots, M^+)$ once in the counterclockwise direction and c_j are arbitrary numbers. Substitution of (2.14) into (2.13) yields the following system of algebraic equations:

$$\sum_{j=1}^{M^+} \frac{c_j}{2\pi i} \int_\gamma B_k(\xi) \xi_n^{j-1} G_+^{-1}(\xi)\, d\xi_n = \Phi_k, \quad (k = 1, \ldots, \mu). \tag{2.15}$$

We assume that $\mu = M^+$, that is, that (2.15) is a square system. By the *Lopatinskij determinant* of (2.1)–(2.3) we mean the determinant of (2.15):

$$\mathscr{L}(\xi_0, \xi'') = \det\left(\frac{1}{2\pi i}\int_\gamma B_k(\xi)\xi_n^{j-1}G_+^{-1}(\xi)\,d\xi_n\right)_{k,j=1,\dots,\mu}$$

We can define a more general determinant $\mathscr{L}(\eta + \tau\theta)$, where $\eta = (\xi_0, \xi'') \in \mathbb{R}^n$ and θ belongs to the hyperbolicity cone $\Gamma(g, N)$ of $P(D)$.

Theorem 1. *For the problem* (2.1)–(2.3) *to be C^∞-well posed, it is necessary that*

$$\mathscr{L}(\eta + \tau N) \neq 0 \quad \forall \eta \in \mathbb{R}^n, \quad \operatorname{Im}\tau < -C_1 \tag{2.16}$$

with a suitable C_1.

In particular, if $P(D) = p(D)$ and $B_k(D) = b_k(D)$ are homogeneous operators, it is necessary that

$$l(\eta + \tau N) \neq 0 \quad \forall \eta \in \mathbb{R}^n, \tau \in \mathbb{C}_-\backslash\mathbb{R}, \tag{2.17}$$

where l is the Lopatinskij determinant for $\{p, b_1, \dots, b_\mu\}$.

For non-homogeneous operators also we can define l by its principal parts. We assume that

$$\exists \eta \in \mathbb{R}^n\colon l(\eta - iN) \neq 0. \tag{2.18}$$

Then l is the principal part of $\mathscr{L}$. (If $l(\cdot - iN) \equiv 0$, then l is defined precisely as the principal part of $\mathscr{L}$.)

Theorem 2. *For the problem* (2.1)–(2.3) *to be C^∞-well posed when* (2.18) *holds, it is necessary that*

$$l(-iN) \neq 0. \tag{2.19}$$

The proofs of Theorem 1 and 2 are based on the construction of asymptotic solutions.

It can easily be shown that (2.16) and (2.19) imply (2.17).

Theorem 3. *If* (2.16) *and* (2.19) *hold and the boundary conditions are normal, that is, if*

$$j > k \Rightarrow m_j > m_k \quad \text{and} \quad b_k(\nu) \neq 0\,\forall k, \tag{2.20}$$

then the boundary value problem (2.1)–(2.3) *is C^∞-well posed.*

Assume that (2.16) and (2.19) hold. Let Γ_1 be a connected component, containing N, of the set $\{\theta \in \Gamma(g, N), l(-i\theta) \neq 0\}$. Then

$$\Gamma_1 \text{ is an open convex cone,} \tag{2.21}$$

$$l(\eta + \tau\theta) \neq 0 \quad \forall \eta \in \mathbb{R}^n, \quad \tau\colon \lim\tau < 0, \tag{2.22}$$

$$\mathscr{L}(\eta + \tau\theta) \neq 0 \quad \exists \eta \in \mathbb{R}, \quad \tau\colon \operatorname{Im}\tau < -C_\theta \tag{2.23}$$

(compare with the similar properties in § 1.3, Chap. 1).

Finally, let

$$\Gamma_+ = \Gamma(g, N) \cap \{\xi + \sigma v, \xi \in \Gamma_1, \sigma \in \mathbb{R}\}$$

and let Γ'_+ be the dual cone and $\Gamma''_+ = \Gamma'_+ \cap \{x_n \geqslant 0\}$.

Theorem 4. *If the hypotheses of Theorem* 3 *are satisfied and* $E_j \in C^\infty(\overline{\mathbb{R}}^+, \mathscr{D}'(\mathbb{R}^n))$ *is a solution of the problem*

$$\begin{gathered} PE_j = 0 \quad \textit{for} \quad x_n > 0, \\ \operatorname{supp} E_j \subset \{x_0 \geqslant 0\}, \\ B_k E_j|_{x_n=0} = \delta_{ij}\delta(x_0)\delta(x'') \quad (k = 1, \ldots, \mu), \end{gathered} \tag{2.24}$$

then supp $E_j \subset \Gamma''_+$.

The proofs of Theorems 3 and 4 are based on the application of Fourier transformation in the complex plane with respect to the variables $(x_0, \ldots, x_{n-1})$ and on proving a Holmgren type uniqueness theorem. In the latter, the fact that the problem is normal plays an important role; this guarantees that the adjoint problem, satisfying all the conditions of Theorem 3, exists.

Theorem 4 and Theorm 7 of Chap. 1 together describe the support of the solution to the problem (2.1)–(2.3).

All these arguments apply also to the problem (2.1′)–(2.3′) with the only difference that the condition (2.20) is replaced by (2.10) and the inequalities $\mathscr{L}(\eta + \tau) \neq 0$ and $l(\eta + \tau\theta) \neq 0$ by the conditions that the row matrices

$$\begin{gathered} B^{co}P(\eta + \tau\theta + \sigma v) \quad (\operatorname{mod} G^+(\eta + \tau\theta + \sigma v)), \\ B^{co}p(\eta + \tau\theta + \sigma v) \quad (\operatorname{mod} g^+(\eta + \tau\theta + \sigma v)) \end{gathered}$$

are, respectively, linearly independent; here ${}^{co}(\cdot)$ denotes the matrix of cofactors.

1.3. Strong L_2-well-posedness of the Mixed Problem (for Equations). As in the preceding section, here also we follow Sakamoto [1970a, b, 1980].

The strong L_2-well-posedness (H-well-posedness in the terminology of Sakamoto) is the strongest form of well-posedness. We denote by H^s various Sobolev spaces. We shall assume in what follows that either $\partial\omega$ is compact or $\partial\omega = \{x_n = 0\}$ and that all the coefficients of P and B_k are constant for $|x'| > R$. For simplicity, we assume that $T_+ < \infty$.

Definition 2. The mixed problem (2.1)–(2.3) is said to be *strongly L_2-well posed* if it is C^∞-well posed and if, for all $u \in C^\infty$, all $\gamma \geqslant 0$, $t_0, t \in [T, T_+]$, $t_0 < t$ and all $s \in \mathbb{Z}^+$, the *a priori* estimate

$$\begin{aligned} &\sqrt{\gamma}\|u; X_{(t_0,t)}\|_{m-1+s,\gamma} + \|u; X_t\|_{m-1+s,\gamma} + \|u; S_{(t_0,t)}\|_{m-1+s,\gamma} \\ &\quad \leqslant C\left[\frac{1}{\sqrt{\gamma}}\|Pu; X_{(t_0,t)}\|_{s,\gamma} + \|u; X_{t_0}\|_{m-1+s,\gamma} + \sum_{k=1}^{\mu} |||B_k u; S_{(t_0,t)}|||_{m-m_k-1+s,\gamma}\right] \end{aligned} \tag{2.25$_s$}$$

holds with the constant independent of u, t_0, t or γ, where

$$\|u; K\|_{q,\gamma} = \sum_{|\alpha|\leqslant q} \gamma^{q-|\alpha|}\|e^{-\gamma x_0}D^\alpha u; L_2(K)\|,$$

and $K_{(t_0,t)} = K \cap \{t_0 \leqslant x_0 \leqslant t\}$. The other norms $|||\cdot; S_t|||_{q,\gamma}$ and $|||\cdot; S_{(t_0,t)}|||_{q,\gamma}$ are defined in a similar fashion, only that the derivatives are now taken only along S.

It is easy to see that if the boundary is non-characteristic, then the estimate (2.25_0) implies (2.25_s) for all $s \in \mathbb{Z}^+$. Moreover, for $s = 0$ analogous estimate is obtained if one employs the norms used by Sakamoto [1970a, b, 1980]. Namely, $\|u; K\|'_{q,\gamma} = \|e^{-\gamma x_0}u; K\|_q$, and similarly, for $|||u; K|||'_{q,\gamma}$; the index $\gamma = 1$ is omitted.

Let $\mathscr{F}$ be the set of infinitely smooth right-hand sides and initial and boundary data having compact supports and satisfying the compatibility conditions of infinite order. Let $\mathscr{F}_s$, the closure of $\mathscr{F}$ in the space

$$H^s(X_T^+) \times \prod_{j=0}^{m-1} H^{s+m-1-j}(X_T) \times \prod_{k=1}^{\mu} H^{s+m-1-m_k}(S_T^+)$$

be the set of right-hand sides, initial and boundary data having smoothness as mentioned and satisfying the *compatibility conditions of order s*. It can be shown that if S is non-characteristic and the boundary-value problem is normal, then the compatibility conditions are of local nature, and when $\partial\omega = \{x_n = 0\}$, the compatibility conditions of order s are the $\left(\mu(s + m - 1) - \sum_k m_k\right)$ conditions that connect the values on S_T of the function $D_0^{\alpha_0}D_n^{\alpha_n}f(\alpha_0 + \alpha_n \leqslant s - 1)$, $D_n^{\alpha_n}g_j(\alpha_n \leqslant s + m - 2 - j)$, $D_0^{\alpha_0}\varphi_k(\alpha_0 \leqslant s + m - 2 - m_k)$ and their derivatives along S_T. These conditions enable us to express $D_0^{\alpha_0}\varphi_k|_{S_T}$ in terms of the remaining functions and their derivatives along S_T.

It follows from the definition that if (2.1)–(2.3) is strongly L_2-well posed and if

$$\{f, g_j(j = 0, \ldots, m - 1), \varphi_k(k = 1, \ldots, \mu)\} \in \mathscr{F}_s,$$

then the problem (2.1)–(2.3) has a solution

$$u \in \bigcap_{j=0}^{s+m-1} C^j([T, T_+], \qquad H^{s+m-1-j}(\omega)) \subset H^{s+m-1}(X_T^+);$$

moreover, $D^\alpha u \in H^{s+m-1-|\alpha|}(S_T^+)$ for all α, $|\alpha| \leqslant s + m - 1$.

Theorem 5. *Let the estimate* (2.25_s) *hold for some s. Then*

(i) *P is strongly x_0-hyperbolic in X_T^+.*

(ii) *Either at no point is S_T^+ non-characteristic or else $\mu = m$, $s = 0$.*

(iii) *$\mu \geqslant M^+$.*

(iv) *If $\mu = M^+$, then the* ***uniform Lopatinskij condition*** *holds*:

$$l(x, \xi + \tau N) \neq 0 \tag{2.26}$$

$$\forall x \in S_T^+, \qquad \xi \in \mathbb{R}^{n+1}, \qquad \tau \in \mathbb{C}_-\backslash\mathbb{R},$$
$$l(x, \xi) \neq 0 \quad \forall x \in S_{T^+}, \qquad \xi \in \mathbb{R}^{n+1}\backslash\mathbb{R}\cdot\nu, \tag{2.27}$$

where $l(x, \xi) \equiv l(x, \xi - 0iN)$.

Conversely, we have

Theorem 6. *Let the operator P be strictly x_0-hyperbolic in X, S be non-characteristic and let $\mu = M^+$. If the uniform Lopatinskij condition holds on S, then the estimate* (2.25) *is true.*

The proof of Theorem 6 uses the method of the divisor operator, which in the case of a boundary-value problem turns out to be much more complicated than for the Cauchy problem.

On account of (2.27) and (2.22), N in condition (2.26) may be replaced by any vector $\theta \in \Gamma(g, N)$. What is more, if the boundary-value problem is normal and if the hypotheses of Theorem 6 are satisfied, then the same conditions are fulfilled for the adjoint problem. Therefore, by means of functional-analytic arguments, we can establish the following

Theorem 7. *Let the operator P be strictly x_0-hyperbolic in X, S be non-characteristic, and let $\mu = M^+$. Assume that the uniform Lopatinskij condition holds on S and that the boundary-value problem is normal. Then the mixed problem* (2.1)–(2.3) *is strongly L_2-well posed. Furthermore, let $X_T^+ \supset X'$, $\partial X' = X'_T \cup X'_{T_+} \cup S' \cup S''$, where $S'' \subset S$, $S' \in C^\infty$. If at each point $x \in S'$ the outward normal $\nu'(x)$ to S' lies in $\Gamma(g, N)$, then the facts that $u \in H^m(X')$, $Pu = 0$ in X' and*

$$D_0^j u|_{X'_T \setminus S'_T} = 0 \quad (j = 0, \dots, m-1),$$

$$B_k u|_{S' \setminus S'_T} = 0 \quad (k = 1, \dots, \mu)$$

imply that $u = 0$ in X'.

In conclusion, we note that we could have required that the estimate

$$\|u; X_t\|_{m-1+s} + \|u; S_{(t_0,t)}\|_{m-1+s}$$
$$\leqslant C\left[\|u; X_{t_0}\|_{m-1+s} + \int_{t_0}^{t} \|Pu; X_{t'}\|_s \, dt' + \sum_{k=1}^{\mu} |||B_k u; S_{(t_0,t)}|||_{m-m_k-1+s}\right], \tag{2.28$_s$}$$

instead of (2.25$_s$), holds for all $t_0, t \in [T, T_+]$, $t_0 < t$, $u \in C^\infty$ and $s \in \mathbb{Z}^+$ with a constant that is independent of t_0, t or u.

Using the Duhamel's principle, we can easily show that the estimates (2.25$_0$) and (2.28$_0$) are equivalent; moreover, (2.28$_0$) implies (2.28$_s$) for an arbitrary $s \in \mathbb{Z}^+$. Using (2.28$_s$), we find that the smoothness condition on the right-hand side can be relaxed:

$$D_0^j f \in L_1([T, T_+], H^{s-j}(\omega)) \quad (j = 0, \dots, s).$$

1.4. Strong L_2-well-posedness of the Mixed Problem for Systems of First Order. In a series of works (Friedrichs [1958], Lax and Phillips [1960]) devoted to the symmetric systems with "non-negative" boundary conditions, there appeared L_2-estimates in two forms leading to the notions of strong and weak

L_2-well-posedness. The first of these is of the form

$$\sqrt{\gamma}\|u: X_{(t_0,t)}\|_{s,\gamma} + \|u; X_t\|_{s,\gamma} + \|u; S_{(t_0,t)}\|_{s,\gamma}$$
$$\leqslant C\left[\frac{1}{\sqrt{\gamma}}\|Pu; X_{(t_0,t)}\|_{s,\gamma} + \|u; X_{t_0}\|_{s,\gamma} + |||Bu; S_{(t_0,t)}|||_{s,\gamma}\right]$$

or

$$\|u; X_t\|_s + \|u; S_{(t_0,t)}\|_s \leqslant C[\|u; X_{t_0}\|_s + \int_{t_0}^{t} \|Pu; X_{t'}\|_s \, dt' + |||Bu; S_{(t_0,t)}|||_s] \tag{2.28$'_s$}$$

for all $t_0, t \in [T, T_+]$, $t_0 < t$, $\gamma > 0$, $u \in C^\infty$ and $s \in \mathbb{Z}^+$, with constants independent of u, t_0, t or γ.

Theorem 8. *Let the estimate* (2.25$_s$) *hold for some s. Then*

(i) *The operator P is x_0-hyperbolic in X_T^+ and the matrix $p(x, 0, \xi')$ is uniformly diagonalizable in X_T^+ (it is assumed throughout this section that $p(x, N) = I$).*

(ii) *Either at no point is S_T^+ non-characteristic or else $\mu = d$ and $s = 0$.*

(iii) $\mu \geqslant M^+$.

(iv) *If $\mu = M^+$, then the* ***uniform Lopatinskij condition*** *holds*:

a) *For all $x \in S_T^+$, $\xi \in \mathbb{R}^{n+1}$, and $\tau \in \mathbb{C}_-\backslash\mathbb{R}$, the rows of the matrix $B^{co}p(x, \xi + \tau N + \sigma v)$ are linearly independent* (mod $g_+(x, \xi + \tau N + \sigma v)$).

b) *For all $x \in S_T^+$, $\xi \in \mathbb{R}^{n+1}\backslash\mathbb{R}\cdot v$, the rows of the matrix $B^{co}p(x, \xi + \sigma v)$ are linearly independent* (mod $g_+(x, \xi + \sigma v)$), *where $g_+(x, \eta) \equiv g_+(x, \eta - 0iN)$.*

Conversely, we have

Theorem 9. *Let the operator P be x_0-hyperbolic and the matrix $p(x, 0, \xi')$ be smoothly diagonalizable in X. Suppose that S is non-characteristic, $\mu = M^+$, and that on S the condition* (2.10) *and the uniform Lopatinskij conditions* (a) *and* b)) *hold. Then the mixed problem* (2.1)–(2.3) *is strongly L_2-well posed and, moreover, the conclusion of Theorem 7 regarding the uniqueness holds.*

The proof of this theorem rests on deducing the energy estimate by the symmetrizer method (Kreiss [1970]) and then applying functional-analytic arguments. Note that the symmetrizer method is much more complicated for the mixed problem than it is for the Cauchy problem. We also note that the problem for strictly hyperbolic equation, examined in the preceding section, can be reduced, after localization, to a problem for a pseudodifferential system of first order.

1.5. Necessary Conditions for C^∞-well-posedness of the Mixed Problem. Using the methods of §2 of Ivrii and Petkov [1974], we can easily prove the following

Theorem 10. *Let $\bar{x} \in S_T^+$, $\mu = M^+(\bar{x})$ and $l(\bar{x}, \xi + \tau N) = 0$ for some $\xi \in \mathbb{R}^{n+1}$ and $\tau \in \mathbb{C}_-\backslash\mathbb{R}$. Assume that $l(\bar{x}, \xi + \cdot N) \not\equiv 0$ on $\mathbb{C}_-$. Then the mixed problem* (2.1)–(2.3) *is not C^∞-well posed.*

Corollary 1. *Let $\bar{x} \in S_T^+$, $\mu = M^+(\bar{x})$ and $l(\bar{x}, N) \neq 0$. Then for the problem* (2.1)–(2.3) *to be C^∞-well posed, it is necessary that the* ***Lopatinskij condition*** (2.26) *holds with $x = \bar{x}$.*

Theorem 11. *Let $\bar{x} \in S_T^+$ and $\mu = M^+(\bar{x})$. Assume that condition* a) *is saitsfied for $x = \bar{x}$, $\xi = \bar{\xi}$ and $\tau = \bar{\tau} \in \mathbb{C}_- \backslash \mathbb{R}$. Then for the C^∞-well-posedness of the mixed problem* (2.1′)–(2.3′) *it is necessary that the* ***Lopatinskij condition*** *holds for $x = \bar{x}$, $\xi = \bar{\xi}$ and for all $\tau \in \mathbb{C}_- \backslash \mathbb{R}$.*

Corollary 2. *Let $\bar{x} \in S_T^+$ and $\mu = M^+(\bar{x})$. Assume that condition* a) *is satisfied for $x = \bar{x}$ and $\xi = N$. Then for* (2.1′)–(2.3′) *to be C^∞-well posed, it is necessary that condition* a) *is satisfied for $x = \bar{x}$ and all $\xi \in \mathbb{R}^{n+1}$ and $\tau \in \mathbb{C}_- \backslash \mathbb{R}$.*

Definition 3. Let $\bar{x} \in S_T^+ \backslash S_T$ and Γ be an open convex cone with vertex $\bar{x}$; $\bar{\Gamma} \backslash \bar{x} \subset \{x_0 < \bar{x}_0\}$. We say that the problem (2.1)–(2.3) satisfies the uniqueness condition (U_Γ) if the facts that $Pu = 0$ in $X_T^+ \cap \Gamma$ and

$$D_0^j u|_{X_T \cap \Gamma} = 0 \quad (j = 0, \ldots, m-1),$$
$$B_k u|_{S_T^+ \cap \Gamma} = 0 \quad (k = 1, \ldots, \mu)$$

imply that $u = 0$ in $X_T^+ \cap \Gamma$.

Theorem 12. *Let $\bar{x} \in S_T^+ \backslash S_T$ and $\mu = M^+(\bar{x})$. Suppose that*

$$l(\bar{x}, \cdot) \not\equiv 0 \quad \text{on} \quad \mathbb{R}^{n+1} - i\Gamma(g, N). \tag{2.29}$$

For the problem (2.1)–(2.3) *to be C^∞-well posed and condition (U_Γ) to be satisfied, it is necessary that*

$$\begin{gathered} l(\bar{x}, \theta) \neq 0 \quad \forall \theta \in \Gamma(g, N): \\ \{\langle \theta, x - \bar{x} \rangle \geqslant 0\} \cap \bar{\Gamma} = \{\bar{x}\}, \end{gathered} \tag{2.30}$$

and, in particular, that

$$l(\bar{x}, N) \neq 0. \tag{2.31}$$

Theorem 13. *Let $\bar{x} \in S_T^+ \backslash S_T$ and $\mu = M^+(\bar{x})$. Suppose that condition* a) *holds for $x = \bar{x}$ and for some $\xi = \bar{\xi} \in \mathbb{R}^{n+1}$ and $\tau = \bar{\tau} \in \mathbb{C}_- \backslash \mathbb{R}$. For* (2.1′)–(2.3′) *to be C^∞-well posed and condition (U_Γ) to be satisfied, it is necessary that condition* b) *holds for $x = \bar{x}$, $\xi = \theta$*

$$\forall \theta \in \Gamma(g, N): \{\langle \theta, x - \bar{x} \rangle \geqslant 0\} \cap \bar{\Gamma} = \{\bar{x}\},$$

and, in particular, for $\theta = N$.

The present author is not aware whether the proofs of these results have been published in their general form somewhere. Corollary 1 of Theorem 10 has been proved by Wakabayashi [1980].

1.6. Weak L_2-well-posedness of the Mixed Problem. The uniform Lopatinskij condition, introduced in § 2.3 and § 2.4, is an extremely severe restriction on the class of problems under discussion because it does not hold, for example, for the wave equation with Neumann boundary conditions. On the other hand, the wave equation with homogeneous Neumann condition possesses the classical energy integral and for it the existence and uniqueness theorems hold. Similar is the case with many problems for the Maxwell system for which there are no strongly L_2-well-posed boundary-value problems at all because the boundary is characteristic. In view of this situation, in their works Agemi and Shirota [1970, 1971] introduced various other definitions for L_2-well-posedness.

Definition 4. The mixed problem (2.1)–(2.3) is said to be *weakly*[1] *L_2-well posed* if it is C^∞-well posed and if for all $t_0, t \in [T, T_+]$, $t_0 < t$, $\gamma \geqslant 0$ and $u \in C^\infty$ such that $B_k u|_S = 0$ $(k = 1, \ldots, \mu)$, $s \in \mathbb{Z}^+$, the *a priori* estimate

$$\sqrt{\gamma}\|u; X_{(t_0,t)}\|_{m-1+s,\gamma} + \|u; X_t\|_{m-1+s,\gamma} \leqslant C\left[\frac{1}{\sqrt{\gamma}}\|Pu; X_{(t_0,t)}\|_{s,\gamma} + \|u; X_{t_0}\|_{m-1+s,\gamma}\right] \tag{2.32}$$

holds with the constant independent of t_0, t, γ or u.

As compared to (2.25_s), the member $\|u; S_{(t_0,t)}\|_{m-1+s,\gamma}$ on the left-hand side is now absent and a condition has been imposed on u that makes the members, absent on the right-hand side, vanish. We shall assume throughout that S is non-characteristic and the boundary-value problem is normal. Then it is not difficult to prove that (2.32_0) implies (2.32_s) for all s and that the problem (2.1)–(2.3) has a solution

$$u \in \bigcap_{j=0}^{s+m-1} C^j([T, T_+], H^{s+m-1-j}(\omega))$$

for any

$$\{f, g_j (j = 0, \ldots, m-1), \varphi_k (k = 1, \ldots, \mu)\} \in \mathscr{F}_s \cap \{\varphi_k \in H^{s+m-m_k-1/2}(S_T^+)\}.$$

Then, by the embedding theorem,

$$D^\alpha u \in H^{s+m-3/2-|\alpha|}(S_T^+)$$

for $|\alpha| \leqslant s + m - 2$. Thus, in comparison to the strong L_2-well-posedness, the required smoothness of boundary data increases by 1/2, while that of $D^\alpha u|_S$ decreases by 1/2.

Moreover, the concept of L_2-well-posedness with the loss in smoothness by $q \in \mathbb{Z}^+$ was also introduced. Apart from the C^∞-well-posedness, the estimate

$$\gamma\|u; X_T^+\|_{m-1+s,\gamma} \leqslant C\|Pu; X_T^+\|_{s+q,\gamma} \tag{2.33}$$

must be satisfied for all $u \in C^\infty$ having the property that $B_k u|_S = 0$ $(k = 1, \ldots, \mu)$ and $u = 0$ for $x_0 < T$. Even for $q = 0$, this definition is weaker than Definition 3, at least formally. When P and B_k are homogeneous operators with constant coefficients and $\omega = \mathbb{R}^{n-1} \times \mathbb{R}^+$, necessary and sufficient conditions for the

[1] Agemi and Shirota termed this problem, in contrast, as strongly L_2-well posed.

L_2-well-posedness with a loss in smoothness by q were given in terms of the Green's function of the problem (2.12)–(2.13), and it was shown that for $q = 0$ the cone of dependence of the mixed problem coincides with the intersection of X_T^+ and the cone of dependence of the Cauchy problem (for $q \geqslant 1$ this result is, in general, not true).

Finally, it was shown that if $\omega = \mathbb{R}^{n-1} \times \overline{\mathbb{R}}^+$, the weak L_2-well posedness of (2.1)–(2.3) implies the weak L_2-well-posedness of the problem obtained by freezing the coefficients at any point $\bar{x} \in S_T^+$, and, in particular, implies (2.19). This enables us to obtain conditions that are necessary for the weak L_2-well-posedness, which in many cases turn out to be sufficient as well.

Let us examine in some detail the strictly hyperbolic operators of order $m = 2$. Let $\omega = \mathbb{R}^{n-1} \times \overline{\mathbb{R}}^+$ and S be non-characteristic and non-spatially similar. Then for $x \in S$, $\xi'' \in \mathbb{R}^{n-1}$ and $\xi_0 \in \mathbb{C}_- \setminus \mathbb{R}$, the polynomial $p(x, \xi_0, \xi'', \xi_n)$ has two roots $\xi_n = \lambda^\pm(x, \xi_0, \xi'')$ one with positive and another with negative imaginary parts. Then $\mu = 1$, and

$$l(x, \xi_0, \xi'') = b(x, \xi_0, \xi'', \lambda^+(x, \xi_0, \xi'')),$$

where b is the principal symbol of B. We recall that a necessary condition for the problem (2.1)–(2.3) to be C^∞-well posed is that the Lopatinskij condition holds. Namely, $l(x, \xi_0, \xi'') \neq 0$ for $x \in S_T^+$, $\xi'' \in \mathbb{R}^{n-1}$ and $\xi_0 \in \mathbb{C}_- \setminus \mathbb{R}$ (we assume that (2.19) holds, that is, that $l(x, -i, 0) \neq 0$). A necessary and sufficient condition for the problem to be strongly L_2-well posed is the uniform Lopatinskij condition, namely, $l(x, \xi_0, \xi'') \neq 0$ for $x \in S_T^+$, $\xi'' \in \mathbb{R}^{n-1}$ and $\xi_0 \in \mathbb{C}_-$. This condition is fulfilled, in particular, for the Dirichlet boundary condition with $B = I$; it is assumed that the boundary condition is normal.

Theorem 14 (Miyatake [1975]). *Let*

$$B = \sum_{j=0}^{n} b_j(x) D_j + b'(x),$$

where $b_n = 1$.

(i) *If P and B have constant coefficients, then a necessary and sufficient condition for the problem* (2.1)–(2.3) *to be weakly L_2-well posed is the inequality*

$$\begin{gathered} |\operatorname{Im} \xi_0| \leqslant C |l(x, \xi_0, \xi'')| \cdot |\operatorname{Im} \lambda(x, \xi_0, \xi'')| \\ \forall x \in S_{T^+}, \qquad \xi'' \in \mathbb{R}^{n-1}, \qquad \xi_0 \in \mathbb{C}_- \end{gathered} \tag{2.34}$$

(*of course, in this case l and λ^+ do not depend on x*).

(ii) *If P and B have variable coefficients, then* (2.34) *is a necessary condition for the problem to be weakly L_2-well posed.*

(iii) *Assume that the coefficients of P and B are constant for $|x'| > R$ and that* (2.34) *as well as the following additional condition hold:*

If

$$\bar{x} \in S_T^+, \qquad c(\bar{x}) = 0, \tag{2.35}$$

then either $c \equiv 0$ *or* $\operatorname{Im} b_0 \equiv \cdots \equiv \operatorname{Im} b_{n-1} \equiv 0$ *in some neighbourhood of* $\bar{x}$, *where*

$$c = \operatorname{Re} b_0 - P_{\xi_n \xi_0} \cdot P^{-1}_{\xi_n \xi_n}.$$

Then the problem (2.1)–(2.3) *is weakly* L_2*-well posed.*

We note that if (2.35) is violated at some point $\bar{x} \in S_T^+$, we can impose other technical conditions that guarantee weak L_2-well-posedness (see Miyatake [1975], where (2.34) is given in various other forms). It turns out that if conditions (2.34) and (2.35) or some of the above-mentioned technical conditions are satisfied, then the conoid of dependence of the mixed problem (2.1)–(2.3) will be the intersection of X_T^+ and the conoid of dependence of the Cauchy problem. We also note that (2.34) implies that the Lopatinskij determinant l does not vanish at hyperbolic points of $T^*S_T^+ \backslash 0$ but may vanish at tangent points and at elliptic points (the classification of points of $T^*S \backslash 0$ is given in the next section).

We have already discussed the question of the loss in smoothness in the case of weak L_2-well-posedness. But in some cases the actual loss is smaller. Let $P = \Box$ be the wave operator and $X = \mathbb{R} \times \omega$. Consider the maps:

$$v \in C^\infty(s), \qquad \operatorname{supp} v \subset S_0^+ \to u \in C^\infty(X): Pu = 0,$$

$$\operatorname{supp} u \subset X_0^+, \qquad u|_s = v \to \Lambda v = \left.\frac{\partial u}{\partial n}\right|_s.$$

If ω is strictly concave (that is, if ω is an "exterior" domain with a non-zero Gaussian curvature of boundary), then Λ is an Airy type pseudodifferential operator of class $L^1_{1/3,2/3}$ (Melrose [1978]). This is a classical pseudodifferential operator outside the tangent set $\mathscr{G}$ and the inverse operator Λ^{-1} is an Airy type pseudodifferential operator of class $L^{-2/3}_{1/3,2/3}$. What is more, Λ^{-1} is a classical pseudodifferential operator of order (-1) outside $\mathscr{G}$. Therefore for a solution of the problem

$$Pu = 0, \qquad \operatorname{supp} u \subset X_0^+, \qquad \left.\frac{\partial u}{\partial n}\right|_S = w$$

the estimate

$$\|u; X_t^-\|_{k+1} + \|u; S_t^-\|_{k+1} \leqslant C\|w; S_t^-\|_{k+1/3}$$

holds for all $k \in \mathbb{Z}^+$. This estimate cannot be improved upon.

1.7. C^∞-well Posed Mixed Problem for Strictly Hyperbolic Equations of Second Order. Let us begin with the results of Ikawa and Soga for the wave equation. Let $\omega \subset \mathbb{R}^{n+1}$, $\partial\omega$ be compact and v the inward unit normal to $\partial\omega$. Let $P = \Box$ and $B = \sum_{j=0}^n b_j D_j + b'$, where $\sum_{j=1}^n b_j v_j = 1$ and b_j are real. A necessary and sufficient condition for strong (weak) L_2-well-posedness is the inequality $b_0 + \beta < 0$ $(b_0 + \beta \leqslant 0)$, where $\beta = (\sum_{j=1}^n b_j^2 - 1)^{1/2}$. By Theorem 10, a necessary condition for the C^∞-well-posedness is that $b_0 \leqslant 1$ for $n \geqslant 3$ and $b_0 \leqslant (1 + \beta^2)^{1/2}$ for $n = 2$.

Theorem 15 (Ikawa [1981]). *Let $n = 3$. Let ω be a strictly concave domain. For the mixed problem (2.1)–(2.3) to be C^∞-well posed and to have a finite propagation speed, it is necessary and sufficient that $b_0 < 1$.*

Now let $n = 2$. Let $b_0 = 0$, b_j be independent of x_0, $\tau = \tau(s) = b_1 v_2 - b_2 v_1$, $\lambda = \lambda(s) = b_1 v_1 + b_2 v_2$ and $x' = x'(s)$, where s is a parameter on $\partial\omega$. We drop the earlier assumption that $\lambda \equiv 1$.

Theorem 16 (Ikawa [1981]). *Let $\lambda \not\equiv 1$. Let ω be a strictly concave domain. For the problem (2.1)–(2.3) to be C^∞-well posed, it is necessary and sufficient that τ and $d\tau/dS$ do not both vanish simultaneously.*

Theorem 17 (Soga [1980]). *For the problem (2.1)–(2.3) to be C^∞-well posed, it is necessary that at no point of $\partial\omega$ does $\lambda(s)$ change its sign.*

Theorem 18 (Soga [1980]). *Assume that $\tau(s)$ does not vanish and $\lambda(s)$ does not change its sign on $\partial\omega$. If $[\lambda(s)]^{1/2}$ is a smooth function, then the mixed problem (2.1)–(2.3) is C^∞-well posed but does not have, in general, a finite propagation speed.*

We now turn our attention to far more general results of Eskin. Let P be a strictly x_0-hyperbolic operator of second order and B a differential operator of an arbitrary order M. We examine the mixed problem in a somewhat different form:

$$Pu = 0, \qquad \operatorname{supp} u \subset X_T^+, \qquad Bu|_S = \varphi. \tag{2.36}$$

Suppose that S is non-characteristic and non-spatially similar. Then locally, in a suitable coordinate system,

$$X = \{x_n \geqslant 0\}, \qquad p(x, \xi) = \xi_n^2 - q(x, \xi_0, \xi''),$$

where

$$q = 0 \Rightarrow q_{\xi_0} \neq 0. \tag{2.37}$$

Let $T^*S \backslash 0 \supset \mathscr{H} = \{q > 0\}$, $\mathscr{E} = \{q < 0\}$ and $\mathscr{G} = \{q = 0\}$ be, respectively, a *hyperbolic*, *elliptic* and *tangent set* (or set of *hyperbolic*, *elliptic* and *tangent points*, respectively). Let

$$l(x, \xi_0, \xi'') = b(x, \xi_0, \xi'', \lambda^+(x, \xi_0, \xi''))$$

be the Lopatinskij determinant and $\lambda^\pm = \pm\sqrt{q(x, \xi_0, \xi'')}$, where $\operatorname{Im} \lambda^\pm \gtrless 0$ for $\operatorname{Im} \xi_0 < 0$.

Assume that the Lopatinskij condition

$$l(x, \xi_0, \xi'') \neq 0 \quad \text{for} \quad \operatorname{Im} \xi_0 < 0 \tag{2.26'}$$

holds and that

$$l(x, \xi_0, \xi'') = 0, \quad (x, \xi_0, \xi'') \in \mathscr{H} \cup \mathscr{E} \Rightarrow \frac{\partial l}{\partial \xi_0}(x, \xi_0, \xi'') \neq 0, \tag{2.38}$$

$$l(x, \xi_0, \xi'') = 0, \quad (x, \xi_0, \xi'') \in \mathscr{G} \Rightarrow \frac{\partial b}{\partial \xi_n}(x, \xi_0, \xi'') \neq 0. \tag{2.39}$$

A domain X is said to be *bicharacteristically strictly convex* (*concave*) if

$$x_n = 0, \qquad \{p, x_n\} = 0 \Rightarrow H_p^2 x_n \gtrless 0,$$

that is, if $\left.\frac{\partial q}{\partial x_n}\right|_{\mathcal{G}} \gtrless 0$. The behaviour of bicharacteristics in bicharacteristically convex and concave domains is depicted in Fig. 2 and Fig. 3, respectively.

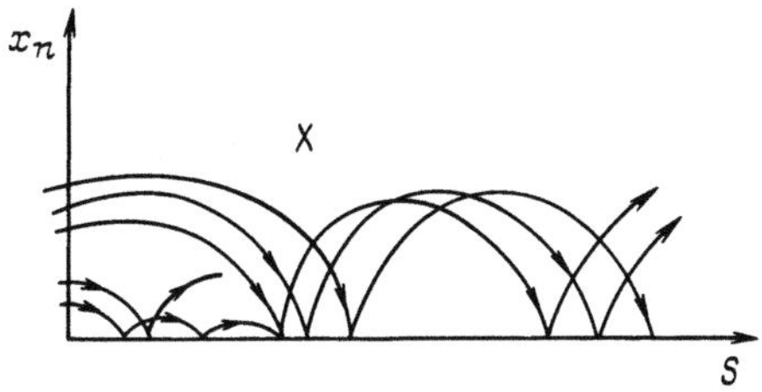

Fig. 2

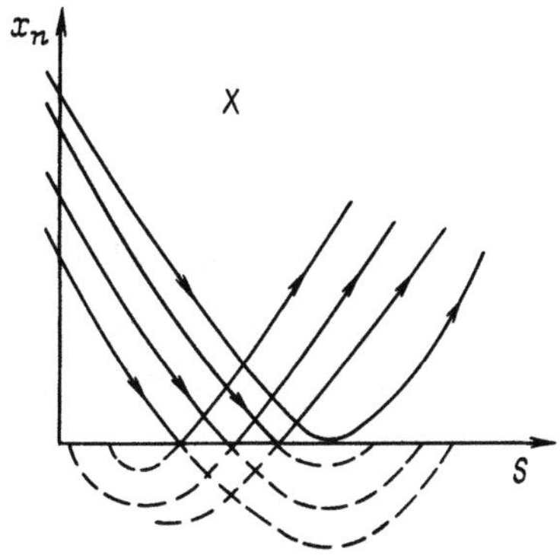

Fig. 3

Theorem 19 (Eskin [1979/80, 1981]). *If the domain X is bicharacteristically strictly concave and if conditions (2.26′), (2.38) and (2.39) hold, then the mixed problem is C^∞-well posed. Furthermore, if $\varphi \in H_{loc}^{s-M+1}(S)$ and $\varphi = 0$ for $x_0 < T$, then $u \in H_{loc}^s(X)$.*

Theorem 20 (Eskin [1979/80]). *Let the domain X be bicharacteristically strictly convex. Let conditions (2.26′), (2.38) and (2.39) be satisfied. Assume that in a neighbourhood of any point $\bar{\rho} \in \mathcal{G} \cap \{l = 0\}$ the condition*

$$|\operatorname{Re} \gamma(\rho)| \ln^{-1} |\operatorname{Re} \gamma(\rho)| \operatorname{sign} q_{\xi_0}(\rho) \leqslant C |\operatorname{Im} \gamma(\rho)|^2$$
$$\forall \rho = (x, \xi_0, \xi'') \in \mathcal{G} \cap \{|\xi_0| + |\xi''| = 1\} \tag{2.40}$$

is fulfilled; here $\xi_n = \gamma(\rho)$ is a smooth root of the polynomial $b(\rho, \xi_n)$ such that $\gamma(\bar{\rho}) = 0$. Then the mixed problem (2.36) is C^∞-well posed. Furhermore, if $\varphi \in H_{loc}^{s-M+1}(S)$, $\varphi = 0$ for $x_0 < T$ and $0 < t - T$ is sufficiently small, then $u \in H_{loc}^s(X_t^-)$.

Theorem 21 (Eskin [1979/80]). *Let the domain X be bicharacteristically strictly convex. Let conditions (2.26), (2.38) and (2.39) be satisfied. Assume that $\{\gamma, q\}(\bar{\rho}) \neq 0$ at any point $\bar{\rho} \in \mathcal{G} \cap \{l = 0\}$ in a neighbourhood of which (2.40) is not satisfied. Then the mixed problem (2.36) is C^∞-well posed. Furthermore, if $\varphi \in H_{loc}^{s-M+L+1}(S)$, $\varphi = 0$ for $x_0 < T$ and $0 < t - T$ is sufficiently small, then $u \in H_{loc}^s(X_t^-)$, where*

$$L > \max\left(-\frac{1}{3}\{\xi_n - \gamma, q\} |\{\gamma, q\}|^{-1}\right)$$

and the maximum is taken over all the points $\bar{\rho} \in \mathscr{G} \cap \{l = 0\}$ in a neighbourhood of which the condition (2.40) *fails to hold.*

On the other hand, we have

Theorem 22 (Eskin [1979/80]). *Let the domain X be bicharacteristically strictly convex. Assume that there exists a sequence of points $\rho_k \in \mathscr{G} \cap \{|\xi_0| + |\xi''| = 1\}$, $\rho_k \to \bar{\rho}$, such that $\pm q_{\xi_0}(\bar{\rho}) > 0$, $\pm \operatorname{Re} \gamma(\rho_k) < 0$,*

$$(\operatorname{Im} \gamma(\rho_k))^2 \ln |\operatorname{Re} \gamma(\rho_k)| (\operatorname{Re} \gamma(\rho_k))^{-1} \to 0 \quad \text{as} \quad k \to \infty$$

and

$$|\{\gamma, q\}(\rho_k)| \leqslant C |\operatorname{Re} \gamma(\rho_k)|^{\delta}, \quad \delta > 0.$$

Then the mixed problem (2.36) *is not C^∞-well posed.*

We conclude the present section with the remark that for non-strictly hyperbolic equations lower terms also figure in the well-posedness conditions. Thus, if $X = \mathbb{R}^2 \times \overline{\mathbb{R}}^+$,

$$P = D_0^2 - D_2^2 - x_2 D_1^2 + c D_1, \qquad c \in \mathbb{C}, \tag{2.41}$$

and $B = D_2 + \alpha D_0$, then the mixed problem is C^∞-well posed for $\operatorname{Re} \alpha < 0$ and for $\operatorname{Re} \alpha = 0$, $c \in \mathbb{R}$; in the remaining cases it is not well posed. When $B = I$, a necessary and sufficient condition for the well-posedness is that $c \in \mathbb{R}$ (Zajtseva [1983]). Similar, though more complicated, well-posedness conditions have been obtained by Zajtseva [1987] for the operator

$$P = D_0^2 - D_2^2 - x_2^2 D_1 + c D_1, \qquad c \in \mathbb{C}. \tag{2.42}$$

1.8. Mixed Problem in the Classes of Analytic Functions. We formulate Duff's theorem [1959], which is an analogue of the Cauchy-Kovalevskaya theorem for the Cauchy problem.

Let $\omega = \{x_n > 0\}$ and $T = 0$. Let P and B_k ($k = 1, \ldots, \mu$) be scalar differential operators of order m and m_k, respectively, whose coefficients are analytic in a neighbourhood of 0. Assume that

$$p(x, \xi_0, 0, \xi_n) = \prod_{j=1}^{m} (\xi_0 - \lambda_j(x)\xi_n), \tag{2.43}$$

where $\lambda_j(x)$ are real, $\lambda_j \neq \lambda_k$ for $j \neq k$, $\lambda_j < 0$ for $j \leqslant M^+$, and $\lambda_j > 0$ for $j \geqslant M^+ + 1$.

We also assume that

$\mu = M^+$, the boundary-value problem is normal, and the polynomials

$b_k(x, \xi_0, 0, \xi_n)$ are linearly independent (mod $p_+(x, \xi_0, \xi_n)$),

where

$$p_+(x, \xi_0, \xi_n) = \prod_{j=1}^{M^+} (\xi_0 - \lambda_j(x)\xi_n). \tag{2.44}$$

These assumptions ensure strong L_2-well-posedness for the mixed problem in the quadrant $\{x_0 \geqslant 0, x_n \geqslant 0\}$ that is obtained from the original problem by removing all differentiation operators other than D_0 and D_n and the variables $x_1, \ldots, x_{n-1}$ are regarded as parameters.

Under these assumptions, we can construct P-characteristic surfaces Λ_k that contain S_0 and are orthogonal to the vectors $(\lambda_k(x), 0, \ldots, 0, 1)$ at points of S_0. When $k \leqslant M^+$, these surfaces lie in X_0^+ for $x_0 \geqslant 0$ and divide X_0^+ into $M^+ + 1$ sectors $T_k (k = 0, 1, \ldots, M^+)$ (see Fig. 4).

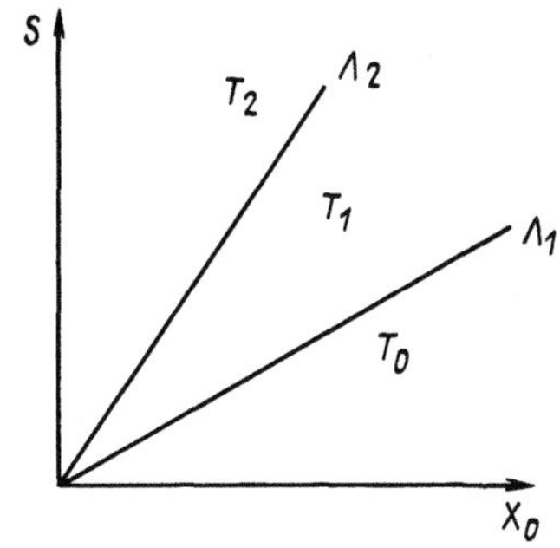

Fig. 4

Theorem 23. *Let conditions* (2.43) *and* (2.44) *be satisfied. Let the functions* f, g_j $(j = 0, \ldots, m-1)$ *and* φ_k $(k = 1, \ldots, \mu)$ *be analytic in a neighbourhood of* 0. *Then in the neighbourhood of* 0 *there exists a function* $u \in L_\infty$ *satisfying* (2.1) *in* $\mathring{X}_0^+$ *in the sense of the theory of generalized functions and the boundary conditions* (2.2) *on* $S_0^+ \backslash S_0$ *and the initial conditions* (2.3) *on* $X_0 \backslash S_0$. *Moreover, its restrictions to* T_k *can be continued to functions that are analytic in a neighbourhood of* 0. *Further, if* f, g_j *and* φ_k *satisfy the compatibility conditions of order* s, *then* $u \in C^{s+m-2}$ *and the solution is unique.*

Note that the compatibility conditions of infinite order enable us to determine the analytic functions φ_k uniquely, and in this case the solution u can be obtained by the Cauchy–Kovalevskaya theorem.

We also note that if the mixed problem (2.1)–(2.3) is C^∞-well posed (in general, the coefficients of P and B_k are no longer analytic), conditions (2.43) and (2.44) hold, and the functions f, g_j and φ_k belong to C^∞, then in a neighbourhood of S_0 the solution is infinitely smooth in each of the sectors T_k up to its boundary and belongs to C^{s+m-2} if f, g_j and φ_k satisfy the compatibility conditions of order s.

§ 2. Propagation of C^∞-Singularities

In this section we examine the propagation of C^∞-wave fronts of solutions to the boundary-value problems for hyperbolic equations and systems. However, we shall not discuss the well-posedness question for these problems.

2.1. The Wave Fronts. Let $X = X' \times \mathbb{R} = \mathbb{R}^n \times \overline{\mathbb{R}}^+ \ni (x_0, \ldots, x_{n-1}, x_n)$, $S = X' \times 0$ and $\mathring{X} = X \backslash S$. Let $\imath$: $T^*X|_S \to T^*S$ be the natural map. We denote by $L^m(X')$ the space of classical pseudodifferential operators of order m on X' and set $L^{m'}(X) = C^\infty(\mathbb{R}^+, L^m(X'))$. We shall discuss only operators with compact supports. We introduce the function space

$$\mathscr{H}^s(X) = \{u, D_0^j u \in L_2(\overline{\mathbb{R}}^+, H^{s-j}(X')) \quad \forall j \in \mathbb{Z}^+\}.$$

Definition 5. By the *boundary wave front* (*boundary wave front of order s*) of a distribution $u \in \mathscr{D}'(\mathring{X})$ we mean that set $\mathrm{WF}_b(u) \subset T^*S\backslash 0$ ($\mathrm{WF}_b^s(u) \subset T^*S\backslash 0$): $\rho \notin \mathrm{WF}_b(u)$ ($\rho \notin \mathrm{WF}_b^s(u)$) if we can find an operator $a(x, D') \in L^{0'}(X)$ which is elliptic at the point ρ and is such that $au \in C^\infty(X)$ ($au \in \mathscr{H}^s(X)$).

Unfortunately, the invariance of such a definition under a change of the variable x_n is very doubtful. However, in many cases this invariance can be established for solutions of systems with smooth right-hand sides. Another definition of the boundary wave front can be found in Melrose [1981].

We also introduce the *complete wave fronts*

$$\mathrm{WF}_f(u) = \mathrm{WF}(u) \cup \imath^{-1}\mathrm{WF}_b(u),$$

$$\mathrm{WF}_f^s = \mathrm{WF}^s(u) \cup \imath^{-1}\mathrm{WF}_b^s(u).$$

Thus "gluing" of T^*X on S appears already in the definition: two points θ and $\theta' \in T^*X|_S$ are identified if $\imath\theta = \imath\theta'$.

2.2. Propagation of Singularities of Solutions to Dissipative Boundary-Value Problems for Symmetric Hyperbolic Systems. We consider a $d \times d$ matrix operator

$$P = \sum_{k=0}^{m} A_k D_n^{m-k}, \tag{2.45}$$

where $A_k \in L_d^{k'}(X)$. Here the notation "$d_1 \times d_2$" stands for the size, and "$d \times d$" will be written as "d". Let a_k and

$$p = \sum_{k=0}^{m} a_k \xi_n^{m-k}$$

be the principal symbols of A_k and P, respectively.

Let $T^*S\backslash 0 \ni \rho^*$ be the point in a neighbourhood of which the discussions will be carried out. We assume that

$$\operatorname{rank} a_0(\rho^*) = d \tag{2.46}$$

(that is, S is non-characteristic at ρ^*), and that

$$p^* = p \tag{2.47}$$

(that is, the system is principally symmetric).

Then for all $u, v \in \mathscr{S}(\mathbb{R}^+, \mathbb{C}^d)$ and all $\rho \in T^*S\backslash 0$ lying in a neighbourhood of ρ^*, we have

$$\langle p(\rho, D_n)u, v\rangle_+ - \langle u, p(\rho, D_n)v\rangle_+ = i\langle T(\rho)\gamma u, \gamma v\rangle_D,$$

where $\langle\cdot,\cdot\rangle_+$ denotes the scalar product in $L_2(\mathbb{R}^+, \mathbb{C}^d)$, $\langle\cdot,\cdot\rangle_L$ the Hermitian product in $\mathbb{C}^L$ and $\gamma u = (u, D_n u, \ldots, D_n^{m-1}u)|_{x_n=0}$ are the Cauchy data of u on S, $D = md$, and $T(\rho)$ is a $D \times D$ smooth Hermitian matrix, with rank $T(\rho^*) = D$, that can be expressed in an obvious manner in terms of $a_k(\rho)$. Let D_+ and d_+ denote, respectively, the number of positive eigenvalues of the matrices $T(\rho^*)$ and $a_0(\rho^*)$. Then $D_+ = md/2$ for even m and $D_+ = (m-1)d/2 + d_+$ for odd m.

Let $B = (B_0, \ldots, B_{m-1})$ be the boundary operator and $B_k \in L^{-k}_{D_+\times d}(S)$. Let b_k, $b = (b_0, \ldots, b_{m-1})$, the principal symbols of B_k and B, respectively, be $D_+ \times d$ and $D_+ \times D$ matrices. We assume that

$$\operatorname{rank} b(\rho^*) = D_+, \tag{2.48}$$

$$\langle T(\rho)U, U\rangle_D \leqslant 0 \quad \forall U \in \operatorname{Ker} b(\rho) \tag{2.49}$$

in some neighbourhood of ρ^*, that is, there the boundary operator B is non-degenerate and is *dissipative*.

Then there exists a $D \times D_+$ smooth matrix $\beta(\rho)$ such that

$$-T(\rho) + \beta(\rho)b(\rho) + b^*(\rho)\beta^*(\rho)$$

is a non-negative definite matrix in the neighbourhood of ρ^*.

We already know that for $u \in \mathscr{D}'(\mathring{X})$

$$\mathrm{WF}^s(u) \subset \mathrm{WF}^{s-m}(Pu) \cup \Sigma,$$

where $\Sigma = \operatorname{Char} P = \{\theta \in T^*X, g \equiv \det p = 0\}$ is the *characteristic set of operator P*; outside this set P is *elliptic*.

Let $E(\rho)$ denote the solution space of the system $(\rho, D_n)v = 0$ on $\mathbb{R}^+$, the solutions decaying as $x_n \to +\infty$.

Definition 6. The problem (P, B), satisfying conditions (2.46) and (2.47), is *elliptic at a point* $\rho \in T^*S\backslash 0$ *if* $\rho \notin \iota\Sigma$ *and the map* $b(\rho)\gamma\colon E(\rho) \to \mathbb{C}^\mu$ *is a bijection.* Here $\mu = \dim E$; $\mu = D_+$ if (2.47) holds and $\rho \notin \iota\Sigma$.

If Σ_b denotes the set of points where the problem (P, B) is non-elliptic, Σ_b is known as a *characteristic set of the problem* (P, B). By the standard methods of the theory of elliptic boundary-value problems, we can easily show that if (2.46) and (2.48) hold and $u \in C^\infty(\overline{\mathbb{R}}^+, \mathscr{D}'(X'))$, then

$$\mathrm{WF}^s_b(u) \subset \mathrm{WF}^{s-m}_b(Pu) \cup \mathrm{WF}^{s-1/2}(B\gamma u) \cup \Sigma_b. \tag{2.50}$$

Finally, let $\xi_n = z_k$ $(k = 1, \ldots, N)$ be all the real and distinct roots of the characteristic equation $g(\rho^*, \xi_n) = 0$.

Theorem 24 (Ivrii [1986]). *Assume that conditions* (2.46), (2.47), (2.48) *and* (2.49) *hold. Let* $\varphi_k \in C^\infty(\overline{\mathbb{R}}^+ \times (T^*X'\backslash 0))$, $\varphi \in C^\infty(T^*S\backslash 0)$ *be real, positive homogeneous functions in* ξ' *of degree* 0 *such that* $\varphi_1 = \cdots = \varphi_N = \varphi$ *on* T^*S, $\varphi(\rho^*) = 0$ *and*

$$\begin{gathered}\langle\{p, \varphi_k\}(\rho^*, z_k)v, v\rangle > 0\\ \forall v \in \operatorname{Ker} p(\rho^*, z_k)\backslash 0, \quad k = 1, \ldots, N,\end{gathered} \tag{2.51}$$

$$\begin{gathered}\operatorname{Re}\langle\{p, \varphi\}(\rho^*, D_n)v, v\rangle_+ - \operatorname{Re} i\langle\beta(\rho^*)\{b, \varphi\}(\rho^*)\gamma v, \gamma v\rangle_D > 0\\ \forall v \in E(\rho^*) \cap \operatorname{Ker} b(\rho^*)\gamma\backslash 0.\end{gathered} \tag{2.52}$$

If $u \in C^\infty(\overline{\mathbb{R}}^+, \mathscr{D}'(X'))$ *and*

$$\rho^* \notin \mathrm{WF}_b^{s-m+1}(Pu) \cup \mathrm{WF}^{s+1/2}(B\gamma u), \tag{2.53}$$

$$\mathrm{WF}^s(u) \cap \{\varphi_k < 0\} \cap W_k = \varnothing \quad \forall k = 1, \ldots, N, \tag{2.54}$$

$$\mathrm{WF}_b^s(u) \cap \{\varphi < 0\} \cap W = \varnothing, \tag{2.55}$$

where W_k *and* W *are sufficiently small neighbourhoods of the points* (ρ^*, z_k) *and* ρ^* *in* T^*X *and* T^*S, *respectively, then* $\rho^* \notin \mathrm{WF}_b^s(u)$. *The same assertions remain valid for the wave fronts of infinite order also.*

In analogy with the established terminology of § 3.1, it is natural to call P, satisfying (2.47) and (2.51) with a given k, a *microhyperbolic operator at the point* (ρ^*, z_k) *in the direction* $H_{\varphi_k}(\rho^*)$ and the problem (P, B), satisfying (2.46)–(2.49), (2.51) and (2.52), *microhyperbolic at the point* ρ^* *in the multi-direction*

$$\left(H_\varphi, -\frac{\partial \varphi_1}{\partial x_n}, \ldots, -\frac{\partial \varphi_N}{\partial x_n}\right)(\rho^*).$$

Remark 1. If the non-degeneracy condition

$$E(\rho^*) \cap \operatorname{Ker} b(\rho^*)\gamma = 0, \tag{2.56}$$

is satisfied, then (2.52) of Theorem 24 is empty and then the propagation of singularities does not depend on any concrete choice of the boundary operator B (though for some B there may be abnormally poor propagation of singularities). This theorem has been proved roughly under such an assumption by Ivrii [1979d, e, 1980a].

If $\rho^* \notin \imath\Sigma$, then the study of the propagation of singularities of solutions to boundary-value problems reduces to that of propagation of singularities for a certain operator $\mathscr{B} \in L^0_{d/2}(S)$. Its characteristic set is Σ_b. The Rayleigh waves are a phenomenon of this very type in which $\mathscr{B}$ is a symmetric operator having characteristics of constant multiplicity 2.

In the general case, there is a complicated combination of the singularity propagation. A detailed examination in the particular case of the wave equation can be found in § 2.6.

Remark 2. If the condition of strict dissipativity

$$\begin{gathered}\langle T(\rho^*)U, U\rangle_D < 0 \\ \forall U \in \operatorname{Ker} b(\rho^*)\backslash 0,\end{gathered} \tag{2.57}$$

holds at the point ρ^*, then (2.56) also holds. What is more, in this case lower smoothness is required from the boundary data if (2.53) is replaced by

$$\rho^* \notin \mathrm{WF}_b^{s-m+1}(Pu) \cup \mathrm{WF}^s(B\gamma u), \tag{2.53$'$}$$

yielding, in addition, that

$$\rho^* \notin \bigcup_{k=0}^{\infty} \mathrm{WF}^{s-k}(D_n^k u|_s).$$

Remark 3. Assume that (2.46)–(2.48) hold. Let $\dim \operatorname{Ker} p(\rho^*, z_k) = 1$ for all $k = 1, \ldots, N$. Let r_k denote the multiplicity of z_k, a root of the characteristic equation. Assume that E_k is the space of generalized-exponential solutions on $\overline{\mathbb{R}}^+$ of the system $p(\rho^*, D_n)v = 0$ that correspond to the characteristic root z_k and to the following growth condition as $x_n \to +\infty$:

$$\begin{aligned} v &= 0(x_n^{(r_k-2)/2}) \quad \text{for even } r_k, \\ v &= 0(x_n^{(r_k-3)/2}) \quad \text{for odd } r_k \end{aligned} \tag{2.58}$$

and for

$$\alpha_k \equiv (-1)^{(r_k-1)/2}\left(\left(\frac{\partial}{\partial \xi_n}\right)^{r_k} g\right)(\rho^*, z_k)\{g, \varphi_k\}(\rho^*, z_k) < 0, \tag{2.58'}$$

while

$$v = 0(x_n^{(r_k-1)/2}) \quad \text{for odd } r_k \text{ and } \alpha_k > 0, \tag{2.58''}$$

where φ_k satisfies (2.51). We set $\tilde{E} = E_1 \oplus E_2 \oplus \cdots \oplus E_N \oplus E(\rho^*)$. If the strict dissipativity condition is satisfied, then

$$\tilde{E} \cap \operatorname{Ker} b(\rho^*)\gamma = 0. \tag{2.59}$$

Conversely, if (2.46)–(2.48) hold and

$$\dim \operatorname{Ker} p(\rho^*, z_k) = 1 \quad \text{for all } k = 1, \ldots, N,$$

then the problem satisfying (2.59) can be reduced to a strictly dissipative problem for a symmetric system of first order.

2.3. The Geometry of the Propagation of Singularities. Assuming that the non-degeneracy condition (2.48) holds, we list a number of concrete results. We start with the assumption that P has simple characteristics. Analogous results also hold for operators with characteristics of constant multiplicity, but in this case g is not the characteristic polynomial of P but a polynomial of principal type whose zeros coincide with those of the characteristic polynomial. We shall assume throughout, without stating it each time, that the right-hand side and the boundary data are infinitely smooth in the desired domain $V \subset T^*X\backslash 0$ of the phase space, that is, it does not intersect $\mathrm{WF}_f(Pu)$ or $\imath^{-1}\mathrm{WF}(Bu|_S)$.

We immediately find that if the only bicharacteristics are those transversal to S, then the wave fronts of the solutions arrive on S along the incoming bicharacteristics and move away from S along the outgoing ones. In this way they are propagated along *bicharacteristic billiards* having branches, in general, on S. We can easily drop the transversal bicharacteristics from the microlocal analysis, which indeed we shall be doing for the convenience in formulation, without stating it each time.

Suppose that over the point $\rho^* \in T^*S\backslash 0$ there lies exactly one tangent bicharacteristic, that is, the characteristic polynomial $g(\rho, \xi_n)$ has for $\rho = \rho^*$ exactly one multiple real root $\xi_n = z_1$. Let $\theta^* = (\rho^*, z_1)$. We recall that since the characteristics of P are simple, it follows that $g_{\xi_0}(\theta^*) \neq 0$. We also assume that

z_1 is a root of multiplicity 2 of the characteristic polynomial, that is, $g(\theta^*) = (H_{x_n}g)(\theta^*) = 0$, $H^2_{x_n}g(\theta^*) \neq 0$. To be definite, we assume that $H^2_{x_n}g(\theta^*) > 0$. If the domain X is bicharacteristically strictly convex at θ^*, that is, if $(H^2_g x_n)(\theta^*) < 0$, then in the neighbourhood of $\iota^{-1}\rho^*$ the wave fronts are propagated along bicharacteristic billiards of g as well as along *boundary bicharacteristics*, namely, along bicharacteristics, defined on $T^*S\backslash 0$, of the *boundary symbol* $g_b(\rho) = g(\rho, \gamma(\rho))$, where $\xi_n = \gamma(\rho)$ is a root of the polynomial $\{g, x_n\}\{\rho, \xi_n\}$ such that $\gamma(\rho^*) = z_1$. A boundary bicharacteristic is the ι-projection of a curve, lying in $T^*X|_S$, along which

$$\frac{d\rho}{dt} = H_g(\rho) + \varkappa H_{x_n}(\rho),$$
$$g = \{g, x_n\} = 0; \tag{2.60}$$

this curve will also be referred to as a boundary bicharacteristic. Note that $\varkappa = (H^2_g x_n)(H^2_{x_n}g)^{-1}$. A boundary bicharacteristic will be termed a *C^∞-boundary bicharacteristic* if $H^2_g x_n \leqslant 0$ on it. Such bicharacteristics are limits of those bicharacteristic billiards which press against the boundary more and more tightly. In particular, for the wave equation bicharacteristic billiards are the usual billiard trajectories, while boundary bicharacteristics are the geodesic boundaries.

If X is *bicharacteristically concave*, that is, if $H^2_g x_n \geqslant 0$, then C^∞-singularities are, perhaps, not propagated along boundary bicharacteristics (see the next two sections). Therefore, in this case Theorem 24 does not furnish a precise description of the propagation of C^∞-singularities.

If there are several tangent bicharacteristics over ρ^*, each satisfying the assumptions made above, and if X is a bicharacteristically strictly convex domain relative to each of them, then there arise several boundary symbols $g_{b,k}$ and there may be different types of propagation depending on their "interaction". The simplest case is the one where all the boundary symbols $g_{b,k}$ ($k = 1, \ldots, N$) are in involution and the vector fields $H_{g_{b,1}}, \ldots, H_{g_{b,N}}, \xi' \frac{\partial}{\partial \xi'}$ are linearly independent. Then there arises the propagation of singularities along N-dimensional bicharacteristic sheets that are constructed in accordance with the aggregate of boundary symbols in exactly the same way as was done in § 3.2, Chap. 1, in accordance with the aggregate of factors.

Ivrii [1979d, e, 1980a] has also analysed the propagation of singularities for the case where one tangent bicharacteristic lies over ρ^* and the root $\xi_n = z_1$ of the characteristic polynomial is of multiplicity 3.

The propagation of singularities of solutions to boundary-value problems for systems with characteristics of variable multiplicity has been studied to a lesser extent, and for concrete results the reader is referred to Ivrii [1979d, e, 1980a].

2.4. The Propagation of Singularities of Solutions to Strictly Dissipative Boundary-Value Problems for Symmetric Hyperbolic Systems. Theorem 24, though fairly general, does not enable us to study conclusively the propagation of singularities in bicharacteristically concave domains near the points of tangency. This is due to the requirement that the function φ should not change when the bicharacteristic is reflected from S, that is, that this function should not depend on ξ_n. True, some implicit and weak dependence was assumed for $N \geqslant 2$ because we had considered a collection of functions φ_k. We now drop this requirement and impose a more natural condition that φ does not decrease when the bicharacteristic is reflected from S. However, we are forced to examine a considerably narrow class of systems and only strictly dissipative boundary-value problems for them.

We assume that the following conditions are satisfied.

a) In a neighbourhood of $L^{-1}\rho^*$ in $T^*X|_S$, the characteristic symbol can be factorized

$$g = e \prod_{k=1}^{N} h_k^{r_k}, \tag{2.61}$$

where e is an elliptic symbol and h_k are real polynomials in ξ_n of degree $m_k = 1$, 2 with the stipulation that h_j and h_k, $j \neq k$, do not both vanish simultaneously and that

$$h_k = 0 \Rightarrow \{h_k, x_0\} > 0.$$

b) In a neighbourhood of ρ^* there exists a real, positive homogeneous symbol η of degree 0 such that $\eta(\rho^*) = 0$, $\{\eta, x_0\}(\rho^*) \neq 0$ and for $\eta(\rho) \geqslant 0$ all the roots ξ_n of $h_k(\rho, \xi_n)$ are real, whereas for $\eta(\rho) < 0, m_k = 2$, all the roots of $h_k(\rho, \xi_n)$ are non-real.

Then both the roots of $h_k(\rho, \xi_n)$ coincide if $\eta(\rho) = 0$ and $m_k = 2$. Condition b) clearly rules out the presence of bicharacteristic boundary sheets of dimensuion greater than 1.

Theorem 25 (Ivrii [1980b]). *Suppose that conditions* a) *and* b), *conditions* b) *of Theorem* 37, *Chap.* 1, *and conditions* (2.46), (2.48) *and* (2.49) *hold in a neighbourhood of* $\imath^{-1}\rho^*$. *Let* $\varphi \in C^\infty(T^*X\backslash 0)$ *be a real positive homogeneous function of degree* 0 *that satisfies* (1.38) (*of Theorem* 37, *Chap.* 1) *and the conditions*

$$\begin{gathered} h_k(\theta') = h_j(\theta'') = 0, \\ \{h_k, x_n\}(\theta') \geqslant 0 \geqslant \{h_j, x_n\}(\theta''), \end{gathered} \tag{2.62}$$

$$\begin{gathered} \imath\theta' = \imath\theta'' \Rightarrow \varphi(\theta') \geqslant \varphi(\theta''); \\ \varphi = 0 \quad \text{on} \quad \imath^{-1}\rho^* \cap \{g = 0\} \end{gathered} \tag{2.63}$$

in the neighbourhood of $\imath^{-1}\rho^*$. *If* $u \in C^\infty(\overline{\mathbb{R}}^+, \mathscr{D}'(X'))$ *and*

$$\begin{gathered} \rho^* \notin \mathrm{WF}_b^s(Pu) \cup \mathrm{WF}^s(B\gamma u), \\ \mathrm{WF}_f^s(u) \cap \{g = 0\} \cap \{\varphi < 0\} \cap W = \varnothing, \end{gathered} \tag{2.64}$$

where W is a neighbourhood of $\imath^{-1}\rho^*$ in T^*X, then

$$\rho^* \notin \mathrm{WF}_b^s(u) \cup \bigcup_{k=0}^{\infty} \mathrm{WF}^{s-k}(D_n^k u|_S).$$

We do not know whether this theorem remains valid for wave fronts of infinite order. However, applying it to concrete cases, we obtain results for wave fronts of finite as well as infinite order. For $N = 1$, we can easily obtain a global version of this theorem that is also true for infinite-order wave fronts.

Theorem 25 is applicable to strictly hyperbolic operators of second order. Namely, to operators of the form

$$P = -D_n^2 + Q, \qquad Q \in L^{2'}(X), \tag{2.65}$$

that are such that

a) q, the principal symbol of Q, is real and $q \geqslant 0 \Rightarrow \varepsilon\{q, x_0\} > 0$, $\varepsilon = \pm 1$ fixed, if the boundary operator $B = \gamma^0 D_n + \gamma^1$, $\gamma^j \in L^j(S)$, satisfies (2.59). This condition in this case is equivalent to the two uniform Lopatinskij-type conditions

$$\begin{gathered} \theta \in T^*X|_s, \qquad p(\theta) = 0, \\ \varepsilon\{p, x_n\}(\theta) \geqslant 0 \Rightarrow b(\theta) \neq 0, \end{gathered} \tag{2.66}$$

$$\begin{gathered} \theta \in T^*X|_s, \qquad p(\theta + \lambda i H_{x_n}) = 0, \\ \lambda > 0 \Rightarrow b(\theta + \lambda i H_{x_n}) \neq 0, \end{gathered} \tag{2.67}$$

where p and b are principal symbols of P and B, respectively. As we have already noted, such problems lead to strictly dissipative boundary-value problems for first-order systems satisfying the conditions (2.46)–(2.48) and condition b) of Theorem 37, Chap. 1. When $\gamma^0 = 1$ and $\gamma^1 \in L^0(S)$, such a reduction is not possible in the neighbourhood of points of tangency. However, in this case Melrose and Sjöstrand [1978, 1982] have obtained similar results, as we describe below.

2.5. The Geometry of the Propagation of Singularities (the Concluding Part). We assume in this section that the operators P and B satisfy the hypotheses of Theorem 25.

We start by describing the propagation of singularities when the characteristic polynomial is of the form $g = -\xi_n^2 + q$, where the symbol $q = q(x, \xi')$ satisfies condition a) of Theorem 25. Under these assumptions

$$\mathrm{WF}_b(u) \subset \mathrm{WF}_b(Pu) \cap \mathrm{WF}(Bu|_s) \cup \mathscr{H} \cup \mathscr{G}, \tag{2.68}$$

where, we recall, $\mathscr{H}$, $\mathscr{E}$, and $\mathscr{G}$ are, respectively, hyperbolic, elliptic and tangent sets. We shall consider the restriction of complete sets. We shall examine the restriction of the complete wave fronts to the manifold $\Sigma = \{g = 0\}$, identifying the equivalent points of $\Sigma|_S$.

Definition 7. By a C^∞-*ray* we mean a Lipschitz curve $\gamma = \{\theta = \theta(t)\}$, with parameter $t = x_0$, lying in Σ that in the neighbourhood of almost all its points is either a bicharacteristic or a C^∞-boundary bicharacteristic.

If $\bar{\theta} \in \Sigma$ and $\iota^{-1}\iota V \subset V$ is a neighbourhood of $\bar{\theta}$, we denote by $K^{\pm}(\bar{\theta}, V)$ the union of all C^∞-rays that issue from $\bar{\theta}$ in the direction of increasing (decreasing) x_0 and lie in $\bar{V}$. It can be shown that $K^{\pm}(\bar{\theta}, V)$ have the properties (1.43)–(1.46). Further, if the domain X is bicharacteristically concave or its boundary S is *bicharacteristically non-degenerate*, that is, if the set

$$\mathscr{G}_\infty = \{\rho \in \mathscr{G}, (H_g^k x_n)(\rho) = 0 \quad \forall k \in \mathbb{Z}^+\}$$

is empty or, more generally, each point $\rho \in \mathscr{G}_\infty$ is an interior point of some bicharacteristic of g lying in $\mathscr{G}$ (in particular, if g is an analytic function), then through each point $\bar{\theta} \in \Sigma$ there passes a unique maximally extended C^∞-ray $\gamma(\bar{\theta})$. On this ray the exceptional points (the points where the bicharacteritic is reflected from S and the points where transition takes place from the bicharacteristic to the C^∞-boundary bicharacteristic and vice-versa) constitute a locally finite set. In the general case, however, C^∞-rays have branching on $\mathscr{G}_\infty$, and together with them the wave fronts too have branching.

Theorem 26. *Let the above-mentioned assumptions be fulfilled. Let $\iota^{-1}\iota V \subset V$ be an open set, $V \cap \Sigma \Subset \Sigma$ and $\bar{\theta} \in \Sigma \cap V$. If*

$$(\mathrm{WF}_f(Pu) \cup \iota^{-1}\mathrm{WF}(Bu|_s)) \cap K^-(\bar{\theta}, V) = \varnothing, \tag{2.69}$$

$$\mathrm{WF}_f(u) \cap K^-(\bar{\theta}, V) \cap \partial V = \varnothing, \tag{2.70}$$

then

$$\mathrm{WF}_f(u) \cap K^-(\bar{\theta}, V) = \varnothing. \tag{2.71}$$

We now assume that the hypotheses of Theorem 25 hold and that the decomposition (2.61) holds with $N \geqslant 2$ in a neighbourhood of $\iota^{-1}\rho^*$ in T^*X. Assume that the domain X is bicharacteristically strictly concave with respect to h_k ($k = 1, \ldots, M$) and $M = N - 1$ or else $M \leqslant N - 1$ and X is bicharacteristically strictly convex with respect to h_k ($k = M + 1, \ldots, N$). In the first case, the singularities are propagated along billiards that consist of bicharacteristics h_k ($k = 1, \ldots, M - 1$) and C^∞-rays h_N. In the second case, singularities are propagated along billiards consisting of bicharacteristics h_k ($k = 1, \ldots, N$) and boundary bicharacteristics, that is, bicharacteristics of the boundary symbol. For $M = 0$ Theorem 24 can also be applied to the present situation. Therefore in this case (2.57) can be replaced by (2.56).

We observe that for bicharacteristically strictly concave domains there is, in addition to the implicit study of propagation of singularities, an explicit construction of a parametrix (Eskin [1976], Melrose [1975], Taylor [1976a, b, 1979]) that is based on Ludwig's constructions and uses Airy functions as the components of oscillatory integrals. In bicharacteristically strictly convex domains there is "semi-explicit" construction (Eskin [1977]). There are also several works devoted to the reflection and refraction problems that fit into the given scheme.

2.6. The Propagation of Singularities of Solutions to Non-classical Problems. In this section we shall examine the propagation of singularities of solutions to the non-classical problems. That is, to these boundary-value problems for strictly hyperbolic equations of second order that do not satisfy the conditions (2.66) and (2.67). Let P be an operator of the form (2.65), $B = D_n + M$, $M \in L^1(S)$ and μ its principal symbol. Then the sets of points where (2.66) and (2.67) are violated are, respectively, the sets $\Lambda^- = \{(\varepsilon\sqrt{q} + \mu = 0\} \subset \mathscr{H} \cup \mathscr{G}$ and $\Lambda^+ \setminus \Lambda^0$, where

$$\Lambda^+ = \{i\sqrt{-q} + \mu = 0\} \subset \mathscr{E} \cup \mathscr{G},$$

$$\Lambda^0 = \{q = \mu = 0\} = \Lambda^- \cap \mathscr{G} = \mathscr{H} \cap \mathscr{G}$$

(we identify the points of $T^*X_S|_S$ with their $\imath$-projections).

Evidently,

$$\Lambda^- \cup \Lambda^+ \subset \Lambda = \{q = \mu^2\}.$$

If $\rho \in \mathscr{E}$ and $\rho \notin \mathrm{WF}_b(Pu)$, we can easily show that $\rho \notin \mathrm{WF}((D_n - \alpha)u|_S)$, where $L^1(S) \ni \alpha$ is an operator with principal symbol $\alpha_1 = i\sqrt{-q}$ in a neighbourhood of ρ. If, in addition, $\rho \notin \mathrm{WF}(Bu|_S)$, then $\rho \notin \mathrm{WF}(av)$, where $v = u|_S$ and $a = i(\alpha + M) \in L^1(S)$ is an operator with principal symbol $-\sqrt{-q} + i\mu$ and with characteristic manifold $\Lambda^+ \setminus \Lambda^0$.

It is also easy to show that if $\rho \in \mathscr{H}$, $\rho \notin \mathrm{WF}_b(Pu)$ and $K^-(\rho, V) \cap \mathrm{WF}(u) = \phi$, where V is a sufficiently small nighbourhood of the set $\imath^{-1}\rho \cap \Sigma$, then $\rho \notin \mathrm{WF}((D_n - \beta)u|_S)$, where $L^1(S) \ni \beta$ is an operator with principal symbol $\beta_1 = -\varepsilon\sqrt{q}$ in the neighbourhood of ρ. If, in addition, $\rho \notin \mathrm{WF}(Bu|_S)$, then $\rho \notin \mathrm{WF}(bv)$, where $b = \beta + M$ is an operator with principal symbol $-\varepsilon\sqrt{q} + \mu$ and characteristic manifold $\Lambda^- \setminus \Lambda^0$.

Thus outside the tangent set $\mathscr{G}$ the study of the propagation of singularities of solutions to boundary-value problems reduces to that of solutions to pseudodifferential equations on S. On the other hand, B satisfies (2.66) and (2.67) outside $\Lambda^+ \cup \Lambda^-$. Therefore our main interest must lie in the propagation of singularities only near the set Λ^0. It is evident that there a and b are operators of principal type.

Definition 8. By a *generalized* C^∞*-ray* we mean a Lipschitz curve $\gamma = \{\theta = \theta(t)\} \subset \Sigma \cup \Lambda^+ \cup \Lambda^-$, with parameter $t = x_0$, that in the neighbourhood of almost all its points $\theta \in \Sigma \cup \Lambda^- \setminus \Lambda^0$ is either a C^∞-ray or a bicharacteristic of b and in the neighbourhood of each of its points $\theta \in \Lambda^+ \setminus \Lambda^0$ it is a bicharacteristic of a, and, finally, if at each of its points $\theta \in \Lambda^0$ it satisfies the equation $\dfrac{d\theta}{dt} = \eta H_q(\theta)$, $\eta \neq 0$.

Let $\bar{\theta} \in V \cap (\Sigma \cup \Lambda^+)$ and $T^*X \setminus 0 \supset V$ be an open set such that $\imath^{-1}\imath V \subset V$. We now denote by $K^\pm(\bar{\theta}, V)$ the union of all generalized C^∞-rays issuing from $\bar{\theta}$ in the direction of increasing (decreasing) x_0 and lying in $\bar{V}$. It can be shown that $K^\pm(\theta, V)$ have the properties (1.43)–(1.46) if, at least, one of the following conditions is satisfied:

$$\operatorname{Re}\mu = \omega \operatorname{Im}\mu, \qquad \omega \neq 0 \quad \text{on} \quad \Lambda^0 \tag{2.72}$$

(then the operators a and b are elliptic near Λ^0);

$$\{q, \operatorname{Re} \mu\} > 0, \qquad \varepsilon\{q, \operatorname{Im} \mu\} < 0 \quad \text{on} \quad \Lambda^0 \tag{2.73}$$

(then a and b are subelliptic operators near Λ^0);

$$\operatorname{Im} \mu \equiv 0, \qquad \{q, \operatorname{Re} \mu\} > 0 \quad \text{on} \quad \Lambda^0 \tag{2.74}$$

(then the operator a is elliptic and b is an operator of real principal type near Λ^0, Λ is a C^∞-manifold and Λ^- is a C^∞-manifold with edge Λ^0; the bicharacteristics of b, parametrized by x_0, start from Λ^0);

$$\varepsilon \operatorname{Re} \mu \leqslant 0 \quad \text{if} \quad q + (\operatorname{Im} \mu)^2 = 0 \tag{2.75}$$

(then the operator b is elliptic).

The inclusion (2.67) remains valid if one of the conditions (2.72) to (2.74) holds. In the general case the inclusion

$$\mathrm{WF}_b(u) \subset \mathrm{WF}_b(Pu) \cup \mathrm{WF}(Bu|_s) \cup \mathscr{H} \cup \mathscr{G} \cup \Lambda^+ \tag{2.76}$$

holds.

Theorem 27. *Assume that* (2.65) *and one of the conditions* (2.72)–(2.75) *hold. Let* $\bigvee$ *be a sufficiently close neighbourhood of* $V \cap \Lambda^0$ *such that* $\imath^{-1}\imath\bigvee \subset \bigvee$, $\theta \in V$. *Then* (2.69) *and* (2.70) *imply* (2.71).

Under assumptions (2.72)–(2.74) this theorem has been established by Ivrii [1981] while under (2.75) by Melrose [1975]. Thus under assumptiions (2.72) and (2.73) the propagation of singularities remains classical, but under (2.74) there appears additional propagation of singularities along bicharacteristics of b and additional branching of singularities at points of Λ^-. If (2.75) is assumed to hold, there appears additional propagation of singularities along bicharacteristics of a and branching at points of Λ^0, though, possibly, not at all points. We note that under (2.75) a does not always have a real principal symbol a_1. Therefore by a bicharacteristic in this case we mean the curve along which

$$a = 0, \qquad \frac{d\rho}{dt} = z(\rho) H_a(\rho),$$

where $z(\rho)$ is a non-vanishing complex-valued function.

§ 3. The Propagation of Analytic Singularities

We devote a separate section to the study of propagation of analytic singularities because in the case of the mixed problem it differs significantly from the propagation of C^∞-singularities. At the same time, we shall use the notation of the preceding section. We basically follow the works of Sjöstrand [1980a, b, c, 1981].

3.1. The General Theory. Let $\tilde{X}$, containing 0, be an open domain in $\mathbb{R}^{n+1}$. Put $X = \tilde{X} \cap \{x_n \geqslant 0\}$, $S = \tilde{X} \cap \{x_n = 0\}$, and $\mathring{X} = X \backslash S$. Let $\mathscr{A}(X)$ and $\mathscr{A}(S)$ denote, respectively, the space of functions that are real-analytic in X and on S. We introduce the analytic wave fronts

$$\mathrm{WF}_a(u) = \mathrm{WF}_a(u|_{\mathring{X}}) \subset T^*\mathring{X}\backslash 0,$$

$$\mathrm{WF}_{ba}(u) = \bigcup_{k \in \mathbb{Z}^+} \mathrm{WF}_a(D_n^k u|_s) \subset T^*S\backslash 0,$$

$$\mathrm{WF}_{fa}(u) = \mathrm{WF}_a(u) \cup \imath^{-1} \mathrm{WF}_{ba}(u).$$

Let P be a $d \times d$ matrix differential operator of order m with coefficients in $\mathscr{A}(X)$. Put $B = (B_1, \ldots, B_\mu)$, where B_k are $1 \times d$ matrix differential operators of order m_k with coefficients in $\mathscr{A}(S)$. Let p, b_k and b be the principal symbol of P, B_k and B, respectively, and let $g = \det p$ be the characteristic symbol of P. We assume that S is a non-characteristic surface.

Definition 9. *The operator P is said to be elliptic at a point $\theta \in T^*X\backslash 0$ if $g(\theta) \neq 0$. The problem (P, B) is elliptic at a point $\rho \in T^*S\backslash 0$ if P is elliptic on $\imath^{-1}\rho$ and the map*

$$\gamma b(\rho, D_n)\colon \operatorname{Ker} p(\rho, D_n) \cap \varphi(\mathbb{R}^+) \to \mathbb{C}^\mu \tag{2.77}$$

is a bijection; in particular it is necessary that

$$\dim(\operatorname{Ker} p(\rho, D_n) \cap \varphi(\mathbb{R}^+) = \mu.$$

Here, and in what follows, γ denotes the restriction operator to $\{x_n = 0\}$.

Let $T^*X\backslash 0 \supset \Sigma$ and $T^*S\backslash 0 \supset \Sigma_b$ be the set of points where P and (P, B) are, respectively, non-elliptic. We shall call Σ and Σ_b the *characteristic set* of P and (P, B), respectively.

Theorem 28. (i) *If $Pu \in \mathscr{A}(\mathring{X})$, then* $\mathrm{WF}_a(u) \subset \Sigma$.
(ii) *If $Pu \in \mathscr{A}(X)$ and $\gamma Bu \in \mathscr{A}(S)$, then*

$$\mathrm{WF}_{ba}(u) \subset \Sigma_b.$$

Definition 10. *The operator P is said to be microhyperbolic at a point $\theta^* \in T^*X\backslash 0$ in the direction $\upsilon = (\upsilon_x, \upsilon_\xi)$ if there exists a neighbourhood V of this point in $\mathbb{R}^{2n+2}$ and an $\varepsilon > 0$ such that*

$$g(\theta + it\upsilon) \neq 0 \quad \forall \theta \in V, \qquad t \in (0, \varepsilon). \tag{2.78}$$

Let $\rho^* \in T^*S\backslash 0$. Let $z_k (k = 1, \ldots, N)$ be the distinct roots of the polynomial $g(\rho^*, \xi_n)$, where $z_1, \ldots, z_M \in \mathbb{R}$ and $z_{M+1}, \ldots, z_N \in \mathbb{C}\backslash\mathbb{R}$. Let $\upsilon = (\upsilon'_x, \upsilon'_\xi) \in \mathbb{R}^{2n}$ and $\zeta_k \in \mathbb{R}$ $(k = 1, \ldots, M)$. Assume that P is microhyperbolic at the points

$$\begin{gathered}\theta_k^* = (\rho^*, z_k) \text{ in the directions} \\ \upsilon_k = (\upsilon'_x, 0, \upsilon'_\xi, \zeta_k) \quad (k = 1, \ldots, M).\end{gathered} \tag{2.79}$$

If V is a sufficiently small neighbourhood of ρ^* in $T^*S\backslash 0$ and $0 < \varepsilon$ is sufficiently small, and if $\rho \in V$, $t \in (0, \varepsilon)$ and z is a root of the characteristic polynomial $g(\rho + itv, z)$ that is close to z_k, then $\operatorname{Im}(z - it\zeta_k) \neq 0$ ($k = 1, \ldots, M$), by virtue of (2.31). Now suppose that $R'(\rho, t)$ is the set of roots z of this polynomial that are close to z_k with some $k = 1, \ldots, M$ and with $\operatorname{Im}(z - it\zeta_k) > 0$ and $R''(\rho, t)$ is the set of roots z of the same polynomial that are close to z_k with some $k = M + 1, \ldots, N$ and with $\operatorname{Im} z_k > 0$. We denote by $E(\rho, t)$ the space spanned over generalized exponential solutions of the system $p(\rho + itv, D_n)v = 0$ corresponding to $R'(\rho, t) \cup R''(\rho, t)$, the set of roots of the characteristic polynomial.

Definition 11. If the condition (2.79) holds and if the map

$$\gamma b(\rho + itv, D_n): E(\rho, t) \to \mathbb{C}^\mu \tag{2.80}$$

is a bijection for sufficiently small V and $\varepsilon > 0$, then *the problem* (P, B) *is microhyperbolic at the point* ρ^* *in the multi-direction* $(v, \zeta_1, \ldots, \zeta_M)$.

Theorem 29. *Let* $\rho^* \in T^*S\backslash 0$. *Let* z_k ($k = 1, \ldots, M$) *be distinct real roots of the polynomial* $g(\rho^*, z)$. *Suppose that* $\theta_k^* = (\rho^*, z_k) \in \imath^{-1}\rho^*$ *and* V *and* V_k *are neighbourhoods of* ρ^* *and* θ^* *in* T^*S *and* T^*X, *respectively. Let* $\varphi_k = \varphi_k(x, \xi') \in \mathscr{A}(V_k)$ ($k = 1, \ldots, M$), $\varphi = \varphi(x', \xi') \in \mathscr{A}(V)$ *be real, positive homogeneous functions of degree* 0 *such that* $\varphi_1 = \cdots = \varphi_M = \varphi$ *on* T^*S, $\varphi(\rho^*) = 0$ *and the problem* (P, B) *is microhyperbolic at* ρ^* *in the multi-direction* $\left(H_\varphi, -\frac{\partial \varphi_1}{\partial x_n}, \ldots, -\frac{\partial \varphi_M}{\partial x_n}\right)(\rho^*)$. *If* $Pu \in \mathscr{A}(X)$, $\gamma Bu \in \mathscr{A}(S)$ *and*

$$\mathrm{WF}_a(u) \cap V_k \cap \{\varphi_k < 0\} = \varnothing \quad \forall k = 1, \ldots, M, \tag{2.81}$$

$$\mathrm{WF}_{ba}(u) \cap V \cap \{\varphi < 0\} = \varnothing, \tag{2.82}$$

then $\rho^* \in \mathrm{WF}_{ba}(u)$.

We note the closeness between Theorems 24 and 29.

3.2. The Wave Equation. We examine the wave equation in some more detail. More generally, we shall consider a differential operator P of second order of the form (2.65) with analytic coefficients. Suppose that either $B = I$ or B is a differential operator of first order satisfying conditions (2.66) and (2.67), with analytic coefficients, or, finally, $B = D_n + \beta(x')$, with $\beta \in \mathscr{A}(S)$. Let us consider the problem

$$Pu \in \mathscr{A}(X), \qquad \gamma Bu \in \mathscr{A}(S). \tag{2.83}$$

Since it is elliptic in $\mathscr{E}$, its solutions satisfy the inclusion $\mathrm{WF}_{fa}(u) \subset \Sigma$.

Definition 12. By an *analytic ray* we mean a Lipschitz curve $\gamma = \{\theta = \theta(t)\}$, with parameter $t = x_0$, that lies in Σ and is either a bicharacteristic or a boundary bicharacteristic in the neighbourhood of all its points.

The exceptional points then constitute a locally finite set on each analytic ray. For $\bar{\theta} \in V \cap \Sigma$ and an open set $\imath^{-1}\imath V \subset V$, we denote by $K_a^{\pm}(\bar{\theta}, V)$ the union of all analytic rays issuing from $\bar{\theta}$ in the direction of increasing (decreasing) x_0 and lying in $\bar{V}$. It can be shown that $K_a^{\pm}(\bar{\theta}, V)$ possess properties (1.43)–(1.46). From Theorem 29, we have

Theorem 30. *Let* $u \in \mathscr{D}'(\tilde{X})$ *satisfy* (2.83), $\imath^{-1}\imath V \subset V$ *be an open set, and* $\bar{\theta} \in V \cap \Sigma \Subset \Sigma$. *If*

$$\mathrm{WF}_{fa}(u) \cap K_a^-(\bar{\theta}, V) \cap \partial V = \varnothing,$$

then

$$\mathrm{WF}_{fa}(u) \cap K_a^-(\bar{\theta}, V) = \varnothing.$$

What is more, with the aid of Theorems 28 and 29 we can investigate the propagation of analytic singularities of solutions to non-classical problems.

Let us discuss the distinction between the C^∞-rays and the analytic rays. A boundary bicharacteristic is always an analytic ray and it is a C^∞-ray only outside the *diffractive set* $\mathscr{G}_+ = \{\rho \in \mathscr{G}, q_{x_n}(\rho) > 0\}$. Therefore through every point of this set there passes (locally) only one C^∞-ray but several analytic rays, and the branching of analytic rays takes place at points of $\mathscr{G}_+$ and at some points of its boundary $\partial\mathscr{G}_+$; at these points transition takes place from tangent bicharacteristic to S to boundary bicharacteristics and vice-versa. To see the effects of the fact that an analytic ray through $\bar{\rho} \in \mathscr{G}_+$ is not unique, we let pass through this point a bicharacteristic γ and a boundary bicharacteristic α. Put $\gamma_\pm = \gamma \cap \{\pm x_0 > 0\}$ and $\alpha_\pm = \alpha \cap \{\pm x_0 > 0\}$ (we assume that $x_0(\bar{\rho}) = 0$); everything is depicted in Fig. 5 for the wave equation where S is not straight. Then $K^+(\bar{\rho}) = \bar{\rho} \cup \gamma_+$ but at the same time $K_a^+(\bar{\rho})$ has a more complicated structure (see Fig. 6).

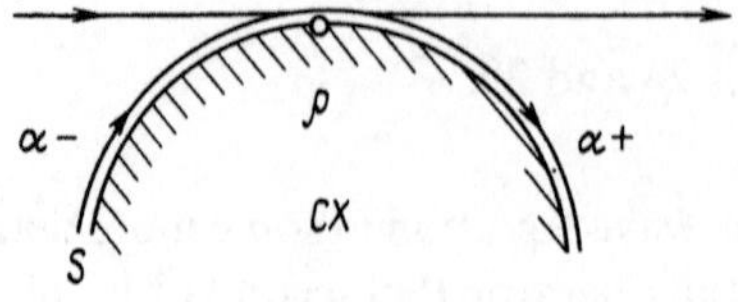

Fig. 5

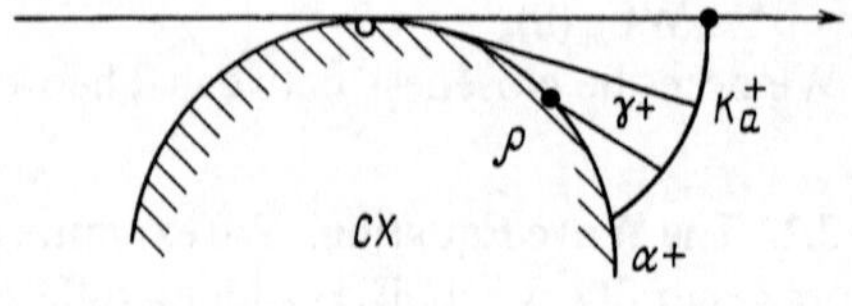

Fig. 6

In accordance with the customary terminology, γ_- is an *incident ray*, γ_+ a *reflected ray*, α_+ a *crawling ray*, and for any $\rho \in \alpha_+$ $\gamma'_+ = \gamma_+(\rho)$ is a *glancing ray*. Therefore

$$K_a^+(\bar{\rho}) = \bar{\rho} \cup \gamma_+ \cup \alpha_+ \cup \bigcup_{\rho \in \alpha_+} \gamma_+(\rho).$$

Instead of showing $K^+(\bar{\rho})$ and $K_a^+(\bar{\rho})$, Fig. 6 shows their intersection with the hyperplane $\{x_0 = t\}$, $t > 0$.

We discuss the propagation of singularities by assuming that either $B = I$ or $B = D_n + \beta$. If u is a solution of (2.83), then, by Theorem 30,

$$(\gamma_{\mp} \cup \alpha_{\mp}) \cap \mathrm{WF}_{fa}(u) = \varnothing \Rightarrow \bar{\rho} \notin \mathrm{WF}_{fa}(u). \tag{2.84}$$

We recall that, by virtue of Theorem 25,

$$\gamma_{\mp} \cap \mathrm{WF}_{f}(u) = \varnothing \Rightarrow \bar{\rho} \notin \mathrm{WF}_{f}(u). \tag{2.85}$$

As shown by a direct computation by Rauch [1977] for the wave equation and by Friedlander and Melrose [1977] for the model equation, the implication (2.85) is not valid for analytic wave fronts.

We also have the following results.

Theorem 31 (Kataoka [1981]). *If* $Pu \in \mathscr{A}(X)$, *then*

$$(\gamma_{-} \cup \gamma_{+}) \cap \mathrm{WF}_{fa}(u) = \varnothing \Rightarrow \bar{\rho} \notin \mathrm{WF}_{fa}(u).$$

We note that here we do not require that boundary conditions be satisfied; we also note that the assertion is not valid for C^∞-wave fronts.

Theorem 32 (Sjöstrand [1980b, c, 1981]). *The problem* (2.83) *has a solution* u *such that* $\mathrm{WF}_f(u) = \mathrm{WF}_{fa}(u) = \gamma_c$. *In particular*, $(\alpha_{-} \cup \alpha_{+}) \cap \mathrm{WF}_{fa}(u) = \varnothing$ *does not imply that* $\bar{\rho} \notin \mathrm{WF}_{fa}(u)$.

Theorem 33 (Sjöstrand [1980b, c, 1981]). *If* u *is a solution of* (2.83), *then*

$$(\gamma_{\mp} \cup \alpha_{\pm}) \cup \mathrm{WF}_{fa}(u) = \varnothing \Rightarrow \bar{\rho} \notin \mathrm{WF}_{fa}(u).$$

In conclusion, we consider precisely the wave equation $\Box u = 0$ with n space variables. Let ω be an open domain in $\mathbb{R}^n$ with analytic boundary $\partial\omega$ and convex complement, that is, ω is an "exterior" domain. Let $y' \in \omega$ and u a solution of the problem

$$\Box u = 0 \text{ in } \mathbb{R} \times \omega, \qquad u|_{\mathbb{R}\times\partial\omega} = 0 \tag{2.86}$$

$$u|_{x_0=0} = 0, \qquad D_0 u|_{x_0=0} = -i\delta(x' - y'). \tag{2.87}$$

Theorem 34 (Rauch and Sjöstrand [1981]). *Suppose that either* $n = 2$ *and* $\partial\omega$ *is convex or* $n \geqslant 3$ *and* $\partial\omega$ *is strictly convex. If* u *is a solution of* (2.86)–(2.87), *then* $\mathrm{WF}_{fa}(u)$ *is precisely the union of all the analytic rays issuing from the points* $(0, y', 1, \xi')$, *where* ξ' *runs through* $\mathbb{R}^n \cap \{|\xi'| = 1\}$. *At the same time*, $\mathrm{WF}_f(u)$ *is the union of all* C^∞*-rays issuing from these very points.*

In particular, $\operatorname{sing\,supp}_a u = \pi_x \mathrm{WF}_{fa}(u)$ *contains the frontal boundary of* $\operatorname{supp} u$.

Problems

In conclusion, we should like to formulate a number of problems whose solution should prove, in our view, extremely useful. Among these problems there are problems of most diverse degree of difficulties.

1. Prove the Ivrii-Petkov conjecture (§ 1.4, Chap. 1) in the general case (that is, when $S_+ \cup S_- \neq \varnothing$).

2. Generalize Nishitani's results (§ 1.9, Chap. 1) to the case when $S_+ \cup S_- = \varnothing$ and to equations of arbitrary order.

3. Is it true that if P is an x_0-hyperbolic operator, then an operator Q of order not exceeding $m-1$ can be found such that the Cauchy problem for $P+Q$ is C^∞-well posed?

4. Is it true that if the Cauchy problem for P is C^∞-well posed and r is the maximum multiplicity of characteristics (or of characteristic roots), then the Cauchy problem for $P+Q$ is also C^∞-well posed provided that the order of Q does not exceed $m-r$?

5. Apart from the strong formulation of the Cauchy problem, which we have used, weak formulation for it is also possible under which we seek the solutions in the spaces $\mathscr{D}'$ and $\mathscr{D}'_{\{\varkappa\}}$. To prove the necessity of the well-posedness conditions, we often start from the weak formulation while to prove the sufficiency from the strong formulation, and thereby establishing their equivalence. We should like to have a more general and direct proof of the equivalence of these two formulations.

6. Give a general and direct proof of the fact that under suitable regularity conditions on the coefficients of P the following implications hold:

$$C^\infty\text{-well-posedness} \Rightarrow G^{(\varkappa)}\text{-well-posedness} \Rightarrow$$

$$G^{\langle\varkappa\rangle}\text{-well-posedness} \Rightarrow G^{(\varkappa_1)}\text{-well-posedness} \quad (\text{for } \varkappa_1 < \varkappa)$$

and

$$\text{local } G^{(\varkappa)}\text{-well-posedness} \Rightarrow G^{\langle\varkappa\rangle}\text{-well-posedness in some neighbourhood of the set } X_T.$$

This problem will be solved if one solves the preceding problem.

7. Prove Theorem 2, Chap. 1, under the assumption that $S_+ \cup S_- = \varnothing$.

8. Let ρ be a critical point of the principal symbol p of a scalar x_0-hyperbolic operator P. Assume that the fundamental matrix $F_p(\rho)$ does not have non-zero real eigenvalues and that one of the conditions a) or b) of Theorem 18, Chap. 1, is violated. Then it can easily be shown that the Cauchy problem is not $G^{\{\varkappa\}}$-well posed for sufficiently large $\varkappa$. Find the (best possible) value of $\varkappa$. Does $\varkappa$ depend on the structure of the fundamental matrix or on the fact which of the conditions a) or b) is violated? What, in particular, is the value of $\varkappa$ for the operators $P_l + cD_n$ ($l = 2, 3$) of Example 3?

9. Is the Cauchy problem for the operator

$$P = D_0^2 - x_1^2 D_1^2 + aD_1$$

$G^{(\varkappa)}$-well posed? If not, establish a Theorem 33 type (Chap. 1) general result that will imply this result.

10. Is it true that all the well-posedness conditions, except the x_0-hyperbolicity condition, for the Cauchy problem can be written in a symplectically invariant form? Is it true, in particular, that if the Cauchy problem is well posed

in some spatial type lens, it will remain so in a smaller spatial type lens whose base is a piece of an arbitrary spatially similar surface, and not necessarily the hyperplane $\{x_0 = T\}$?

11. Prove Theorems 36, 37, 38, 39 and 42, Chap. 1, for all those operators for which it is known that the Cauchy problem is suitably well posed.

12. More generally, give a direct proof of the fact that the well-posedness of the Cauchy problem implies the validity of these theorems.

13. If

$$P = \sum_{j=0}^{n} a_j(x)D_j + b(x)$$

is a symmetric hyperbolic matrix operator, then prove that there exists a nowhere dense closed set Λ in the space of C^∞-matrices that depends on p and is such that the inclusion (1.51) turns into an equality when $b \notin \Lambda$.

14. Similarly, show that if P is x_0-hyperbolic, we can find an operator Q of order not exceeding $m - 1$ such that the Cauchy problem for $P + Q$ is C^∞-well posed and that for this operator equality (1.51) holds.

15. Obtain necessary and sufficient conditions for the weak L_2-well-posedness of the normal mixed problem for equations of an arbitrary order.

16. Define the notions of a C^∞-ray and an analytic ray for the general mixed problem and reformulate in terms of these Theorems 24, 25 and 29, Chap. 2.

17. Investigate the propagation of singularities of solutions of the mixed problem in the Gevrey classes for the wave equation, first of all, and before this for the Freidlander's model problem [1976], and show that for $\varkappa \geqslant \varkappa^*$ the $G^{(\varkappa)}$-singularities and for $\varkappa > \varkappa^*$ the $G^{\langle\varkappa\rangle}$-singularities are propagated in the same manner as the C^∞-singularities are. Show further that for other values of $\varkappa$, the singularities in Gevrey classes are propagated just as analytic singularities are. And that in this case $\varkappa^* = 3$, a fact dictated by the order of decay of solution to the Helmholtz equation $(\Delta + k^2)v = \delta(x' - y')$ in the shade zone as the frequency k increases.

Bibliographical Remarks

In preparing the bibliography the author was faced with the daunting task of selecting the titles to be included in the list. It suffices to mention that the first version of the list of references was four times as large as the final version, and yet it was far from complete. In selecting the titles, the author was guided by a number of principles. An attempt was made to include the most recent works, or those which remained undeservedly, in the author's view, less known or those whose results were used by the author or, finally, the ones which had a rich bibliography. But at the same time, some works which proved to be useful for the author in writing this paper could not be included in the list.

The author does not consider himself to be competent enough to comment on the historical themes. Therefore these remarks should not be taken either as a historical survey or as an investigation into the question of priority.

We note, first of all, the fundamental works of Hörmander [1963, 1983, 1985] devoted to the general theory of partial differential operators, the works of Duistermaat and Hörmander [1972], Hörmander [1971, 1983, 1985], Taylor [1981] and Trèves [1980] devoted to the theory of pseudo-differential and Fourier integral operators, and the works of Maslov [1965, 1983] and Maslov and Fedoryuk [1973, 1976] where the theory of canonical operators has been discussed.

A large number of works are devoted to the C^∞-well posed Cauchy problem. Here one must mention, first of all, the classical work of Hadamard [1932]. Petrovskij [1938], Leray [1954], and Gårding [1957] have established the L_2-well-posedness of the Cauchy problem for strictly hyperbolic operators. The necessary conditions (but not the sufficient conditions for a matrix operator) for the L_2-well-posedness have been obtained by Ivrii and Petkov [1974], and Strang [1966, 1967]. That hyperbolicity is necessary for the C^∞-well-posedness has been proved by Lax [1957], Mizohata [1961], and Ivrii and Petkov [1974]. Necessary and sufficient conditions for the Cauchy problem to be C^∞-well posed for equations with characteristics of constant multiplicity have been obtained by Flaschka and Strang [1971], Ivrii and Petkov [1974], and Chazarain [1974], respectively. Operators with constant coefficients have been studied by Gårding [1951] and Hörmander [1963, Chap. 5]. These work as well as those by Bronshtejn [1974], Atiyah, Bott and Gårding [1970, 1973], John [1978], Nuij [1968], and Svensson [1970] discuss the algebraic properties of hyperbolic polynomials. The necessary conditions for the regular and completely regular hyperbolicity can be found in the work of Ivrii and Petkov [1974], while the sufficient conditions in the works of Ivrii [1976a], Iwasaki [1983, 1984], and Melrose [1983]. The necessary conditions for the C^∞-well-posedness of the Cauchy problem for equations with characteristics of variable multiplicity were obtained by Ivrii and Petkov [1974], Hörmander [1977], and Yagdzhyan [1980]; the last work mentioned deals with degenerate equations having gluing of characteristic roots of infinite order. Sufficient conditions under which the Cauchy problem for degenerate equations is well posed have been obtained by Nuij [1968] and Yagdzhyan [1980], and for fairly general irregular hyperbolic equations by Ivrii [1977] and Hörmander [1977] by the method of energy integral. A complete analysis of equations of second order in two independent variables with analytic coefficients in an open domain has been carried out by Nishitani [1984]; some model equations with characteristics of variable multiplicity 3 or 4 have been investigated by Zajtseva [1980]. Petkov [1975, 1978] and Demay [1977] have obtained necessary and sufficient conditions for the C^∞-well-posedness of the Cauchy problem for systems of first order with characteristics of constant multiplicity. A necessary condition for the regular hyperbolicity of systems in two independent variables has been obtained by Petkov and Kutev [1976]. A rich bibliography on the C^∞-well posed Cauchy problem can be found in the works of Ivrii and Petkov [1974], Gårding [1982], and Iwasaki [1983].

Not too many works have been devoted to the well-posedness of the Cauchy problem in Gevrey classes. The necessity of the hyperbolicity has been established by Ivrii [1976b], Komatsu [1977], and Nishitani [1978], while sufficiency for the well-posedness in arbitrary Gevrey classes by Ivrii [1975], Bronshtejn [1980], and Kajitani [1983]. The necessary and sufficient conditions for well-posedness in Gevrey class with suitable exponents for equations with characteristics of constant multiplicity have been obtained by Ivrii [1976b], and Komatsu [1977], while the paper by Ivrii [1976c] contains the necessary conditions for the well-posedness of equations with characteristics of variable multiplicity. Some degenerate equations have been dealt with by Ivrii [1978], and Yagdzhyan [1978]. Ivrii [1975, 1976b, 1978] and Nishitani [1978] have obtained the energy estimates in Gevrey classes. For a discussion of operators with constant coefficients, we once again refer the reader to Hörmander [1963, Chap. 5].

A large number of works have been devoted to the study of the propagation of singularities of solutions to the Cauchy problem and the related propagation in the whole space. One of the earliest works in this direction is by Babich [1960], where a fundamental solution for the strictly hyperbolic equation has been constructed, and by Lax [1975], where the propagation of oscillations has been described; a still earlier work of Courant and Lax has been overshadowed by Babich [1960]. The remaining works in our list are recent. Atiyah, Bott and Gårding [1970, 1973] discuss the singularities of the fundamental solutions of equations with constant coefficients. Duistermaat and Hörmander [1972] describe the propagation of singularities of principal type equations by means of the Fourier integral operators. Parametrices for the Cauchy problem for operators with characteristics of constant multiplicity have been constructed by Chazarain [1974] and by Petkov [1978], respectively, for equations and for first-order systems. Alinhac [1978], Yoshikawa [1977, 1978a,b, 1980], and Kucherenko and Osipov [1983] have constructed parametrices for various classes of effectively hyperbolic operators. Nishitani [1983] describes the propagation of singularities for general effectively hyperbolic operators; close results have been obtained by Ivrii [1979a]. The propagation of the singularities of the solutions to various classes of equations has been discussed by Ivrii [1979a], R. Lascar [1981], Melrose and Uhlmann [1979a,b], and Uhlmann [1982]. For a description of the propagation of singularities of solutions to symmetric hyperbolic systems of first order, the reader is referred to Ivrii [1979c]; all these results have been reformulated by Wakabayashi [1980] in the language of generalized bicharacteristics. A detailed discussion of hyperbolic pseudodifferential systems that are principally diagonal can be found in Kucherenko [1974], R. Lascar [1981], Lax [1957] and Ralston [1976]. All these works concern the propagation of C^∞-singularities.

The works dealing with the propagation of singularities in Gevrey classes are few, the work by Wakabayashi [1983] being the most important. From among a significant number of works devoted to the propagation of analytic singularities, we mention only Wakabayashi [1983], where the results of Sato, Kashiwara and Kawai have been reformulated in terms of the generalized bicharacteristics.

Of works concerning the well-posedness of the mixed problem, we must mention, first of all, the works of Schauder, of Krzyzanski and of Ladyzhenskaya that are devoted to strictly hyperbolic equations of second order, and the works of Friedrichs [1958] and of his students Lax and Phillips [1960] that are devoted to positive (that is, strictly dissipative) problems for symmetric systems. Of the more recent works, we note Sakamoto [1970a,b, 1980], and Kreiss [1970], Agranovich [1969, 1972], Ralston [1971] and Rauch [1972], where strong L_2-well-posedness of the mixed problem has been investigated, respectively, for equations and first-order systems.

The weak L_2-well-posedness has been dealt with by Agemi and Shirota [1970, 1971], and by Miyatake [1975, 1977]. Miyatake's papers contain almost exhaustive results for second-order equations. A necessary condition for the C^∞-well-posedness under very general condition can be found in Wakabayashi [1980].

The C^∞-well-posedness of the mixed problem for the wave equation has been analysed by Hörmander [1983, 1985] and Soga [1979, 1980]. The general problems for strictly hyperbolic equations of second order have been studied by Eskin [1979/80, 1981a,b]. A non-strictly model hyperbolic problem has been discussed by Zajtseva [1983]. Under fairly weak assumptions, it has been shown by Duff [1959] that the mixed problem is analytically well posed. Gårding's [1982] survey paper should be mentioned once again.

Povzner and Sukharevskij inagurated the study of the propagation of C^∞-singularities of the solutions to the mixed problem. Chazarain [1977] and Nirenberg dealt with the same question in the language of wave fronts, but they examined only transversal reflection of waves from the boundary. For this case, a parametrix has been constructed by Fedoryuk [1977]. The propagation of singularities in bicharacteristically strictly concave domains has been investigated by Friedlander [1976], Melrose [1975, 1978], Taylor [1976a,b, 1979], Eskin [1976], while in bicharacteristically convex domains by Ivrii [1979d,e, 1980], Eskin [1977]; the domains of an arbitrary nature have been dealt with by Ivrii [1980b], Melrose and Sjöstrand [1978, 1982]; for investigations concerning very general symmetric systems, see Ivrii [1979d,e 1980a,b, 1986]. The propagation of singularities of solutions to the non-classical problems for strictly hyperbolic equations of second order has been examined by Ivrii [1981], Eskin [1979/80, 1981a,b], Melrose and Sjöstrand [1978, 1982].

Very few works deal with the question of the propagation of analytic singularities. In this direction, one must mention, first of all, a very general work of Sjöstrand [1980]. Friedlander and Melrose [1977], Kataoka [1981], Rauch and Sjöstrand [1981] and Sjöstrand [1980b,c, 1981] have investigated the propagation of analytic singularities for strictly hyperbolic equations of second order. The maximum attention in these works is directed to bicharacteristically concave domains where the propagation of C^∞-singularities and that of analytic singularities are significantly different. In this very situation, Rauch [1977] examined the propagation of the frontal boundary of the support of the solution, the propagation being closely related to that of analytic singularities.

A Survey of Recent Results

The main part of the paper was written in 1984; however, the works published during the past three years[1] have made this supplement necessary.

We mention, first of all, the surveys by Gårding [1984] and by Mizohata [1986a]. The book Mizohata [1986b], where the latter survey has been published, contains a number of important works some of which have been included in our reference list. We note that at last we have in the Russian language the main works of Petrovskij; in particular, the works on the hyperbolic equations and systems (see Petrovksij [1986]).

A number of works have been devoted to the C^∞-well-posedness of the Cauchy problem. Nishitani [1985, 1986] has obtained the energy estimates for effectively hyperbolic as well as for some other operators with double characteristics, and has given an alternative proof for the C^∞-well-posedness of the Cauchy problem for effectively hyperbolic operators. Yagdzhyan [1985, 1986] has found necessary and sufficient conditions for the Cauchy problem to be C^∞-well posed for degenerate operators having gluing of characteristic roots of infinite order, and has also constructed a parametrix for the Cauchy problem. Zajtseva [1987b] has shown that in order that the Cauchy problem for the operator

$$P = D_t^2 - D_x^2 - x^{2n}D_y^2 + x^{n-1}D_y$$

be C^∞-well posed for odd n, it is necessary and sufficient that $a \in \mathbb{R}$, $|a| \leqslant n$. It has been recently discovered that the condition $a \in \mathbb{R} \cup i\mathbb{R}$ is necessary and sufficient for the problem to be C^∞-well posed for even n provided that $|a|$ is sufficiently small. Nishitani [1985] has studied the first-order systems and has shown that if P is a matrix operator of first order with no more than double characteristics, then for the strong (regular) hyperbolicity of P, it is necessary that in a neighbourhood of every point ρ either the principal symbol p is uniformly diagonalizable or $g = \det p$ is effectively hyperbolic. If both these conditions are violated at $\rho \in T^*\mathbb{R}^{n+1}\backslash 0$, then for the Cauchy problem to be C^∞-well posed, it is necessary that there exists an $\alpha \in [-1, 1]$ and a matrix B such that

$$\left(p^{s\,\mathrm{co}}p - \frac{i}{2}\{p, {}^{\mathrm{co}}p\} + \alpha \operatorname{Tr}^+ g - pB\right)(\rho) = 0.$$

Conversely, if in the neighbourhood of every point either g is effectively hyperbolic or p is uniformly diagonalizable and if $\Sigma = \{g = dg = 0\}$ is a C^∞-manifold of codimension $d \leqslant 3$, $T_\rho\Sigma = \operatorname{Ker} F_g(\rho)$, then P is a regularly hyperbolic operator.

A number of works deal with the well-posedness of the Cauchy problem in Gevrey classes. Kajitani [1986] has established that under suitable regularity conditions on the coefficients, the Cauchy problem for a uniformly diagonalizable system of first order is $G^{(\varkappa)}$-well posed for $\varkappa \leqslant 2$. The results of Bronshtejn [1980] have been generalized by Ohya and Tarama [1986] to the case where the coefficients are not very smooth with respect to x_0. Namely, they have shown that if the coefficients of the principal part of the scalar operator P lie in $C^\sigma([0, T], G^{(\varkappa)}(\mathbb{R}^n))$, with $\sigma \in (0, 2]$, and the lower terms lie in $C^0([0, T], G^{(\varkappa)}(\mathbb{R}^n))$, then the Cauchy problem will be $G^{(\varkappa)}$-well posed if $1 < \varkappa < \min(1 + \sigma/r, r/(r-1))$, where r is the maximum multiplicity of the characteristic roots. The Italian mathematicians have obtained considerably more precise results for second-order operators, and very often they have examined operators without lower terms and with coefficients depending only on x_0. They have succeeded in discovering a large number of illuminating counter examples. In particular, it has been shown that for a strictly hyperbolic operator, with $\sigma < 1$, $\varkappa > 1/(1-\sigma)$, the Cauchy problem, in general, is neither $G^{(\varkappa)}$-well posed nor C^∞-well posed. Moreover, for $\sigma < 1$, even the uniqueness theorem may fail to hold. A survey and bibliography can be found in Spagnolo [1986] and Colombini, Jannelli and Spagnolo [1986].

[1] The original Russian edition of the book was submitted in 1987 and published in 1988.— Translator

Of the works on the propagation of singularities of solutions to the Cauchy problem, we mention the one by Melrose [1986]. This paper examines the propagation of singularities for the equation $u_{tt} + Au = 0$, where A is a self-adjoint positive definite differential operator of the second order and is hypoelliptic with a loss of $2 - \delta$ derivatives, with $\delta > 0$.

A number of works deal with the propagation of the polarization front. Namely, if $u \in \mathscr{D}'(X, \mathbb{C}^N)$, then $(x, \xi, w) \in T^*X \times \mathbb{C}^N$ lies in $\mathrm{WF}^s_{pol}(u)$ provided that for any pseudodifferential operator A of order 0 the inclusion $Au \in H^s$ implies that $a(x, \xi)w = 0$, where a is the principal symbol of A. The wave front is a projection onto T^*X of the polarization front. The concept of the polarization front was introduced by Dencker [1982], and his investigation concerns the propagation of polarization fronts of solutions to systems of real principal type with characteristics of constant multiplicity. It turns out that for such systems the polarization fronts are propagated along Hamiltonian orbits, namely, curves lying in $T^*X \times \mathbb{C}^N$ and determined by a system of ordinary differential equations; their projections onto T^*X are bicharacteristics. Dencker [1986] studies the propagation of the polarization fronts for systems with conical refraction, and in this study the distinction between the real polarization and the complex polarization plays a significant role. Some results also exist concerning the propagation of the polarization fronts near the boundary, namely, concerning their reflection under transversality of bicharacteristics and even diffraction. Some results regarding the analytic polarization fronts have also been obtained (see the bibliography in Esser [1986]).

Many works study mixed problems. Eskin [1985] has investigated the well-posedness of the mixed problem for a strictly hyperbolic operator of second order in a bicharacteristically convex domain and has generalized his results of [1981b]. For this problem, a parametrix has been constructed by Kubota [1986] under certain conditions imposed on the boundary operator and the standard results on the propagation of singularities have been obtained. Laubin [1986] has given new proofs for Theorems 32 to 34 of Chapter 2 of the present survey. Zajtseva [1987a] has obtained a condition on $(\alpha, a) \in (\mathbb{C} \cup \infty) \times \mathbb{C}$ that is necessary and sufficient for the mixed problem

$$\{D_t^2 - D_x^2 - x^2 D_y^2 + aD_y, D_x + \alpha D_t\}$$

to be well posed.

Finally, Lebeau [1984, 1986] has shown that for the Dirichlet problem for a strictly hyperbolic operator of second order with analytic coefficients in a domain having analytic boundary, the $G^{(\varkappa)}$-singularities, for $\varkappa \geqslant 3$, are propagated in the same manner as the C^∞-singularities are, while for $1 < \varkappa < 3$ in the same way as the analytic singularities are (recall that analytic singularities can be propagated along crawling rays and glancing rays but C^∞-singularities cannot). Thus, Problem 17, formulated above, has been partially solved.

References*

Agemi, R., Shirota, T.

[1970] On necessary and sufficient conditions for L^2-well-posedness of mixed problems for hyperbolic equations I. J. Fac. Sci., Hokkaido Univ., Ser. I, *21*, No. 2, 133–151, Zbl.207,403

[1971] On necessary and sufficient conditions for L^2-well-posedness of mixed problems for hyperbolic equations II. ibid. *22*, No. 3, 137–149, Zbl.241.35055

Agranovich, M.S.

[1969] Boundary-value problems for systems of first order pseudodifferential operators. Usp. Mat. Nauk *24*, No. 1, 61–125. English transl.: Russ. Math. Surv. *24*, No. 1, 59–126 (1969), Zbl.175,108

* For the convenience of the reader, references to reviews in Zentralblatt für Mathematik (Zbl.), compiled using the MATH database, have, as far as possible, been included in this bibliography.

[1972] A theorem on matrices depending on parameters and its applications to hyperbolic systems. Funkts. Anal. Prilozh. *6*, No. 2, 1–11. English transl.: Funct. Anal. Appl. *6*, 85–93 (1972), Zbl.253.35060

Alinhac, S.

[1978] Branching of singularities for a class of hyperbolic operators. Indiana Univ. Math. J. *27*, No. 6, 1027–1037, Zbl.502.35058

Atiyah, M.F., Bott, R., Gårding, L.

[1970] Lacunas for hyperbolic differential operators with constant coefficients. I. Acta Math. *124*, 109–189, Zbl.191,112

[1973] Lacunas for hyperbolic differential operators with constant coefficients. II. ibid. *131*, 145–206, Zbl.266.35045

Babich, V.M.

[1960] Fundamental solutions of hyperbolic equations with variable coefficients. Mat. Sb. Nov. Ser. *52*, No. 2, 709–738, Zbl.96,67

Bronshtejn, M.D.

[1979] Smoothness of roots of polynomials depending on parameters. Sib. Mat. Zh. *20*, No. 3, 493–501. English transl.: Sib. Math. J. *20*, 347–352 (1980), Zbl.415.30003

[1980] The Cauchy problem for hyperbolic operators with characteristics of variable multiplicity. Tr. Mosk. Mat. O.-va *41*, 83–99. English transl.: Trans. Mosc. Math. Soc. 1982, No. 1, 87–103 (1982), Zbl.468.35062

Chazarain, J.

[1974] Operateurs hyperboliques à caracteristiques de multiplicité constante. Ann. Inst. Fourier. *24*, No. 1, 173–202, Zbl.274.35045

[1977] Reflection of C^∞-singularities for a class of operators with multiple characteristics. Publ. Res. Inst. Math. Sci., Kyoto Univ. *12*, Suppl., 39–52, Zbl.365.35050

Colombini, F., Jannelli, E., Spagnolo, S.

[1986] Non-uniqueness examples for second-order hyperbolic equations. Current Topics in Partial Differential Equations. Pap. dedic. S. Mizohata, 309–326, Zbl.627.35053

Demay, Y.

[1977] Paramétrix pour des systèmes hyperboliques du premier ordre à multiplicité constante. J. Math. Pures Appl. *56*, No. 4, 393–422, Zbl.379.35068

Dencker, N.

[1982] On the propagation of polarization sets for systems of real principal type. J. Funct. Anal. *46*, No. 3, 351–372, Zbl.487.58028

[1986] On the propagation of polarization in conical refraction. Preprint, Univ. Lund, *3*, 88 pp. Appeared in: Duke Math. J. *57*, No. 1, 85–134 (1988), Zbl.669.35116

Duff, G.F.D.

[1959] Mixed problems for hyperbolic equations of general order. Can. J. Math. *9*, 195–221, Zbl.86,76

Duistermaat, J.J., Hörmander, L.

[1972] Fourier integral operators. II. Acta Math. *128*, No. 3–4, 183–269, Zbl.232,47055

Eskin, G.

[1976] A parametrix for mixed problems for strictly hyperbolic equations of arbitrary order. Commun. Partial Differ. Equations *1*, No. 6, 521–560, Zbl.355.35053

[1977] Parametrix and propagation of singularities for the interior mixed hyperbolic problem. J. Anal. Math. *32*, 17–62, Zbl.375.35037

[1979/80] Well-posedness and propagation of singularities for initial boundary value problems for second order hyperbolic equations with general boundary conditions. Semin. Goulaouic-Schwartz, Exp. No. 11, 14 pp.

[1981a] Initial boundary value problem for second order hyperbolic equations with general boundary conditions. I. J. Anal. Math. *40*, 43–89, Zbl.492.35042

[1981b] General initial-boundary problems for second order hyperbolic equations. Singularities in boundary value problems. Proc. NATO Adv. Stud. Inst., Maratea/Italy 1980, 19–54, Zbl.475.35061

[1985] Initial-boundary value problems for second order hyperbolic equations with general boundary conditions. II. Commun. Partial Differ. Equations *10*, No. 10, 1117–1212, Zbl.585.35059

Esser, P.

[1986] Polarisation analytique dans les problèmes aux limites non-diffractifs de multiplicité constante. Bull. Soc. R. Sci. Liege *55*, No. 3, 371–465, Zbl.603.35089

Fedoryuk, M.V.

[1977] Singularities of the kernels of Fourier integral operators and asymptotics of the solutions of mixed problems. Usp Mat. Nauk *32*, No. 6, 67–115. English transl.: Russ. Math. Surv. *32*, No. 6, 67–120 (1977), Zbl.376.35001

Flaschka, H., Strang, G.

[1971] The correctness of the Cauchy problem. Adv. Math. *6*, No. 3, 347–379, Zbl.213,373

Friedlander, F.G.

[1976] The wave front set of the solution of a simple initial-boundary value problem with glancing rays. I. Math. Proc. Camb. Philos. Soc. *79*, No. 1, 145–159, Zbl.319.35053

Friedlander, F.G., Melrose, R.B.

[1977] The wave front set of solutions of a simple initial-boundary value problem with glancing rays. II. Math. Proc. Camb. Philos. Soc. *81*, No. 1, 97–120, Zbl.356.35056

Friedrichs, K.O.

[1958] Symmetric positive linear differential equations. Commun. Pure Appl. Math. *11*, 333–418, Zbl.83,318

Gårding, L.

[1951] Linear hyperbolic partial differential equations with constant coefficients. Acta Math. *85*, No. 1, 1–62, Zbl.45,202

[1957] Cauchy's problem for hyperbolic equations. Univ. Chicago, Zbl.91,269

[1982] Hyperbolic boundary problems. Differential Geometry and Differential Equations, Proc. 1980 Beijing Symp., Vol. 1, 371–465, Zbl.527.35051

[1984] Hyperbolic differential operators. Perspective in Math., Anniv. Oberwolfach 1984, 215–247, Zbl.572.35012

Hadamard, J.

[1932] Le problème de Cauchy et les équations aux dérives partielles lineaires hyperboliques. Paris, Zbl.6,205

Hörmander, L.

[1963] Linear Partial Differential Operators. Berlin-Heidelberg-New York: Springer-Verlag, Zbl.108,93

[1971] Fourier integral operators. I. Acta Math. *127*, No. 1–2, 79–183, Zbl.212,466

[1977] The Cauchy problem for differential equations with double characteristics. J. Anal. Math. *32*, 118–196, Zbl.367.35054

[1983] The Analysis of Linear Partial Differential Operators. Vols. I and II. Berlin-Heidelberg-New York: Springer-Verlag, Zbl.521.35001 and Zbl.521.35002

[1985] The Analysis of Linear Partial Differential Operators. Vols. III and IV. Berlin-Heidelberg-New York: Springer-Verlag, Zbl.601.35001 and Zbl.612.35001

Ikawa, M.

[1981] Mixed problems for the wave equation. Singularities in Boundary-Value Problems. Proc. NATO Adv. Stud. Inst., Marateo/Italy 1980, 97–119. Zbl.473.35047

Ivrii, V.Ya. (= Ivrij, V.Ya.)

[1975] Well-posedness in Gevrey classes of the Cauchy problem for non-strictly hyperbolic operators. Mat. Sb., Nov. Ser. *96*, No. 3, 390–413. English transl.: Math. USSR, Sb. *25* (1975), 365–387 (1976). Zbl.319.35054

[1976a] Sufficient conditions for regular and completely regular hyperbolicity. Tr. Mosk. Mat. O.-va *33*, 3–65. English transl.: Trans. Mosc. Math. Soc. *33*, 1–65 (1978), Zbl.355.35052

[1976b] Well-posedness conditions in Gevrey classes for the Cauchy problem for non-strictly hyperbolic operators. Sib. Mat. Zh. *17*, No. 3, 547–563. English transl.: Sib. Math. J. *17* (1976), 422–435 (1977), Zbl.338.35088

[1976c] Well-posedness conditions in Gevrey classes for the Cauchy problem for hyperbolic operators with characteristics of variable multiplicity. Sib. Mat. Zh. *17*, No. 6, 1256–1270. English transl.: Sib. Math. J. *17*, 921–931 (1977), Zbl.404.35068

[1977] The well-posed Cauchy problem for non-strictly hyperbolic operators. III. The energy integral. Tr. Mosk. Mat. O.-va *34*, 151–170. English transl.: Trans. Mosc. Math. Soc. *34*, 149–168 (1978), Zbl.403.35060

[1978] Well-posedness in Gevrey classes of the Cauchy problem for non-strictly hyperbolic operators. Izv. Vyssh. Uchebn. Zaved. Mat. 1978, No. 2, 26–35, Zbl.395.35055

[1979a] Wave fronts of solutions of some pseudodifferential equations. Tr. Mosk. Mat. O.-va *39*, 49–82. English transl.: Trans. Mosc. Math. Soc. 1981, No. 1, 49–86 (1981), Zbl. 433.35077

[1979b] Wave fronts of solutions of some hyperbolic pseudodifferential equations. Tr. Mosk. Mat. O.-va *39*, 83–112. English transl.: Trans. Mosc. Math. Soc. 1981, No. 1, 87–119 (1981), Zbl.433.35078

[1979c] Wave fronts of solutions of symmetric pseudodifferential systems. Sib. Mat. Zh. *20*, No. 3, 557–578. English transl.: Sib. Math. J. *20*, 390–405 (1980), Zbl.417.35080

[1979d] Wave fronts of solutions of boundary-value problems for symmetric hyperbolic systems. I. Sib. Mat. Zh. *20*, No. 4, 741–755. English transl.: Sib. Math. J. *20*, 516–524 (1980), Zbl.454.35055

[1979e] Wave fronts of solutions of boundary-value problems for symmetric hyperbolic systems. II. Sib. Mat. Zh. *20*, No. 5, 1022–1038. English transl.: Sib. Math. J. *20*, 722–734 (1980), Zbl.454.35056

[1980a] Wave fronts of solutions of boundary-value problems for symmetric hyperbolic systems. III. Sib. Mat. Zh. *21*, No. 1, 74–81. English transl.: Sib. Math. J. *21*, 54–60 (1980), Zbl.454.35057

[1980b] Wave fronts of solutions of boundary-value problems for a class of symmetric hyperbolic systems. Sib. Mat. Zh. *21*, No. 4, 62–71. English transl.: Sib. Math. J. *21*, 527–534 (1981), Zbl.447.35055

[1981] The propagation of singularities of solutions of non-classical boundary-value problems for hyperbolic equations of second order. Tr. Mosk. Mat. O-va *43*, 81–91. English transl.: Trans. Mosc. Math. Soc. 1983, No. 1, 87–99 (1983), Zbl.507.35059

[1986] The propagation of singularities of solutions of dissipative boundary-value problems for symmetric hyperbolic systems. Dokl. Akad. Nauk SSSR *287*, No. 6, 1299–1301. English transl.: Sov. Math., Dokl. *33*, 526–529 (1986), Zbl.626.35057

Ivrii, V.Ya; Petkov, V.M. (= Ivrij, V.Ya.; Petkov, V.M.)

[1974] Necessary conditions for the well-posedness of the Cauchy problem for non-strictly hyperbolic equations. Usp Mat. Nauk *29*, No. 5, 3–70. English transl.: Russ. Math. Surv. *29*, No. 5, 1–70 (1974), Zbl.312.35049

Iwasaki, N.

[1983a] The Cauchy problem for hyperbolic equations with double characteristics. Publ. Res. Inst. Math. Sci. *19*, No. 3, 927–942, Zbl.571.35067

[1983b] The Cauchy problem for effectively hyperbolic equations (A special case). J. Math. Kyoto Univ. *23*, 503–562, Zbl.546.35037

[1984] Cauchy problems for effectively hyperbolic equations (A standard type). Publ. Res. Inst. Math. Sci. *20*, No. 3, 543–584, Zbl.565.35065

John, F.

[1978] Algebraic conditions for hyperbolicity of systems of partial differential equations. Commun. Pure Appl. Math. *31*, No. 1, 89–106, Zbl.381.35059

Kajitani, K.

[1983] Cauchy problem for nonstrictly hyperbolic systems in Gevrey classes. J. Math. Kyoto Univ. *23*, No. 3, 599–616, Zbl.544.35063

[1986] The Cauchy problem for uniformly diagonalizable hyperbolic systems in Gevrey classes. In: Mizohata (Ed.): Hyperbolic Equations and Related Topics, Proc. Taniguchi Int. Symp. Katata and Kyoto 1984, 101–123, Zbl.669.35065

Kataoka, K.

[1981] Microlocal analysis of boundary-value problems with application to diffraction. Singularities in Boundary-Value Problems. Proc. NATO Adv. Stud. Inst., Marateo/Italy 1980, 121–131, Zbl.461.35088

Komatsu, H.

[1977] Irregularity of characteristic elements and hyperbolicity. Publ. Res. Inst. Math. Sci. *12*, Suppl., 233–245, Zbl.375.35044

Kreiss, H.O.

[1970] Initial-boundary value problems for hyperbolic systems. Commun. Pure Appl. Math. *23*, 277–298, Zbl.193,69

Kubota, K.

[1986] Microlocal parametrices and propagation of singularities near gliding points for hyperbolic mixed problems. I. Hokkaido Math J. *15*, No. 2, 243–308, Zbl.619.35071

Kucherenko, V.V.

[1974] Asymptotics, as $h \to 0$, of solutions of the system $A\left(x, -h\frac{\partial}{\partial x}\right)u = 0$ in the case of characteristics of variable multiplicity. Izv. Akad. Nauk SSSR, Ser. Mat *38*, 625–662. English transl.: Math. USSR, Izvestija *8* (1974), 631–666 (1975), Zbl.308.35080

Kucherenko, V.V., Osipov, Yu.V.

[1983] The Cauchy problem for non-strictly hyperbolic equations. Mat. Sb., Nov. Ser. *120*, No. 1, 84–111. English transl.: Math. USSR, Sb. *48*, 81–109 (1984), Zbl.519.35048

Kumano-go, H.

[1979] Fundamental solution for a hyperbolic system with diagonal principal part. Commun. Partial Differ. Equations *4*, No. 9, 959–1015, Zbl.431.35062

Kumano-go, H., Taniguchi, K.

[1979] Fourier integral operators of multiphase and the fundamental solution to a hyperbolic system. Funkc. Ekvacioj, Ser. Int. *22*, No. 2, 161–196, Zbl.568.35092

Lascar, R.

[1981] Propagation des singularités des solutions d'équations pseudodifferentielles à caracteristiques de multiplicités variables. Lect. Notes Math. *856*, Zbl.482.58031

Laubin, P.

[1986] Asymptotic solutions of hyperbolic boundary value problems with diffraction. Advances in Microlocal Analysis. NATO ASI Ser., Ser. C *168*, 165–202, Zbl.601.35113

Lax, P.D.

[1957] Asymptotic solutions of oscillatory initial value problems. Duke Math. J. *24*, No. 4, 627–646, Zbl.83,318

Lax, P.D., Phillips, R.S.

[1960] Local boundary conditions for dissipative symmetric linear differential operators. Commun. Pure Appl. Math. *13*, 427–455, Zbl.94,75

Lebeau, G.

[1984] Regularité Gevrey 3 pour la diffraction. Commun. Partial Differ. Equations *9*, No. 15, 1437–1494, Zbl.559.35019

[1986] Propagation des singularités Gevrey pour le problème de Dirichlet. Advances in Microlocal Analysis. NATO ASI Ser., Ser. C *168*, 203–223, Zbl.603.35015

Leray, J.

[1954] Hyperbolic Differential Equations. Princeton: Inst. Adv. Study, Zb.1.75,98

Maslov, V.P.

[1965] Theory of Perturbations and Asymptotic Methods. Moscow: Moskov. Gos. Univ. French translation: Paris: Dunod (1972), Zbl.247.47010

[1983] Non-standard characteristics in asymptotic problems. Usp. Mat. Nauk *38*, No. 6, 3–36. English transl.: Russ. Math. Surv. *38*, 1–42 (1983), Zbl.562.35007

Maslov, V.P., Fedoryuk, M.V.

[1973] The canonical operator (real case). Itogi Nauki Tek. Ser., Sovrem. Probl. Mat. *1*, 85–167. English transl.: J. Sov. Math. *3*, 217–279 (1975), Zbl.303.35069

[1976] Quasi-Classical Approximation to Equations of Quantum Mechanics. Moscow: Nauka. English transl.: Dordrecht-Boston-London: D. Reidel (1981), Zbl.449.58002

Melrose, R.B.

[1975] Local Fourier-Airy integral operators. Microlocal parametrices for diffractive boundary-value problems. Duke Math. J. *42*, No. 4, 583–604, Zbl.368.35054, 605–635, Zbl.368.35055

[1978] Airy operators. Commun. Partial Differ. Equations *3*, No. 1, 1–76, Zbl.384.35052

[1981] Transformation of boundary value problems. Acta Math. *147*, 149–236, Zbl.492.58023

[1983] The Cauchy problem for effectively hyperbolic operators. Hokkaido Math. J. *12*, 371–391, Zbl.544.35094

[1986] Propagation for the wave group of positive subelliptic second-order differential operators. In: Mizohata (Ed.): Hyperbolic Equations and Related Problems, Proc. Taniguchi Int. Symp., Katata and Kyoto 1984, 181–192, Zbl.696.35064

Melrose, R.B., Sjöstrand, J.

[1978] Singularities of boundary value problems. I. Commun. Pure Appl. Math. *31*, 593–617, Zbl.368.35020

[1982] Singularities of boundary value problems. II. ibid. *35*, 129–168, Zbl.546.35083

Melrose, R.B., Uhlmann, G.A.

[1979a] Lagrangian intersection and the Cauchy problem. Commun. Pure Appl. Math. *32*, 483–519, Zbl.396.58006

[1979b] Microlocal structure of involutive conical refraction. Duke Math. J. *46*, No. 3, 571–582, Zbl.422.58026

Miyatake, S.

[1975] Mixed problems for hyperbolic equations of second order with first order complex boundary operators. Japan J. Math. New Ser. *1*, No. 1, 111–158, Zbl.337.35047

[1977] A sharp form of the existence theorem for hyperbolic mixed problems of second order. J. Math. Kyoto Univ., *17*, No. 2, 199–223, Zbl.374.35028

Mizohata, S.

[1961] Some remarks on the Cauchy problem. J. Math. Kyoto Univ. *1*, No. 1, 109–127, Zbl. 104,319

[1986a] Comments on the development of hyperbolic analysis. In: Mizohata, S. (Ed.): Hyperbolic Equations and Related Topics, Proc. Taniguchi Int. Symp., Katata and Kyoto 1984, IX–XXXIV (1986), Zbl.674.35054

[1986b] (Ed.) Hyperbolic Equations and Related Topics. Proc. Taniguchi Int. Symp., Katata and Kyoto 1984. Boston: Acad. Press, Zbl.643.00010

Nishitani, T.

[1978] On the Lax-Mizohata theorem in the analytic and Gevrey classes. J. Math. Kyoto Univ. *18*, No. 3, 509–521, Zbl.402.35093

[1983a] Energy inequality for non-strictly hyperbolic operators in Gevrey class. J. Math. Kyoto Univ. *23*, No. 4, 739–773, Zbl.552.35051

[1983b] On wave front sets of solutions for effectively hyperbolic operators. Sci. Rep., Osaka *32*, No. 2, 1–7, Zbl.548.35006

[1984a] A necessary and sufficient condiition for the hyperbolicity of second order equations in two independent variables. J. Math. Kyoto Univ. *24*, No. 1, 91–104, Zbl.552.35049

[1984b] Local energy integrals for effectively hyperbolic operators. I, II. J. Math. Kyoto Univ. *24*, No. 4, 623–658, Zbl.582.58035, 659–666, Zbl.589.35078

[1985] On strong hyperbolicity for first order systems. Proc. Japan Acad., Ser. A *61*, No. 7, 193–196, Zbl.579.35046

[1986] Microlocal energy estimates for hyperbolic operators with double characteristics. In: Mizohata (Ed.): Hyperbolic Equations and Related Topics, Proc. Taniguchi Int. Symp., Katata and Kyoto/Jap. 1984, 235–255, Zbl.665.35007

Nuij, W.

[1968] A note on hyperbolic polynomials. Math. Scand. *23*, No. 1, 69–72, Zbl.189,408

Ohya, Yu., Tarama, S.

[1986] La problème de Cauchy à caracteristiques multiples dans la classe de Gevrey—coefficients hölderiens en t. In: Mizohata (Ed.): Hyperbolic Equations and Related Topics, Proc. Taniguchi Int. Symp., Katata and Kyoto/Jap. 1984, 273–306, Zbl.665.35045

Olejnik, O.A.

[1970] On the Cauchy problem for weakly hyperbolic equations. Commun. Pure Appl. Math. *23*, No. 6, 569–586, Zbl.193,386

Petkov, V.M.

[1975] Necessary condition for well-posedness of the Cauchy problem for non-symmetrizable hyperbolic systems. Tr. Semin. I.G. Petrovskogo *1*, 211–236. English transl.: Transl., II. Ser., Am. Math. Soc. *118*, 25–50 (1982), Zbl.317.35055

[1978] A parametrix of the Cauchy problem for non-symmetrizable hyperbolic systems with characteristics of constant multiplicity. Tr. Mosk. Mat. O.-va *37*, 3–47. English transl.: Trans. Mosc. Math. Soc. *1*, 1–47 (1980), Zbl.442.35067

[1981] Propagation des singularites sur le bord pour des systèmes hyperboliques à caracteristiques de multiplicité constante. C.R. Acad. Sci., Paris, Ser. I. *293*, 637–639, Zbl.487.35059

Petkov, V.M., Kutev, N.D.

[1976] On regularly hyperbolic systems of first order. God. Sofij Univ., Fak. Mat. Mekh. *67*, 375–389, Zbl.353.35061

Petrovskij, I.G.

[1938] On the Cauchy problem for systems of linear partial differential equations. Bull. Univ. Mosk., Ser. Int. Mat. Mekh. *1*, No. 7, 1–74, Zbl.24,37

[1986] Selected Works. Systems of Partial Differential Equations. Algebraic Geometry. Moscow: Nauka, Zbl.603.01018

Ralston, J.V.

[1971] Note on a paper of Kreiss. Commun. Pure Appl. Math. *24*, 759–762, Zbl.215,168

[1976] On the propagation of singularities in solutions of symmetric hyperbolic partial differential equations. Commun. Partial Differ. Equations *1*, No. 2, 87–133, Zbl.336.35066

Rauch, J.

[1972] L_2 is a continuable initial condition for Kreiss' mixed problems. Commun. Pure Appl. Math. *25*, 265–285, Zbl.226.35056

[1977] The leading wave front for hyperbolic mixed problems. Bull. Soc. R. Sci., Liege *46*, 156–161, Zbl.372.35051

Rauch, J., Sjöstrand, J.

[1981] Propagation of analytic singularities along diffracted rays. Indiana Univ. Math. J. *30*, No. 3, 389–401, Zbl.424.35003

Sakamoto, R.

[1970a] Mixed problems for hyperbolic equations. I. J. Math. Kyoto Univ. *10*, No. 2, 349–373, Zbl.203,100

[1970b] Mixed problems for hyperbolic equations. II. ibid, No. 3, 403–417, Zbl.206,401

[1980] Hyperbolic Boundary Value Problems. Cambridge: Univ. Press, Zbl.494.35001

Sjöstrand, J.

[1980a] Analytic singularities and microhyperbolic boundary value problems. Math. Ann. *254*, 211–256, Zbl.459.35007

[1980b] Propagation of analytic singularities for second order Dirichlet problems. I. Commun. Partial Differ. Equations *5*, No. 1, 41–94, Zbl.458.35026

[1980c] Propagation of analytic singularities for second order Dirichlet problems. II. ibid, No. 2, 187–207, Zbl.534.35030

[1981] Propagation of analytic singularities for second order Dirichlet problems. III. ibid. *6*, No. 5, 499–567, Zbl.524.35032

Soga, H.

[1979] Mixed problems in a quarter space for the wave equation with a singular oblique derivative. Publ. Res. Inst. Math. Sci. *15*, No. 3, 631–651, Zbl.433.35044

[1980] Mixed problems for the wave equation with a singular oblique derivative. Osaka J. Math. *17*, No. 1, 199–232, Zbl.429.35045

Spagnolo, S.

[1986] Analytic and Gevrey well-posedness of the Cauchy problem for second order weakly hyperbolic equations with coefficients irregular in time. In: Mizohata (Ed.): Hyperbolic Equations and Related Topics, Proc. Taniguchi Int. Symp., Katata and Kyoto/Jap. 1984, 363–380, Zbl.679.35057

Strang, G.

[1966] Necessary and insufficient conditions for well-posed Cauchy problems. J. Differ. Equations *2*, No. 1, 107–114, Zbl.131,91

[1967] On strong hyperbolicity. J. Math. Kyoto Univ. *6*, No. 3, 397–417, Zbl.174,151

Svensson, S.L.

[1970] Necessary and sufficient conditions for hyperbolicity of polynomials with hyperbolic principal part. Ark. Mat. *8*, 145–162, Zbl.203,409

Taylor, M.E.

[1976a] Grazing rays and reflection of singularities of solutions to wave equations. I. Commun. Pure Appl. Math. *29*, No. 1, 1–38, Zbl.318.35009

[1976b] Grazing rays and reflection of singularities of solutions to wave equations. II. ibid, No. 5, 463–482, Zbl.327.35038

[1979] Grazing rays and reflection of singularities of solutions to wave equations. III. ibid *32*, No. 1, 68–92

[1981] Pseudodifferential Operators. Princeton: Univ. Press, Zbl.453.47026

Trèves, F.

[1980] Introduction to Pseudodifferential and Fourier Integral Operators. Vols. 1 and 2. New York-London: Plenum Press, Zbl.453.47027

Uhlmann, G.A.

[1982] Light intensity distribution in conical refraction. Commun. Pure Appl. Math. *35*, 69–80, Zbl.516.35055

Wakabayashi, S.

[1980] A necessary condition for the mixed problem to be C^∞-well posed. Commun. Partial Differ. Equations *5*, No. 10, 1031–1064, Zbl.445.35084

[1983a] Singularities of solutions of the hyperbolic Cauchy problem in Gevrey classes. Proc. Jap. Acad., Ser. A *59*, No. 5, 182–185, Zbl.527.35048

[1983b] Analytic singularities of solutions of the hyperbolic Cauchy problem. ibid, No. 10, 449–452, Zbl.552.35052

[1984] Singularities of solutions of the Cauchy problem for symmetric hyperbolic systems. Commun. Partial Differ. Equations *9*, No. 12, 1147–1177, Zbl.575.35054

Yagdzhyan, K.A.

[1978] The Cauchy problem for weakly hyperbolic equations in Gevrey classes. Izv. Akad. Nauk Arm. SSR, Mat. *13*, No. 1, 3–22, Zbl.384.35040

[1980] Well-posedness of the Cauchy problem for weakly hyperbolic equations. ibid *15*, No. 6, 475–487. English transl.: Sov. J. Contemp. Math. Anal., Arm. Acad. Sci. *15*, No. 6, 54–65 (1980), Zbl.477.35056

[1985] Necessary and sufficient conditions for the well-posedness of the Cauchy problem for operators with multiple characteristics. ibid *20*, No. 1, 3–25. English transl.: ibid. *20*, No. 1, 1–23 (1985), Zbl.593.35013

[1986] Pseudodifferential operators with a parameter and the fundamental solution of the Cauchy problem for operators with multiple characteristics. ibid *21*, No. 4, 317–344. English transl.: ibid. *21*, No. 4, 1–29 (1986), Zbl.619.35007

Yoshikawa, A.

[1977] Construction of a parametrix for the Cauchy problem of some weakly hyperbolic equation. I. Hokkaido Math. J. *6*, No. 2, 313–344. Zbl.381.35054

[1978a] Construction of a parametrix for the Cauchy problem of some weakly hyperbolic equation. II. ibid 7, No. 1, 1–26, Zbl.423.35063

[1978b] Construction of a parametrix for the Cauchy problem of some weakly hyperbolic equation. III. ibid, 127–141, Zbl.393.35040

[1980] Parametrices for a class of effectively hyperbolic operators. Commun. Partial Differ. Equations 5, No. 11, 1073–1151, Zbl.483.35048

Zajtseva, O.V.

[1980] On the well-posedness of the Cauchy problem for a model equation with quadruple characteristics. Izv. Vyssh. Uchebn. Zaved., Mat. 1980, No. 6, 20–22. English transl.: Sov. Math. *24*, No. 6, 21–23 (1980), Zbl.441.35041

[1983] On the well-posedness of a model non-strictly hyperbolic mixed problem. ibid. 1983, No. 10, 22–25. English transl.: ibid. *27*, No. 10, 28–32 (1983), Zbl.599.35094

[1987a] On the well-posedness of a new model non-strictly hyperbolic mixed problem. ibid. 1987, No. 11, 10–12. English transl.: ibid. *31*, No. 11, 12–15 (1987), Zbl.679.35060

[1987b] The well-posedness conditions for the Cauchy problem for a certain model non-strictly hyperbolic equation. ibid. 1987, No. 12, 43–45. English transl.: ibid. *31*, No. 12, 53–55 (1987), Zbl.672.35038

Author Index

Subject Index

GPSR Compliance

The European Union's (EU) General Product Safety Regulation (GPSR) is a set of rules that requires consumer products to be safe and our obligations to ensure this.

If you have any concerns about our products, you can contact us on ProductSafety@springernature.com

In case Publisher is established outside the EU, the EU authorized representative is:

Springer Nature Customer Service Center GmbH
Europaplatz 3
69115 Heidelberg, Germany

Batch number: 10620141

Printed by Printforce, the Netherlands